Weather and Climate: The M.P. Singh Volume, Part II

Edited by
Sethu Raman
Maithili Sharan

2005

Birkhäuser Verlag
Basel · Boston · Berlin

Reprint from Pure and Applied Geophysics
(PAGEOPH), Volume 162 (2005) No. 10

Editor(s)

Maithili Sharan
Professor, Centre for Atmospheric Sciences
Indian Institute of Technology, Delhi
Hauz Khas New Delhi 11016
India

e-mail: mathilis@cas.iitd.ernet.in

Sethu Raman
Professor and State Climatologist
North Carolina State University
1005 Capability Dr., Suite 213
Campus Box 7236
Raleigh, NC 27695
U.S.A.

e-mail: raman@ncsu.edu

A CIP catalogue record for this book is available from the Library of Congress,
Washington D.C., USA

Bibliographic information published by Die Deutsche Bibliothek:
Die Deutsche Bibliothek lists this publication in the Deutsche Nationalbibliographie; detailed
bibliographic data is available in the internet at <http://dnb.ddb.de>

ISBN 3-7643-7297-4 Birkhäuser Verlag, Basel – Boston – Berlin

© 2005 Birkhäuser Verlag, P.O.Box 133, CH-4010 Basel, Switzerland
Part of Springer Science+Business Media
Printed on acid-free paper produced from chlorine-free pulp TCF ∞
Printed in Germany

ISBN-10: 3-7643-7297-4   e-ISBN 3-7643-7377-6
ISBN-13: 978-3-7643-7297-2

9 8 7 6 5 4 3 2 1   www.birkhauser.ch

PURE AND APPLIED GEOPHYSICS
Vol. 162, No. 10, 2005

Contents

Pure appl. geophys. 162 (2005) 1715–1718
0033–4553/05/101715–4
DOI 10.1007/s00024-005-2791-7

❙ Pure and Applied Geophysics

Weather and Climate: The M.P. Singh Volume
Part II

Preface

Weather and climate are of concern to virtually all countries worldwide. For many countries the economy depends largely on agriculture, which is significantly affected by variations in weather and climate. Many of the South Asian countries, for example, are prone to natural hazards such as tropical cyclones, droughts, and floods. In fact, the Indian economy is dubbed by many as a "Monsoon Gamble." Rapid industrialization and urbanization add to further deterioration of our environment. Changes in weather and climate can be traced to this environmental degradation and changes in land-use patterns brought on by rapid industrialization. Atmospheric and oceanic processes assume significance in understanding all aspects of weather and climate.

A number of field experiments have been conducted to gain an understanding of the physics of the atmosphere and the oceans. Significant advances have taken place in the understanding of atmospheric and oceanic processes and in the development of atmospheric/ oceanic models. Modern technology such as weather and ocean sensing satellites, fast communication systems, improved measurement systems and efficient computational techniques are now used in updating the information base. With increased computational power, finer resolution global / regional models have been developed for weather and climate. This has led to an increase in the predictability of weather-related phenomena at various scales. This special volume on Weather and Climate has provided contributions to the latest developments in this field.

In this volume, peer-reviewed papers related to mathematical techniques/ modeling, numerical simulations, atmospheric and oceanic processes and field experiments are included to gain insight into the weather and climate system. Papers are divided broadly into three groups, namely climate, weather and air quality. The first eight papers are related to climate, the next eleven papers pertain to weather, including land surface processes, and the last seven papers deal with air quality studies. This volume has been divided in to two parts. Part I contains the papers related to climate and weather whereas papers pertaining to boundary layer processes and air quality are included in Part II.

The paper by Marchuk et al. summarizes the "splitting methods" of numerical modeling and illustrates their application to a high resolution model of the Indian Ocean. The article from Mitra et al. describes results from a series of numerical experiments designed to determine the fidelity of predictions of major features of the Indian monsoon season in the period 1987 to 2002, using a state-of-the-art Florida State University coupled ocean-atmospheric general circulation model. The coupled model is also shown to capture the anomalous dry conditions of the monsoon 2002 season. Pielke et al. have discussed in their paper the various aspects of the 2002 Colorado drought to analyze the question whether it is a routine or unprecedented drought. The study of Sanjeeva Rao and Sikka examines several aspects of the intra-seasonal oscillations to understand its mechanisms and how it modulates the performance of the monsoon using data from BOBMEX and ARMEX experiments. The aim of the study undertaken by Mohanty et al. is to understand the climate diagnostics of the Asian summer monsoon and the role of equatorial convection on the summer monsoon activity over India. Le Treut and Bellon show that the direct impact of radiative heating on the development of monsoon may depend in a very complex manner on a combination of strongly interacting radiative and dynamical feedbacks.

Khandekar et al. have presented a review of the present status of global warming science. They have focused on evolution of the earth's atmosphere, and the greenhouse effect. Beniston's paper provides a concise overview of the peculiarities of mountain climates, pointing out the sensitivity of the atmosphere to these systems with a special reference to the European Alps.

There are several papers on weather in this volume dealing with radiation, tornados, tropical cyclones, and land use. The paper by Ramanathan and Ramana discusses the presence of absorbing aerosols over Kathmandu, Nepal and Kanpur, India and their implications on radiation balance. The article by Rao et al. presents the structures of mesocirculations that generated tornadoes associated with Tropical Cyclones Frances (1998) in Texas and in Louisiana. The study (Subramaniam et al.) addresses the thermal, salinity and circulation responses at sea surface due to intense tropical cyclones in 1999 and 2001 in the northern tropical Indian Ocean, based on satellite measurements and model simulations. Dube et al. in their paper describe a depth-averaged storm surge model for simulating the surges along the Orissa coast. This study emphasizes the impact of the Mahanadi River on overall surge development along the Orissa coast.

In the second part of this volume, scientific papers related to the atmospheric boundary layer and air quality studies are included. The land surface processes and surface characteristics play a dominant role in the development of boundary layer and in turn on weather and climate. For weather prediction and climate modeling, the understanding of the behavior and predictability of the atmospheric boundary layer is crucial. The paper by Raman et al. is a numerical study on convection initiation caused by differences in soil types. Arya provides an overview on micrometeorology and atmospheric boundary layer discussing, recent developments in the field in relation to weather and climate. Orr et al. presents detailed features of mesoscale flows over Greenland, specifically, to understand the influence of high orography on stable flows over and around Greenland. Hozumi and Ueda consider air flow over large mountains in an idealized situation of uniform approach flow, and in the absence of local thermally driven flows on the slopes, e.g. cooling as in the case of Greenland (Orr et al.) and heating as the southern slopes of the Himalayas. The effect of gentle slopes on strong and weak wind nocturnal boundary layer (NBL) is investigated by Gopalakrishnan et al. This study indicates that the wind profiles, temperature profiles and surface layer turbulence characteristics are sensitive to the imposed geostrophic wind when small slopes are present, especially for light winds. Using techniques of nonlinear analysis, Shi et al. examine in their paper the behavior of NBL by analyzing a set of four partial differential equations and this paper explores the role of initial conditions in the final solution and carries out an eigen value analysis to determine whether a pure limit cycle exists. The paper by Vasudev Murthy et al. presents a complete asymptotic analysis of a simple model for the evolution of the nocturnal temperature distribution on bare soil in calm and clear conditions. This simple model is able to explain the occurrence of minima in the vertical temperature profile during the calm/clear night and thus describes the structure and dynamics of the Ramdas layer.

Pollutants emitted into the atmosphere from various sources such as industrial chimneys, vehicular exhaust, power plants, nuclear and chemical plants etc. degrade the environment and in turn influence weather and climate. Sharan and Modani describe an analytical model for the dispersion of pollutants in a finite layer in low wind conditions. The proposed model is validated with the data from the Hanford diffusion experiment in stable conditions and IIT diffusion experiment in unstable conditions. The paper by Rao addresses the analysis and quantification of various types of uncertainties associated with the prediction of concentrations from atmospheric dispersion models. Gego et al. address the important problem of consistency between various measurement networks with a particular reference to the particulate nitrate, sulfate and ammonium concentrations. Neophytou and Britter provide a simple model for the variation of smoke and oxygen concentrations in an inclined two-dimensional tunnel that is open at the top.

Following the collapse of the New York World Trade Center Towers (WTC) on September 11, 2001, several studies were initiated to monitor air quality to better understand the impact of emissions from the disaster. The next three papers deal with the various aspects. Urban heat island and roughness length effects, sea breeze structure and behavior are among the features examined by Child and Raman. Gilliam et al. critically examine a dispersion modeling system (CALMET-CAL-PUFF) used to simulate the WTC emissions transport. The first paper focuses on the performance of the diagnostic meteorological model, CALMET during various synoptic flow events whereas in the second, the dispersion patterns of a simulated WTC plume are presented for a three-month period from September 11, 2001.

Earlier research related to meteorology in India was primarily confined to the India Meteorological Department (IMD), the country's National Weather Service. IMD devoted significant concentration on the operational needs of the country. However, efforts on the fundamental research for the understanding of atmospheric and oceanic features were limited in scope. Dr. M. P. Singh, Professor of Applied Mathematics, Indian Institute of Technology (IIT), Delhi (India) took an initiative and played a key role in the development of research and development activities in the areas of atmospheric and oceanic sciences at IIT Delhi. At his initiative, a Centre for Atmospheric Sciences was set-up at IIT Delhi which was cosponsored by the India Meteorological Department on a cost sharing basis. India is among the very few countries where such a joint venture sponsored by the National Weather Service to promote weather research in an academic institute has been established. To recognize the contributions of Professor M.P. Singh in research and developmental activities in the area of atmospheric and oceanic sciences, we honor him by dedicating this issue of Pure and Applied Geophysics as "M.P. Singh volume on Weather and Climate."

We would like to place on record our gratitude to all the authors who have contributed to this special issue. Their association with this special issue will benefit immensely the scientific community at large, especially younger scientists. Late Professor G. V. Rao played a key role in the planning of this special issue. He fell victim to a swimming accident on the Mexican Pacific Coast while he was performing research related to the North American Monsoon Experiment. We wish to thank all the reviewers for their valuable comments/suggestions. We also thank Dr. Renata Dmowska, Topical Editor, Pure and Applied Geophysics for her assistance in issuing out this topical edition in honor of Professor M.P. Singh.

Finally, we express our sincere thanks to the editorial staff of Pure and Applied Geophysics for the meticulous care they have taken in preparing both parts of this special edition.

Editors:

Maithili Sharan
Professor, Centre for Atmospheric Sciences,
Indian Institute of Technology, Delhi
Hauz Khas, New Delhi 110016, India.

Sethu Raman
Professor and State Climatologist,
North Carolina State University,
Raleigh, North Carolina 27695, U.S.A.

C. Boundary Layer Processes

Pure appl. geophys. 162 (2005) 1721–1745
0033–4553/05/101721–25
DOI 10.1007/s00024-005-2690-y

© Birkhäuser Verlag, Basel, 2005

| Pure and Applied Geophysics

Micrometeorology and Atmospheric Boundary Layer

S. Pal Arya[1]

Abstract—Starting with simple definitions and the scope of micrometeorology and the atmospheric planetary boundary layer (PBL), the importance of small-scale turbulent exchange and energy dissipation processes in the PBL to the evolution of local weather and climate is pointed out. The energy budgets of an "ideal" surface and a near-surface canopy layer are briefly described and typical diurnal variations of measured energy fluxes at two (a rural and a suburban) sites are illustrated. Using the conservation equations of thermodynamic energy and moisture, temporal and spatial variations of mean air temperature and specific humidity are described. Diurnal evolutions of potential temperature profiles and the corresponding PBL height are illustrated with some observed data. Typical mean wind profiles in the daytime convective boundary layer (CBL) and the nighttime stable boundary layer (SBL) are also shown and discussed. The strong influence of stability on the PBL height and structure and their diurnal variations are discussed. The widely used flux-profile relations based on the Monin-Obukhov surface-layer similarity theory are presented with a brief discussion of parameterization of surface roughness and fluxes. The statistical description of the PBL turbulence and its commonly used quantitative measures are presented with an illustration of observed time series of turbulence. Some of the better-known similarity theories and scaling of turbulence in the PBL are described. Finally, different types of turbulence closure theories and numerical simulations/models of the PBL are briefly reviewed.

Key words: Atmospheric boundary layer, eddy diffusivity, energy budget and fluxes, micrometeorology, Reynolds stresses, surface layer, and turbulence.

1. Introduction

Micrometeorology is a branch of meteorology which deals with small-scale or microscale atmospheric phenomena and processes occurring in the lowest layer of the atmosphere, commonly called the atmospheric or planetary boundary layer (PBL). The atmospheric PBL is defined as the layer of frictional and thermal influence of the earth's surface over short time scales of a few hours to less than a day (OKE, 1987; ARYA, 2001). Rapid exchanges of momentum, heat, and mass between the earth's surface including vegetation, soil, water, and other materials that constitute the surface, and the atmosphere occur through the PBL.

[1] Department of Marine, Earth and Atmospheric Sciences, North Carolina State University, Raleigh, NC 27695-8208, U.S.A. E-mail: sparya@unity.ncsu.edu

Micrometeorology deals with all these earth-atmospheric exchange processes and the resulting vertical distributions (profiles) of mean wind, temperature, humidity, and carbon dioxide. Micrometeorological measurements aim to capture the whole wide range of eddy motions, ranging from very rapid, random fluctuations to more organized, large eddy structures. They also include the various components of the energy and moisture budgets at or near the surface including the active soil layer and the vegetation canopy layer. The atmospheric PBL thickness varies over a wide range (several tens of meters to several kilometers) and depends on the rate of heating or cooling of the surface, strength of winds, the roughness and topographical characteristics of the surface, large-scale vertical motion, horizontal advection of heat and moisture, the presence of low-level clouds in the PBL, and other factors. In response to the strong diurnal cycle of heating and cooling of land surfaces during fair-weather conditions, the PBL height waxes and wanes between its lowest typical value of the order of 100 m (range $\sim$ 20–500 m) in the morning to the highest value of the order of 1 km (range $\sim$ 0.2–5 km) during late afternoon. Mean winds, temperatures, and other properties, as well as turbulent fluxes also exhibit strong diurnal variations. Diurnal variations of the PBL height and other properties are found to be considerably smaller or almost absent over large lakes, seas and oceans, as well as over high latitude land areas during winter.

The importance of micrometeorology and the atmospheric PBL to local weather and climate has been well recognized (OKE, 1987; GEIGER et al., 1995). Although the PBL comprises only a tiny fraction of the atmosphere, the small-scale processes occurring within the PBL, especially in the shallow surface layer (usually defined as the lowest 10% of the PBL) and its lowest part, called the canopy layer, provide most of the energy and moisture exchanges between the surface and the atmosphere which are vital to the evolutions of mesoscale and large-scale weather. The surface and boundary layer friction is primarily responsible for the low-level convergence and divergence of flow and moisture. Almost one-half of the dissipation of the atmospheric kinetic energy on an annual basis occurs with the PBL. Thus, an accurate parameterization of land-surface and PBL exchange processes is considered an important part of the current weather forecasting and climate models.

This overview of micrometeorology and the atmospheric PBL describes aspects of the near-surface energy budget, air temperature and humidity profiles in the PBL, mean wind distribution in the PBL, and turbulent exchanges of momentum, heat and mass in the PBL. Semi-empirical theories and parameterizations of turbulent exchanges, commonly used in micrometeorology, are also briefly discussed. Some of the surface layer and PBL similarity theories, as well as numerical modeling approaches that are commonly used for studying the atmospheric boundary layer height and structure are also reviewed. More extensive descriptions of micrometeorology and atmospheric PBL are given elsewhere (MUNN, 1966; MONIN and YAGLOM, 1971; BRUTSAERT, 1982; ROSENBERG et al.,

1983; OKE, 1987; STULL, 1988; SORBJAN, 1989; GARRATT, 1992; KAIMAL and FINNEGAN, 1994; ARYA, 1999; 2001).

Energy Budget Near the Surface

Using the principles of the energy conservation at an 'ideal' surface, which is assumed to be flat, bare, extensive and opaque to radiation, the surface energy budget can be expressed as

$$R_N = H + H_L + H_G \tag{1}$$

in which R_N is the net radiation, H and H_L are the sensible and latent heat fluxes to or from the atmosphere, and H_G is the ground heat flux to or from the subsurface medium. Here all the fluxes are normal to the surface and can be expressed in the SI units of W m^{-2}, or J s^{-1}m^{-2}. We use the sign convention that nonradiative fluxes directed away from the surface and the radiation fluxes directed toward the surface are positive and *vice versa*. The net radiation may be considered as the primary energy forcing and Eq. (1) tells us how R_N might be partitioned between other energy exchanges with the atmosphere and the submedium.

All the components of the surface energy budget show strong diurnal variations, in response to the diurnal variation of net radiation and the diurnal cycle of heating and cooling of the surface (OKE, 1987). They also show large seasonal variations. Both the diurnal and seasonal variations are dependent on the latitude, the surface characteristics including thermal properties such as albedo and emissivity, and prevailing weather conditions. The ratios H/R_N, H_L/R_N, and H_G/R_N, which represent relative partitioning of net radiation, are expected to show much less variability than the individual fluxes. Empirical estimates of these ratios for different types of surfaces can be used to parameterize other energy fluxes in terms of R_N. For example, the ground heat flux is often parameterized in this manner. Then, an independent estimate of the Bowen ratio, $B = H/H_L$, would permit an approximate determination of the sensible and latent heat fluxes from a measurement or estimate of R_N only (ARYA, 2001).

In many aspects the earth's surface may not be considered 'ideal,' because it is very rough and may be covered by vegetation, buildings, and water, which are not opaque to radiation. For such vegetation and urban canopies, as well as water layers, which may store or release energy, the following energy budget equation for a near-surface layer is more appropriate:

$$R_N = H + H_L + H_G + \Delta H_s \tag{2}$$

where ΔH_s represents the change in the energy storage in the layer per unit time, per unit horizontal area, which can be expressed in terms of the average rate of warming or cooling of the layer and its thickness (OKE, 1987; ARYA, 2001). An illustration and comparison of the monthly-averaged near-surface energy budgets at two (a rural and a suburban) sites in Greater Vancouver, Canada, for a summer month are shown in Fig. 1. Note that the suburban site has higher R_N, H and ΔH_s and lower H_L than the rural site; here H_G has been combined with ΔH_s. The suburban area is an area of fairly uniform one-or two story houses (36% built and 64% green space), while the rural site represented an extensive area of grassland. For the latter, ΔH_s essentially represents H_G. Some day-to-day variations of the energy budget of the same suburban area are shown and discussed by OKE (1988). GRIMMOND and OKE (1995) compare the summertime energy budgets of the four North American cities (Chicago, Los Angeles, Sacramento, and Tucson), representing different climate conditions. As expected, the magnitudes of individual energy fluxes vary between cities; however the diurnal trends of flux partitioning are similar in terms of the timing of peaks and changes in sign.

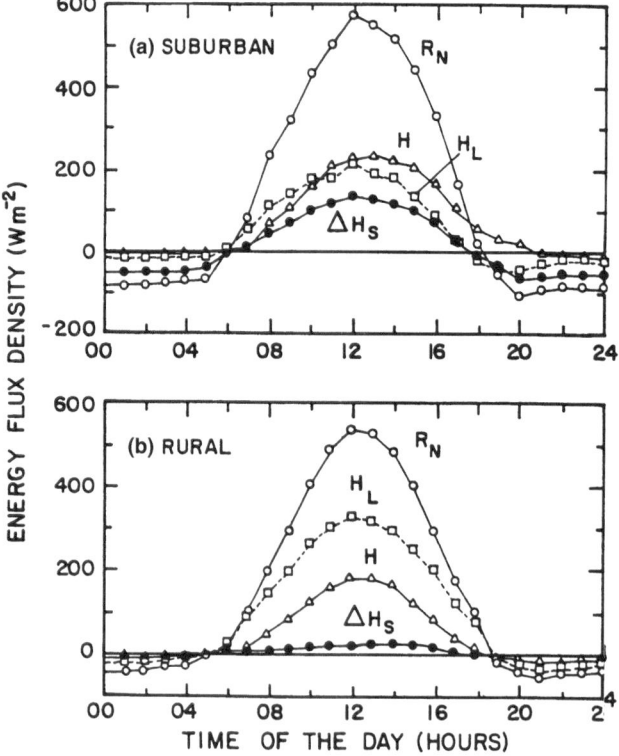

Figure 1
Monthly-averaged near-surface energy budgets at (a) suburban and (b) rural sites in Greater Vancouver, Canada, during summer (from ARYA, 2001, after OKE, 1987).

2. Temperature and Humidity Profiles in the PBL

The vertical distribution of air temperature in the PBL depends on the surface temperature, air temperature at the top of the PBL, the surface heat flux, divergence or vertical gradient of sensible heat flux, divergence of net radiation, horizontal and vertical advections of heat by mean motions, entrainment of free-atmospheric air into the PBL, and local heating or cooling associated with phase changes of H_2O during cloud condensation or evaporation processes.

An 'ideal' surface may be expected to have a well defined surface temperature T_s, which is determined by the surface energy budget, thermal properties of the surface and the subsurface medium, and the air mass characteristics affecting the PBL. The most commonly used force-restore method of predicting the ground surface temperature utilizes a two-layer approximation of the subsurface medium in which a shallow thermally-active layer of soil overlies a much thicker constant-temperature layer (GARRATT, 1992). The resulting prognostic equation for the surface temperature is

$$C_g \frac{\partial T_s}{\partial t} = (R_N - H - H_L) - \mu_g(T_s - T_m), \tag{3}$$

in which T_m is the temperature of the lower soil layer, C_g $\left[\mathrm{J\,m^{-2}K^{-1}}\right]$ is the heat capacity per unit area of the near-ground soil layer, and μ_g $\left[\mathrm{J\,m^{-2}s^{-1}K^{-1}}\right]$ is a heat transfer coefficient, where SI units are indicated in brackets. Both C_g and μ_g can be expressed in terms of better-known soil thermal properties, such as volumetric heat capacity and thermal diffusivity (GARRATT, 1992).

The temporal and spatial variations of mean air temperature in the PBL are given by the thermodynamic energy equation

$$\frac{DT}{Dt} = \frac{1}{\rho c_p} \left(-\frac{\partial H}{\partial z} + \frac{\partial R_N}{\partial z} \right) + S_H, \tag{4}$$

where the total derivative DT/Dt includes the local rate of change of temperature with time, as well as horizontal and vertical advections of temperature (heat) by mean motion, ρ and c_p are density and specific heat of air, and S_H represents the local source or sink of heat associated with condensation or evaporation processes (ARYA, 1999, 2001).

For representing the thermal structure of the PBL and its diurnal variation, it is often more convenient to use the potential temperature Θ, which is a more conservative variable and is related to actual temperature as

$$\Theta = T(P_0/P)^k \tag{5}$$

$$\frac{\partial \Theta}{\partial z} = \frac{\Theta}{T} \left(\frac{\partial T}{\partial z} + \Gamma \right) \simeq \frac{\partial T}{\partial z} + \Gamma, \tag{6}$$

where $P_0 = 100$ kPa is a reference pressure, $k = 0.286$, and Γ is the adiabatic lapse rate (for an unsaturated atmosphere, $\Gamma \cong 0.01$ K m^{-1}).

Figure 2 shows a typical sequence of potential temperature profiles at 3 h intervals during the course of a day and night, under fair-weather conditions (DEARDORFF, 1978; ARYA, 2001). Note that before sunrise, the Θ profiles between 2100 and at 0600 h are characterized by nocturnal inversion. Shortly after sunrise, heating of the surface by solar radiation and exchange of sensible heat with the lowest layer results in its warming. This process progressively erodes the nocturnal

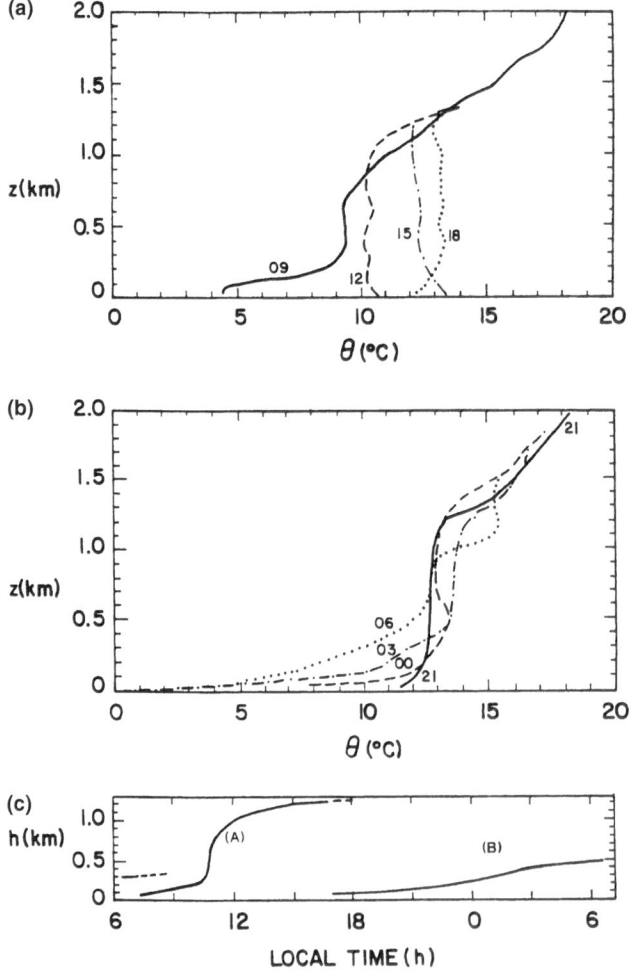

Figure 2

Diurnal variations of potential temperature profiles during (a) day 33 and (b) days 33–34 of the Wangara Experiment in southern Australia. (c) Curve A represents the inversion base height in the daytime CBL and curve B, the surface inversion height (h_i) in the nighttime SBL. (from ARYA, 2001, after DEARDORFF, 1978).

inversion from below and replaces it with an evolving unstable boundary layer. This daytime unstable or convective boundary layer (CBL) has a three-layered structure: (1) a shallow unstable surface layer in which $\partial\Theta/\partial z < 0$; (2) a much deeper mixed layer with nearly uniform Θ; and (3) a shallow transition layer comprising the lower part of the capping inversion in which $\partial\Theta/\partial z > 0$. The height of inversion base, z_i is often used as an approximation for the PBL height (h), although the latter can be substantially (10–30%) larger than the former. A better estimate of h would be the height level in the capping inversion where the potential temperature equals that of the near-surface air, since air parcels from near the surface can rise up to that level under their own buoyancy (ARYA, 1999, 2001), or the height at which the flux of sensible heat becomes negligible.

Shortly before sunset the surface begins to cool and the surface heat flux becomes negative, so that the air near the surface is also cooled. The resulting surface inversion layer in which $\partial\Theta/\partial z > 0$ deepens in the early evening hours and, sometimes, throughout the night. The cooling of air in the nocturnal inversion layer is due to both the radiative and sensible heat flux divergences, while the ground surface cools due to the net radiative loss from the surface. The surface inversion height h_i, may not necessarily correspond to the stable boundary layer (SBL) height, h, in which significant turbulent exchanges occur. There are several different ways of defining and estimating h (GARRATT, 1992). The inversion layer height (h_i) is more easily detected from temperature soundings and is shown in Fig. 2(c). The SBL height can be estimated from remote sensing instruments, such as radar, sodar, and lidar, or from the simultaneous temperature and wind soundings. Such measurements indicate that, unlike the inversion height (h_i), the PBL height (h) generally decreases with time during early evening hours and is usually much less than h_i through most of the night and early morning. The SBL can also be divided into the surface layer, in which $\partial\Theta/\partial z$ is large and turbulent fluxes are nearly constant, and an outer layer of much weaker and intermittent turbulence. It is usually capped by a weaker inversion layer or a remnant of the previous afternoon's mixed layer. Strong shears, if present in this so-called residual layer, may generate intermittent and sporadic turbulence which is not usually connected to the SBL, but may sometimes penetrate the SBL all the way down to the surface (STULL, 1988; MAHRT, 1999).

The vertical distribution of specific humidity in the PBL depends on the type of surface and its moisture content, the rate of evapotranspiration from the surface, specific humidity of air mass above the PBL, advection of moisture by mean motions, and local evaporation and cloud condensation processes within the PBL. The temporal and spatial variations of mean specified humidity Q in the PBL are given by the following equation for the conservation of mass of H_2O in any state:

$$\frac{DQ}{Dt} = -\frac{1}{\rho}\frac{\partial E}{\partial z} + S_w, \tag{7}$$

where E $\left[\text{kg m}^{-2}\text{s}^{-1}\right]$ is the vertical flux of water vapor and S_W represents any local source or sink of water vapor associated with evaporation or condensation processes.

After appropriate parameterizations of turbulent fluxes and source and sink terms, Eqs. (4) and (7) are used to predict temperature and specific humidity distributions in the PBL. Conversely, they can be used to estimate the surface fluxes from the measured temperature and humidity profiles in the PBL. In particular, when advections of heat and moisture can be neglected and fog and clouds are not present in the PBL, the surface fluxes can be expressed through integrations of Eqs. (4) and (7) with respect to z as

$$H_0 = \rho c_p \int_0^h \frac{\partial T}{\partial t}\, dz = \rho c_p h \left(\frac{\partial T}{\partial t}\right)_m , \tag{8}$$

$$E_0 = \rho \int_0^h \frac{\partial Q}{\partial t}\, dz = \rho h \left(\frac{\partial Q}{\partial t}\right)_m , \tag{9}$$

where $(\partial T/\partial t)_m$ and $(\partial Q/\partial t)_m$ are the mean rates of warming and moistening of the PBL. Equations (8) and (9) are especially useful for estimating heat and moisture fluxes during the daytime unstable and convective conditions when $(\partial T/\partial t)$ and $(\partial Q/\partial t)$ are approximately independent of height in the mixed layer and their measurement at one height level might be representative of the whole PBL. These temperature and humidity tendencies generally vary with the height in the stable boundary layer, as well as in the transition layer above the convective mixed layer.

3. Mean Wind Profiles in the PBL

Near-surface winds and their vertical distribution in the PBL are of considerable interest in micrometeorology. The important factors in the determination of wind distribution are the large-scale horizontal pressure and temperature gradients, the surface roughness and topographical characteristics, the earth's rotation and Coriolis effects, thermal stratification caused by the diurnal heating and cooling cycle, horizontal and vertical advections of momentum, and the presence of clouds and precipitation in the PBL (STULL, 1988; ARYA, 2001).

The PBL winds are essentially driven by large-scale horizontal pressure gradients or geostrophic winds, which may vary with height in response to large-scale or mesoscale horizontal temperature gradients. For the given pressure and temperature fields, geostrophic winds are essentially determined by the geostrophic balance between pressure gradient and Coriolis forces. Actual winds in the PBL result from a more complicated balance between pressure gradient, Coriolis, friction and inertia forces on an air parcel, and frequently differ from the geostrophic winds. In the absence of local and advective accelerations, the equations of mean horizontal motion are given by

$$2\rho \, \underset{\sim}{\Omega} \times (\underset{\sim}{V} - \underset{\sim}{G}) = \frac{\partial \underset{\sim}{\tau}}{\partial z},$$ (10)

where Ω is the earth's rotational vector, V and G are the actual and geostrophic wind vectors, and τ $[\mathrm{kg \, m^{-1} \, s^{-2}}]$ is the shear stress vector.

Equation (10) also expresses the balance between the Coriolis force acting normal to V, the pressure gradient force acting normal to G, and the friction force per unit volume of any fluid element. This simpler force balance is valid in an 'idealized' PBL with a quasi-stationary and horizontally-homogeneous flow. It requires that the wind speed and direction change with height above the surface, in response to the decreasing friction force. Directly at the surface, wind speed must vanish due to the effect of viscosity. In the surface layer wind speed increases rapidly with height, but remains subgeostrophic, while wind direction remains nearly constant. Near-surface winds are usually directed toward low pressure at an angle which depends on the surface roughness, geostrophic winds, latitude, and thermal stability and baroclinity of the PBL. Above the surface layer the PBL winds may become supergeostrophic and attain a maximum value, while the wind direction may veer or back with height (ARYA, 2001).

For a given large-scale forcing (geostrophic winds) and uniform surface roughness, the wind distribution in the PBL is strongly dependent on the thermal stability which is determined by the diurnal surface heating and cooling cycle. This strong influence of stability can be seen in the observed wind profiles shown in Figs. 3 and 4 for the typical daytime unstable and nighttime stable conditions over land surfaces (DEARDORFF, 1978). Here U and V are the horizontal wind components in the x and y directions, respectively, in the frequently used surface-layer coordinate system with the x axis taken parallel to the near-surface winds.

Fig. 3 shows the observed wind component profiles as well as potential temperature and specific humidity profiles obtained from pibal and radiosonde soundings during fairly convective conditions. Note that, despite the large

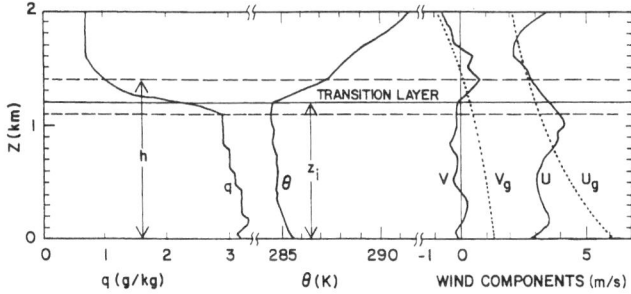

Figure 3

Measured vertical profiles of wind components, potential temperature and specific humidity in the CBL on day 33 of the Wangara Experiment (from ARYA, 2001, after DEARDORFF, 1978).

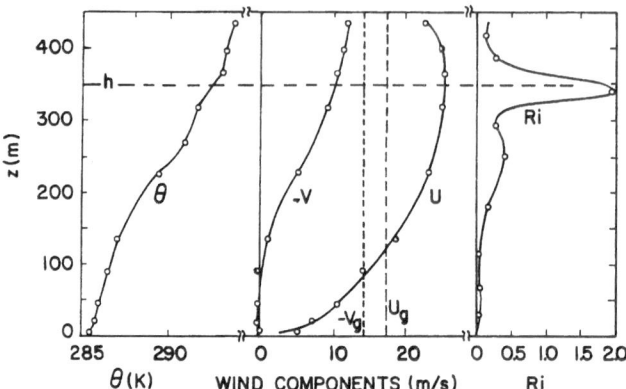

Figure 4
Measured vertical profiles of mean wind components and potential temperature and the calculated Ri-profile in the nocturnal PBL under moderately stable conditions (from ARYA, 2001, after IZUMI and BARAD, 1963).

geostrophic wind shears (U_g and V_g vary with height) present at the time of these observations, the observed profiles display the characteristic features of nearly uniform winds in the convective mixed layer and stronger gradients in the shallow surface layer below and the transition layer above. There is little change, if any, in the wind direction in the CBL, which has a typical mid-day depth of about 1400 m. Similar features have been observed in moderately unstable and convective boundary layers over other homogeneous land and sea surfaces (STULL, 1988; ARYA, 2001).

In contrast to the CBL, the nocturnal stable boundary layer (SBL) is characterized by much shallower depth, stronger wind speed and direction shears throughout the PBL, and much weaker and often sporadic turbulence and mixing. Two broad, but distinct stability regimes have been identified in the SBL: (1) slightly to moderately stable regime in which turbulent exchanges are more or less continuous in time and space through at least the lower half of the SBL; and (2) the very stable regime in which turbulent exchanges occur only intermittently and sporadically in time and space through most of the SBL. The former can exist over land at night only during strong winds and more likely at sea when air is slightly warmer than the sea surface. Fig. 4 illustrates the typical wind and potential temperature profiles in a moderately stable boundary layer. As an indirect measure of turbulence activity, the computed gradient Richardson number

$$\mathrm{Ri} = \frac{g}{T_v}\frac{\partial \Theta_v}{\partial z}\left|\frac{\partial \underset{\sim}{V}}{\partial z}\right|^{-2} \qquad (11)$$

is also shown as a function of height in Fig. 4. When and where Ri exceeds its critical value, estimated to lie between 0.25 and 0.50, turbulence is likely to decay and even disappear temporarily until Ri becomes subcritical again. Such episodes of turbulence decay and regeneration occur frequently, but randomly, in the SBL. Below critical values of Ri in the lower half of the SBL in Fig. 4 are indicative of continuous turbulence there, and considerably larger values at or near the top are indicative of intermittent turbulence there.

The very stable regime is more typical of the nighttime SBL with weak winds and strong radiative cooling over land surfaces. Here, both the wind components attain their maximum values and Ri becomes large (above critical) at relatively low levels. The SBL height based on the maximum in the wind speed profile is substantially smaller than the inversion height. The occurrence of a low-level jet is a common phenomenon in the SBL. The jet can become highly intensified when thermal or slope winds oppose the surface geostrophic wind. The low-level jet is associated with inertial oscillations that occur in a stably stratified atmosphere, deepening of the surface inversion, deceleration of near-surface flow, and reduced mixing and turbulence due to negative buoyancy effects in stably stratified flow. The occurrence of a wind speed maximum, coincident with the top of the inversion, can result in strong wind shears both above and below the jet maximum. These shears can cause turbulence which promotes the upward propagation of surface inversion.

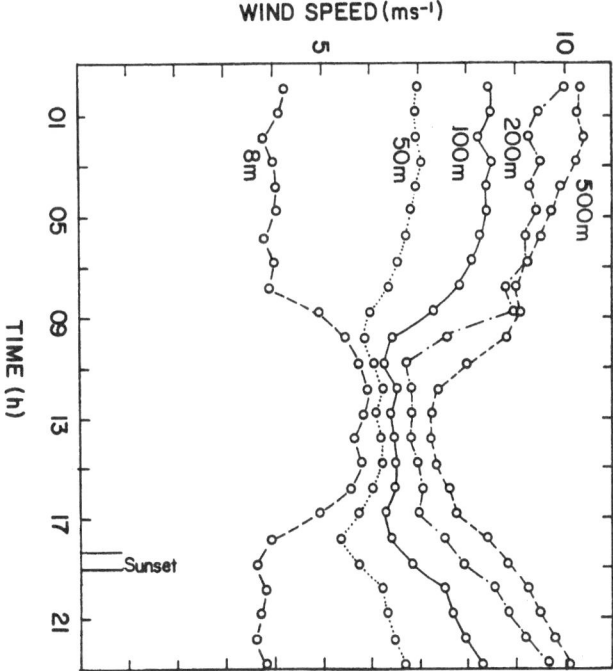

Figure 5

Diurnal variations of 40-day averaged wind speeds at various heights in the PBL during the Wangara Experiment (from ARYA, 2001, after MAHRT, 1981).

The strong influence of stability on wind distribution in the PBL is also reflected in the observed diurnal variations of wind speed at different height levels (Fig. 5). Near the surface, wind speed increases sharply after sunrise, attains a broad maximum during the mid-day period, and decreases sharply just before sunset (MAHRT, 1981; ARYA, 2001). The rapid increase in the morning is due to the increased mixing and efficient transfer of momentum from aloft through the rapidly growing unstable or convective boundary layer. When the surface begins to cool and the CBL collapses in the late afternoon, near-surface winds decelerate rapidly due to greatly reduced mixing in the surface inversion layer. At higher levels the diurnal pattern is reversed, so that wind speed decreases sharply during the morning hours, attains a minimum during mid-day, and increases in the late afternoon and evening hours (see Fig. 5).

4. Flux-profile Relations for the Surface Layer

Measuring or otherwise estimating the surface stress (τ_0) and the vertical fluxes of heat (H_0) and water vapor (E_0) had been one of the primary objectives in micrometeorology. Their direct measurements are rather difficult and require research grade fast-response instrumentation. Indirect estimates from more routine measurements of wind speed, temperature and specific humidity at one or more height levels are usually based on the flux profile relations for the idealized horizontally-homogeneous surface layer in which the vertical fluxes of momentum, heat, and water vapor may be assumed to be independent of height.

4.1. Surface-layer Similarity Theory

For the constant-flux surface layer, the Monin-Obukov (M-O) similarity theory has been widely used to formulate the appropriate scaling parameters, as well as the flux-profile similarity relations based on the same (MONIN and YAGLOM, 1971; STULL, 1988; ARYA, 2001):

$$\frac{kz}{u_*}\frac{\partial U}{\partial z} = \phi_m(\zeta); \quad \frac{kz}{\theta_*}\frac{\partial \Theta}{\partial z} = \phi_h(\zeta); \quad \frac{kz}{q_*}\frac{\partial Q}{\partial z} = \phi_w(\zeta) \qquad (12)$$

$$u_* = \left(\frac{\tau_0}{\rho}\right)^{1/2}; \quad \theta_* = -\frac{H_0}{\rho c_p u_*}; \quad q_* = -\frac{E_0}{\rho u_*} \qquad (13)$$

$$\zeta = \frac{z}{L} = -\frac{kz}{u_*^3}\frac{g}{T_{v0}}\frac{H_{v0}}{\rho c_p}, \qquad (14)$$

where $k \cong 0.40$ is the von Karman constant, u_*, θ_*, and q_* are the velocity, temperature and specific humidity scales, respectively, L is the Obukhov length, and ζ is the M-O similarity or stability parameter. The buoyancy effect of water vapor is considered through the virtual temperature T_{V0} and the virtual heat flux

$$H_{v0} = H_0 + 0.61 c_p \Theta E_0 \cong H_0 \left(1 + 0.07 B^{-1}\right), \tag{15}$$

which are included in the definition of L or ζ in Eq. (14). The M-O similarity relations (12) imply that the dimensionless wind shear and the vertical gradients of Θ and Q in the surface layer are some universal functions of ζ only.

Other parameters involving mean gradients can also be expressed in terms of the same basic M-O similarity functions, e.g.,

$$\frac{K_m}{kzu_*} = \frac{1}{\phi_m(\zeta)}; \quad \frac{K_h}{kzu_*} = \frac{1}{\phi_h(\zeta)}; \quad \frac{K_w}{kzu_*} = \frac{1}{\phi_w(\zeta)} \tag{16}$$

$$\mathrm{Ri} = \zeta \phi_h(\zeta)/\phi_m^2(\zeta), \tag{17}$$

where K_m, K_h, K_w, are the eddy diffusivities or exchange coefficients of momentum, heat and water vapor, respectively; they represent the ratios of turbulent fluxes to mean gradients, and have SI units of $\left[\mathrm{m^2\,s^{-1}}\right]$.

Empirical Forms of Similarity Functions

The empirical forms of the M-O similarity functions have been determined from micrometeorological experiments conducted at various flat and homogeneous sites. There is still no general consensus on the best formulas to be used. The simplest are the Businger-Dyer empirical relations (ARYA, 2001):

$$\begin{aligned}
\phi_w = \phi_h = \phi_m^2 = (1 - 15\zeta)^{-1/2}, \quad \text{for } -5 < \zeta < 0 \\
\phi_w = \phi_h = \phi_m = 1 + 5\zeta, \quad \text{for } 0 \le \zeta < 1
\end{aligned} \tag{18}$$

which imply simple and explicit relations between ζ and Ri

$$\begin{aligned}
\zeta = \mathrm{Ri}, \quad \text{for } -5 < \mathrm{Ri} < 0 \\
\zeta = \mathrm{Ri}/(1 - 5Ri), \quad \text{for } 0 \le \mathrm{Ri} < 0.2.
\end{aligned} \tag{19}$$

The above relations imply that $K_w = K_h$, irrespective of stability conditions. The often assumed equality of these scalar diffusivities has been verified in some experiments.

The mean wind profile in the surface layer can be expressed by the modified log-law:

$$U = \frac{u_*}{k}\left[\ln\frac{z}{z_0} - \Psi_m(\zeta)\right], \quad \text{for } z \gg z_0, \tag{20}$$

in which z_0 is the surface roughness parameter (length) and $\Psi_m(\zeta)$ is a stability-dependent function which is the integral of $\phi_m(\zeta)$ with respect to ζ from z_0/L to z/L (ARYA, 2001), which is uniquely related to $\phi_m(\zeta)$. The deviations in the mean wind profile from the log law increase with increasing stability or instability. Under moderately stable conditions the profile is log-linear and tends to become linear with increasing height or z/L.

There is a question about the validity of the above-mentioned empirical similarity functions under very unstable (approaching free convection) and very stable (approaching critical Ri) conditions. Micrometeorological data used for their determination were limited to a moderate stability range of $-5 < \zeta < 1$. Some additional data taken under more stable conditions indicate that the profile functions become increasingly flatter as ζ increases above 0.5 (HICKS, 1976; HOLTSLAG, 1984).

4.2. Parameterization of Surface Roughness and Fluxes

The roughness parameter is related (proportional) to the average height of surface roughness elements, h_0, as well as the roughness density. Over tall roughness, another surface parameter, called the zero-plane displacement d_0 is also introduced. The reference plane is assumed to be displaced d_0 above the ground surface, so that the appropriate height in the above flux-profile relations is $z = z' - d_0$, where z' is the height measured from the ground level. Both z_0 and d_0 have been determined empirically from wind profile measurements under near-neutral stability conditions over a wide variety of surfaces. The ratios z_0/h_0 and d_0/h_0 largely depend on the roughness element density; for most vegetation canopies, $z_0/h_0 \cong 0.15$ and $d_0/h_0 \cong 0.80$ (ARYA, 2001). For land surfaces, empirical data suggests these ratios to lie within narrow ranges of $z_0/h_0 = 0.03 - 0.2$ and $d_0/h_0 \cong 0.5 - 0.9$ ((GARRATT, 1992; ARYA, 2001).

For parameterizing the surface fluxes in mesoscale and large-scale atmospheric circulation models, the most commonly used are the so-called bulk transfer relations:

$$\tau_0 = \rho C_D U_r^2$$
$$H_0 = \rho c_p C_H U_r (\Theta_s - \Theta_r)$$
$$E_0 = \rho C_W U_r (Q_s - Q_r) \tag{21}$$

in which C_D, C_H and C_W are the drag, heat transfer and water vapor transfer coefficients, U_r, Θ_r, and Q_r are mean variables at a reference height z_r in the surface layer, and Θ_s and Q_s are their values at the surface. The drag and other transfer coefficients can be specified as functions of z_r/z_0 and Ri_b (the bulk Richardson number) using the M-O flux-profile relations (ARYA, 2001). In simpler, but crude parameterizations, C_D and $C_H = C_W$ are specified as constant values (sometimes, different values for land and ocean surfaces). Typical range of their values over ocean surfaces is 0.001 to 0.002, but a considerably wider range of 10^{-4} to 10^{-2} is expected over land surfaces.

Another commonly used method of parameterizing the surface fluxes is the so-called resistance approach in which the fluxes are expressed as

$$\tau_0 = \rho r_M^{-1} U_r$$
$$H_0 = \rho c_p r_H^{-1} (\Theta_s - \Theta_r),$$
$$E_0 = \rho r_W^{-1} (Q_s - Q_r) \tag{22}$$

where r_M, r_H, and r_W are the resistances to the transfer of momentum, heat, and water vapor, respectively, which are inversely related to the surface fluxes. The resistance approach is more commonly used in agricultural meteorology. The main advantage is that the total resistance of a vegetation canopy can be expressed in terms of the resistances of its components (bare soil, leaves, stems, etc.). However, it is not easy to determine and specify some of the component resistances. A comparison of Eqs. (21) and (22) suggests that resistances are inversely related to the product of transfer coefficients and reference wind speed, and these are equivalent relations.

A variety of atmospheric boundary-layers, parameterization schemes have been proposed in the literature for surface fluxes, surface temperature, and surface soil moisture (GARRATT, 1992). Most elaborate models of parameterizing land surface processes in current weather and climate models include subsurface soil layers, snow cover, and simplified vegetation canopy structure and properties (see e.g., DAI et al., 2003).

5. Turbulence in the PBL

Atmospheric motions in the PBL are found to be generally turbulent, although in very stable conditions turbulence is often weak, highly intermittent, and sometimes mixed with internal gravity waves. Turbulence is manifested in the form of very irregular, almost random, three-dimensional fluctuations around some mean motion. The PBL turbulence is also characterized by its efficient mixing, diffusive and dissipative properties, as well as by a wide range of scales and eddies (e.g., length scales from 10^{-3} to 10^4 m).

Using Reynolds' decomposition, any variable in a turbulent flow can be expressed as the sum of its mean and turbulent fluctuation. A statistical description is used to describe the properties of individual fluctuating variables (with zero means) and their correlations with other variables, either at the same time and location in space, or with separations in time and space. All the sophisticated statistical measures, such as probability, correlation and spectrum functions of space and time series of randomly fluctuating variables, are often utilized in the analysis and description of turbulence.

Some observed hour-long time series of velocity, temperature, and humidity fluctuations in an unstable surface layer at a suburban site are shown in Fig. 6 (ROTH, 1991). The corresponding mean values of variables at that time were: $U = 3.66$ m s^{-1}, $V = W = 0$, $\Theta = 294.6$ K; $Q = 8.3$ gm^{-3}. Note the differences in the

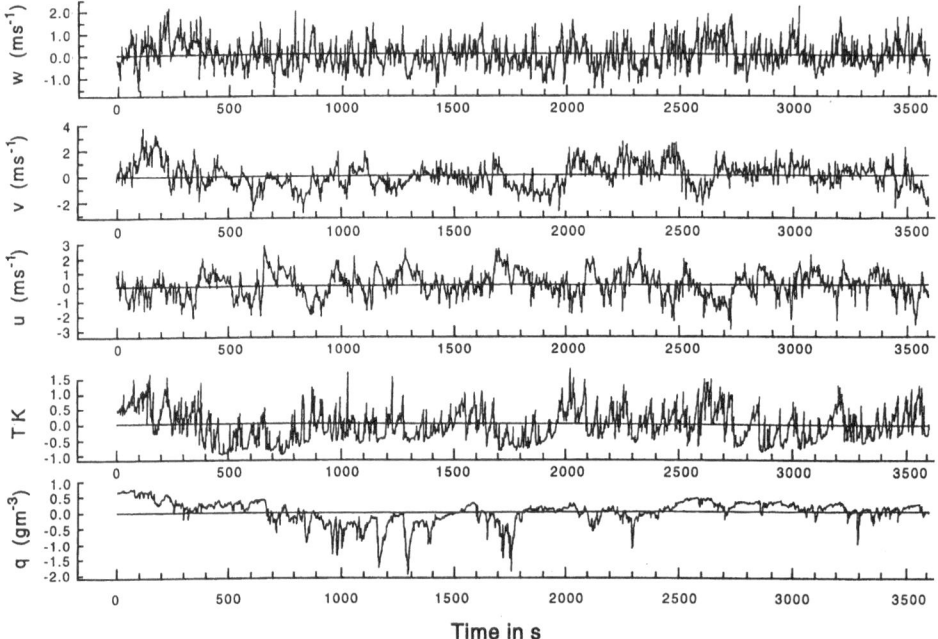

Figure 6
Observed time series of velocity, temperature, and absolute humidity fluctuations in the unstable atmospheric surface layer at a suburban site in Vancouver, Canada (from ARYA, 2001, after ROTH, 1991).

character of traces of horizontal and vertical velocity fluctuations, as well as temperature and humidity fluctuations. In unstable and convective conditions, buoyant plumes and thermals often cause asymmetry between positive and negative fluctuations in vertical velocity, temperature, and humidity (STULL, 1988; ARYA, 2001).

The simplest statistical measures of turbulence (fluctuation levels) are the variances $\overline{u^2}, \overline{v^2}$, etc., the standard deviations $\sigma_u = \left(\overline{u^2}\right)^{1/2}, \sigma_v = \left(\overline{v^2}\right)^{1/2}$, etc., or the turbulent kinetic energy (TKE) per unit mass $E = 0.5(\overline{u^2} + \overline{v^2} + \overline{w^2})$. The relative fluctuation levels in different directions are indicated by turbulence intensities $i_u = \sigma_u \big/ \left|\underset{\sim}{V}\right|$ $i_v = \sigma_v \big/ \left|\underset{\sim}{V}\right|$, and $i_w = \sigma_w \big/ \left|\underset{\sim}{V}\right|$, in which $\left|\underset{\sim}{V}\right|$ represents the mean wind speed.

Measurements of turbulence in the PBL indicate that turbulence intensities are typically less than 10% in the SBL, 10–20% in the near-neutral PBL, and greater than 20% in the CBL. Turbulence intensities are generally largest near the surface and decrease with height as mean wind speed increases and fluctuation levels decrease. However, σ_w is found to increase with height in the lower part of the CBL and attains its maximum value around $z/h = 0.3$, while σ_u and σ_v often show secondary maxima near the inversion base. The TKE generally decreases with height, although not in the same manner under different stability conditions (STULL, 1988). Quite often, in numerical PBL models, the PBL height h is defined as the height level where TKE reduces to 5 or 10% of its maximum value near the surface.

Even more important in micrometeorology are the covariances $\overline{uv}, \overline{vw}, \overline{\theta w}$, etc., which are averages of the products of two fluctuating variables and have magnitudes and signs that depend on the correlations between the two variables. Such negative or positive correlations between u and w, θ and w, and q and w can be visually recognized from their time series in Fig. 6. A more quantitative measure of these correlations is provided by correlation coefficients $r_{uw} = \overline{uw}/\sigma_u\sigma_w$, $r_{\theta w} = \overline{\theta w}/\sigma_\theta\sigma_w$, etc., which always lies between -1 and 1. Covariances can also be interpreted as turbulent fluxes of momentum, heat, etc. Momentum fluxes are directly related to the Reynolds stresses. In particular, one can identify three normal stress components

$$\tau_{xx} = -\rho\overline{u^2}; \quad \tau_{yy} = -\rho\overline{v^2}; \quad \tau_{zz} = -\rho\overline{w^2} \tag{23}$$

and six shear stress components

$$\tau_{xy} = \tau_{yx} = -\rho\overline{uv}; \quad \tau_{xz} = \tau_{zx} = -\rho\overline{uw}; \quad \tau_{yz} = \tau_{zy} = -\rho\overline{vw}, \tag{24}$$

which act on the surface of any cubical fluid element (STULL, 1988). Here the first subscript denotes the direction of normal to the surface and the second subscript denotes the direction of the stress component. In the surface layer, with the x axis along the mean wind, $\tau_{xy} = \tau_{yx} = \tau_{zy} = 0$, and $\tau_{zx} = -\rho\overline{uw}$ is independent of height, so that the surface stress and fluxes of heat and water vapor can be represented as

$$\tau_0 = -\rho\overline{uw}; \quad H_0 = \rho c_p\overline{\theta w}; \quad E_0 = \rho\overline{qw}, \tag{25}$$

which provide the basis for the so-called eddy-correlation method of measuring fluxes. It is more general and direct, but requires sophisticated, fast-response turbulence instrumentation with careful attention to instrument calibration, orientation, and leveling.

6. The PBL Similarity Theories

The observed mean wind, temperature, and specific humidity profiles, as well as turbulence structure (e.g., variances and covariances), in the somewhat idealized quasi-stationary and horizontally-homogeneous PBL are best represented in an appropriate similarity theory framework. We have already described the most widely used Monin-Obukhov similarity theory for the surface layer. The applicability of the M-O similarity theory and scaling is limited to only the homogeneous part of the surface layer, also called the inertial surface layer, which extends from well above (1.5 to 2 h_0) the tops of roughness elements to no more (sometimes, much less) than 10% of the PBL height. For most of the PBL above this shallow surface layer several different similarity theories have been proposed, depending on the prevailing stability and turbulence structure, and the relevant independent variables considered in the formulation of the similarity hypothesis. Some of the more restrictive similarity theories with one or two similarity parameters are found to be more useful in practical applications than the most generalized one with a large number of similarity parameters. Only a few of the simpler PBL similarity theories are reviewed here together with their restrictive assumptions and limitations. In addition to the implied assumptions of quasi-stationarity and horizontal homogeneity, the PBL is assumed to be barotropic and free of active cloud-condensation and precipitation processes. Further restrictions are often placed by the prevailing stability and turbulence regimes for which different similarity theories have been proposed.

6.1. Mixed-layer Similarity Theory

Under moderately unstable and convective conditions, there exists a deep mixed layer within the convective boundary layer (CBL) in which mean potential temperature profile is nearly uniform. Variations in specific humidity and mean wind profiles are also observed to be relatively small over the mixed layer (STULL, 1988; GARRATT, 1992; KAIMAL and FINNIGAN, 1994). Turbulence structure of this convective mixed layer, as well as the related diffusion parameters, are best represented in the framework of the mixed-layer similarity theory, using the mixed-layer height z_i and DEARDORFF's (1972) convective velocity W_* and convective temperature T_*, defined as

$$W_* = \left(\frac{g}{T_{vo}} \frac{H_{vo}}{\rho c_p} z_i \right)^{1/3}; \quad T_* = \frac{H_{vo}}{\rho c_p W_*}. \tag{26}$$

The mixed-layer similarity theory predicts that $\sigma_u/W_*, \sigma_v/W_*$, etc., and other appropriately scaled turbulence statistics must be unique functions of z/z_i only. Actual observations of turbulence in the CBL, including the surface layer, indicate that $\sigma_u/W_* \simeq \sigma_v/W_* \simeq 0.6$, independent of z/z_i. However, the vertical component σ_w/W_* shows a stronger dependence on z/z_i, with a broad maximum in the middle of the CBL (KAIMAL and FINNIGAN, 1994; ARYA, 2001). Approximate empirical relations for the turbulence kinetic energy (E) and its dissipation rate (ε) are:

$$E = 0.36W_*^2 + 0.9 \left(\frac{z}{z_i} \right)^{2/3} \left(1 - 0.8 \frac{z}{z_i} \right)^2 W_*^2, \tag{27}$$

$$\varepsilon = \frac{W_*^3}{z_i} \left(0.8 - 0.3 \frac{z}{z_i} \right). \tag{28}$$

Higher moments of turbulence in the CBL are also found to be well scaled by W_* and T_*. However, the mixed-layer similarity scaling may not be valid in the transition layer above the inversion base, which is influenced by entrainment of free atmospheric air, stable stratification and increased wind shear (STULL, 1988). Statistical moments involving fluctuating temperature, including the downward heat flux at the top of the mixed layer, are also found to depend on the potential temperature gradient in the free atmosphere just above the top of the CBL (SORBJAN, 1996).

6.2. Similarity Theories for the Stratified PBL

An earlier PBL similarity theory proposed by KAZANSKI and MONIN (1961) is based on the addition of the Coriolis parameter f to the list of independent variables considered in the M-O similarity theory. This yielded an additional length scale $u_*/|f|$, which had been recognized as the appropriate height scale for the neutral PBL whose height is given by $h = c\,u_*/|f|$, where $c = 0.2$ to 0.3. The normalized height of the stratified PBL is predicted to be a unique function of the stability parameter $\mu_* = u_*/|f|L$, i.e.,

$$h|f|/u_* = F_h(\mu_*). \tag{29}$$

The similarity theory predicts that the dimensionless ageostrophic wind components $(U - U_g)/u_*$ and $(V - V_g)/u_*$, as well as scaled turbulence parameters $\sigma_u/u_*, \sigma_v/u_*$, etc., must be unique functions of $\xi = |f|z/u_*$ and μ_* only.

Although the Kazanski-Monin (K-M) similarity theory, also known as the Rossby-number similarity theory, was originally proposed for the stratified PBL, regardless of stability conditions, it is only applicable to neutral and stable boundary layers whose heights are not limited by a low-level elevated inversion. For the stable boundary layer (SBL) with continuous turbulent exchange with the surface layer, a simpler explicit form of the function in Eq. (29) has been obtained from scaling considerations (ZILITINKEVICH, 1972) as

$$h|f|/u_* = d\mu_*^{-1/2}$$

or,

$$h = d(u_*L/|f|)^{1/2} \tag{30}$$

in which d is an empirical constant for which estimated values range from 0.2 to 0.6 with an average value of about 0.4. The above similarity theory may not be strictly applicable to an evolving SBL for several hours after the evening transition time (sunset). The SBL evolves rather slowly and rarely reaches a steady state.

Recognizing the importance of the PBL height (h) as an independent height scale and the lack of general validity of Eq. (29), especially in unstable conditions, DEARDORFF (1972) utilized z/h and h/L as dimensionless height and stability parameters. A more detailed formulation and experimental verification of the similarity theory using the above similarity parameters based on the PBL height was presented by ZILITINKEVICH and DEARDORFF (1974). Further consideration of $u_*/|f|$

and baroclinicity added several more similarity parameters. (ARYA and Wyngard, 1975). Thus, a generalized similarity theory for the baroclinic PBL with entrainment in the upper transition layer would have 5 to 6 similarity parameters and may not provide a simple and practically useful framework for representing mean flow and turbulence data.

In practice, the mixed-layer similarity theory is often used for representing the turbulence structure of most unstable and convective boundary layers, while the simpler version of the PBL similarity theory with z/h and h/L as similarity parameters is used for representing mean flow and turbulence variables in slightly unstable to moderately stable boundary layers. For some scaled properties, even the dependence on h/L is found to be weak enough to be ignored. For example, in weakly to moderately stable conditions the turbulence kinetic energy, shear stress and heat flux can be represented by the following similarity expressions:

$$E \simeq 5.75u_*^2\left(1 - \frac{z}{h}\right)^{\alpha_1},$$ (31)

$$\tau = \rho u_*^2\left(1 - \frac{z}{h}\right)^{\alpha_1},$$ (32)

$$H = -\rho c_p u_* \theta_*\left(1 - \frac{z}{h}\right)^{\alpha_2},$$ (33)

in which $\alpha_1 \simeq 1.75$ and $\alpha_2 \simeq 1.5$ (RAO and NAPPO, 1998). The above expressions are also consistent with the local similarity theory, with height-dependent similarity scales, based on the local fluxes of momentum and heat (NIEUWSTADT, 1984).

None of the PBL similarity theories applies to the very stable boundary layer in which turbulence is weak and highly intermittent in both time and space, and upper layers become decoupled from the surface layer.

7. Mathematical Models of the PBL

Most of the mathematical and numerical PBL models are based on the fundamental laws of conservation of mass, momentum, and energy, which are expressed in the form of nonlinear partial differential equations containing velocity, pressure, temperature, etc. (STULL, 1988; ARYA, 2001). These equations are notoriously difficult to solve, even approximately, using standard numerical methods and techniques. The wide variety of PBL models that have been proposed in the literature can be divided into two broad categories, depending on whether any turbulent eddies are explicitly resolved or not; (1) ensemble-averaged turbulence closure models; (2) large-eddy simulation models. There are also different types or orders of turbulence closure models for the PBL.

7.1. First-order Closure PBL Models

Before the advent of digital computers, only the ensemble-averaged conservation equations, utilizing Reynolds' averaging rules, formed the basis of simple analytical PBL models. However, ensemble averaging leads to the turbulence closure problem of more unknowns than the number of equations (STULL, 1988; ARYA, 2001). For example, the equations of mean motion have variances and covariances appearing in them. For the idealized horizontally-homogeneous PBL, these equations can be considerably simplified, but remain unclosed:

$$
\begin{aligned}
\frac{\partial U}{\partial t} - f(V - V_g) &= -\frac{\overline{\partial uw}}{\partial z}, \\
\frac{\partial V}{\partial t} + f(U - U_g) &= -\frac{\overline{\partial vw}}{\partial z},
\end{aligned}
\tag{34}
$$

in which U_g and V_g are the geostrophic wind components in x and y directions, respectively. Similarly, equations for mean potential temperature and specific humidity contain turbulent fluxes of heat and moisture in the vertical direction:

$$
\frac{\partial \Theta}{\partial t} = -\frac{\overline{\partial \theta w}}{\partial z}; \quad \frac{\partial Q}{\partial t} = -\frac{\overline{\partial q w}}{\partial z}.
\tag{35}
$$

In order to close the above set of equations, a variety of local and nonlocal closure hypotheses have been proposed for expressing turbulent fluxes in terms of mean variables. The simplest are the local gradient-transport (K-theory) relations, based on eddy viscosity (diffusivity) and mixing length hypotheses (ARYA, 2001). A comprehensive review of the first-order local closure models of the PBL is given by HOLT and RAMAN (1988). These are based on explicit or implicit specifications of eddy diffusivities or mixing length in the PBL, largely on an *ad hoc* basis. The basic premise of down-gradient transport of flux can, perhaps, be justified in neutral and stable boundary layers, but becomes generally invalid in a convective mixed layer.

For unstable and convective boundary layers, several nonlocal mixing models have been proposed in the literature (STULL, 1988, 1993; PLEIM and CHANG, 1992). Nonlocal formulations for turbulent fluxes are based on the premise that turbulent eddies of different sizes can transport flow properties across finite vertical distances, comparable to their size. Thus, the largest eddies in the CBL can transfer properties all the way from the surface to the top of the CBL and *vice versa*. STULL (1993) has proposed a comprehensive nonlocal mixing model based on his transilient turbulence theory. A vertical column of air is divided into a finite, but large, number of equal size grid boxes. Relative contributions to the turbulent flux at any level from different grid boxes depend on the mean property differences and coefficients of the transilient matrix. The main problem is an unambiguous and accurate specification of those coefficients. Nonlocal mixing models are also computationally more expensive, although certain simplifications aimed at reducing computational cost have been proposed (PLEIM and CHANG, 1992).

By far, the simplest models of the mixed-layer dynamics and structure are the so-called integral or slab models, which are based on the vertically averaged equations of motion and scalars (STULL, 1988). A separate rate equation is used to calculate the boundary layer height. The approach is particularly suited to unstable and convective boundary layers, especially when layer averaged winds are acceptable. With increasingly cheaper and faster computers available for higher resolution numerical models of the PBL, the use of this approach has declined.

7.2. Higher-order Closure Models

Since none of the first-order closure models was deemed to be satisfactory for the entire range of stability conditions encountered in the PBL, a variety of second-order and a few third-order closure models, based on dynamic equations for second and third moments, have been proposed and tested against available PBL data. Higher-order closure models are physically more realistic and have the advantage of computing turbulent variances, covariances, TKE, and dissipation rates, in addition to the mean variables. However, they involve a host of empirical closure constants and are computationally considerably more expensive than the first-order closure models. A good compromise between the two is provided by the so-called TKE models or one and half-order closure models (STULL, 1988; HOLT and RAMAN, 1988; ARYA, 2001). The approach contains more physics than the conventional first-order closure and provides a rational basis for calculating eddy viscosity in terms of the TKE (E) and the rate of energy dissipation (ε) or mixing length (l). Nonetheless, it still employs the local gradient-transport (K-theory) relations for turbulent fluxes. In addition to the mean-field equations, a prognostic equation for E and a prognostic or diagnostic equation for ε or l are carried in the model. HOLT and RAMAN (1988) have reviewed different types of E-ε, E-l, and parameterized length-scale models. The simplest one is the parameterized length-scale model for the idealized PBL, based on the following equations:

$$\frac{DE}{Dt} = K_m \left[\left(\frac{\partial U}{\partial z} \right)^2 + \left(\frac{\partial V}{\partial z} \right)^2 \right] - \frac{g}{T_{vo}} K_h \frac{\partial \Theta_v}{\partial z} + \frac{\partial}{\partial z} \left(K_E \frac{\partial E}{\partial z} \right) - \varepsilon \qquad (36)$$

$$K_m = C l E^{1/2}; \quad \varepsilon = C_\varepsilon E^{3/2} / l_\varepsilon \qquad (37)$$

in which the total derivative includes the advection of the TKE by the mean flow, C and C_ε are empirical constants, and l and l_ε are the mixing and dissipation lengths which are parameterized or specified (HOLT and RAMAN, 1988). The eddy diffusivities of heat (K_h) and the TKE (K_E) are also specified in terms of K_m (e.g., $K_h = \alpha K_m$, where $\alpha = 1$ to 2, depending on stability, and $K_E = K_m$). Equation (36) is a simplified version of the full TKE equation in which pressure transport and turbulent transport are combined and parameterized using the gradient-transport relation.

7.3. Large-Eddy Simulation Models

The large-eddy simulation (LES) approach has led to the most sophisticated numerical models of the PBL to date. More fundamental, direct numerical simulation (DNS) of turbulence is still limited to low Reynolds-number flows and is not considered to be practically and computationally feasible for environmental turbulent flows such as the PBL. In contrast to DNS, which aims to resolve all turbulent eddies, including the microscale motions which dissipate the TKE, LES attempts to resolve and faithfully simulate only the large energy-containing eddies which are responsible for most turbulent transports occurring in the flow. The small subgrid-scale motions are not resolved, but their important contributions to the rate of energy dissipation and somewhat minor contributions to turbulent transports are usually parameterized through the use of simpler subgrid scale (SGS) models (DEARDORFF, 1972; MASON, 1994).

Three-dimensional LES modeling of the PBL started with the pioneering work of DEARDORFF (1972). The usefulness and validity of this computer-intensive numerical modeling has been amply demonstrated for unstable and convective PBLs, even in the presence of moist convection and clouds. A detailed intercomparison study of four different LES models of the CBL, using different SGS models, computer codes and techniques, showed good consistency of their simulation results (NIEUWSTADT et al., 1992). The observed insensitivity of LES results to different SGS models is probably due to the relatively small SGS fluxes and TKE, as compared to the resolved fluxes and TKE in LES models. This is not the case, however, for LES models of neutral and stable boundary layers whose results are found to be more sensitive to initial conditions, time of simulation, the sophistication of the SGS model and its ability to accurately parameterize the observed flux-gradient relations in the surface layer (ANDREN et al., 1994; ARYA, 2001). To date, LES studies of the SBL have been limited to conditions of mild to moderate stability, strong wind shears, and more or less continuous turbulence across the SBL. Further applications of LES to very stable conditions, with frequent episodes of turbulence generation and destruction, should be expected in the near future.

Despite the above-mentioned limitations, LES can be viewed as the best available approach for PBL modeling. It has greatly enhanced our understanding of the complex-turbulent transport and diffusion processes in the PBL. Due to its large computational costs, however, LES will most likely remain a research tool, rather than being routinely used in operational models of weather and air quality.

REFERENCES

ANDREN, A., BROWN, A., GRAF, J., MOENG, C. H., MASON, P. J., NIEUWSTADT, F. T. M., and SCHUMANN, V. (1994), *Large-eddy Simulation of Neutrally-stratified Boundary Layer: A Comparison of Four Computer Codes*, Q. J. Roy. Meteorol. Soc. *120*, 1457–1484.

ARYA, S. P., *Air Pollution Meteorology and Dispersion* (Oxford University Press, New York, 1999).

ARYA, S. P., *Introduction to Micrometeorology*, Second Edition (Academic Press, San Diego, 2001).

ARYA, S. P. S. and WYNGAARD, J. C. (1975), *Effect of Baroclinicity on Wind Profiles and the Geostrophic Drag Law for the Convective Planetary Boundary Layer*, J. Atmos. Sci. *32*, 767–778.

BRUTSAERT, W. H., *Evaporation into the Atmosphere* (Reidel, Dordrecht, The Netherlands, 1982).

DAI, Y., ZENG, X., DICKINSON, R. E., BAKER, I., BONAN, G. B., BOSILOVICH, M. G., DENNING, A. S., DIRMEYER, P. A., HOUSER, P. R., NIU, G-Y., OLESON, K. W., SCHLOSSER, C. A., and YANG, Z.-L. (2003), *The Common Land Model*, Bull. Am. Meteorol. Soc. *84*, 1013–1023.

DEARDORFF, J. W. (1972), *Numerical Investigation of Neutral and Unstable Planetary Boundary Layers*, J. Atmos. Sci. *29*, 91–115.

DEARDORFF, J. W., *Observed characteristics of the outer layer*. In *Short Course on the Planetary Boundary Layer* (A. K. Blackadar, ed.) (American Meteorological Society, Boston, MA, 1978).

GARRATT, J. R., *The Atmospheric Boundary Layer* (Cambridge University Press, Cambridge, UK, 1992).

GEIGER, R., ARON, R. H., and TODHUNTER, P. *The Climate Near the Ground*, 5th Edition, (Viehweg, Weisbaden, Germany, 1995).

GRIMMAND, C. S. B. and OKE, T. R. (1995), *Comparison of Heat Fluxes from Summertime Observations in the Suburbs of four North American Cities*, J. Appl. Meteorol. *34*, 873–889.

HOLT, T. and RAMAN, S. (1988), *A Review and Comprehensive Evaluation of Multilevel Boundary Layer Parameterizations for the First-order and Turbulent Kinetic Energy Closure Schemes*, Rev. Geophys. *26*, 761–780.

HICKS, B. B. (1976), *Wind Profiles Relationships from the 'Wangara' Experiment*, Q. J. Roy. Meterol. Soc. *102*, 535–551.

HOLTSLAG, A. A. M. (1984), *Estimates of Diabatic Wind Speed Profiles from Near-surface Weather Observations*, Boundary-layer Meteorol. *29*, 225–250.

IZUMI, Y. and BARAD, M. L. (1963), *Wind and Temperature Variations during Development of a Low-level Jet*, J. Appl. Meteorol. *2*, 668–673.

KAIMAL, J. C. and FINNIGAN, J. J., *Atmospheric Boundary Layer Flows* (Oxford University Press, New York, 1994).

KAZANSKI, A. B. and MONIN, A. S. (1961), *On the Dynamical Interaction between the Atmosphere and the Earth's Surface*, Izv. Akad, Nauk, SSSR, Geophys. Ser. *5*, 514–515.

MAHRT, L. (1981), *The Early Evening Boundary Layer Transition*, Q. J. Roy, Meteorol. Soc. *107*, 329–343.

MAHRT, L. (1999), *Stratified Atmospheric Boundary Layers*, Bound.-Layer Meteorol. *90*, 375–396.

MASON, P. J. (1994), *Large-eddy Simulation: A Critical Review*, Q. J. Roy. Meteorol. Soc. *120*, 1–26.

MONIN, A. S. and YAGLOM, A. M., *Statistical Fluid Mechanics: Mechanics of Turbulence*, vol. 1 (MIT Press, Cambridge, 1971).

MUNN, R. E., *Descriptive Micrometeorology* (Academic Press, New York, 1966).

NIEUWSTADT, F. T. M. (1984), *Turbulence Structure of the Stable Nocturnal Boundary Layer*, J. Atmos. Sci., *41*, 2202–2216.

NIEUWSTADT, F. T. M., MASON, P. J., MOENG, C. H., and SCHUMANN, U. (1992), *Large-eddy simulation of the convective boundary layer: A comparison of four computer codes*. In *Turbulent Shear Flows* (F. Durst et al., eds.) 8, (Springer-Verlag, Berlin, 1992) pp. 343–367.

OKE, T. R., *Boundary Layer Climates*, Second Edition (Methuen, New York, 1987).

OKE, T. R. (1988), *The Urban Energy Balance*, Progress in Phys. Geog. *12*, 471–508.

PLEIM, J. E. and CHANG, J. S. (1992), *A Non-local Closure Model for Vertical Mixing in Convective Boundary Layer*, Atmos. Environ. *26A*, 965–981.

RAO, K. S. and NAPPO, C. J., *Turbulence and dispersion in the stable atmospheric boundary layer*. In *Dynamics of Atmospheric Flows: Atmospheric Transport and Diffusion Processes* (M. P. Singh and S. Raman, eds) (Computational Fluid Mechanics Publishers, Southampton, U. K., 1998) pp. 39–91.

ROSENBERG, N. J., BLAD, B. L. and VERMA, S. B., *Microclimate: The Biological Environment*, Second Edition (Wiley Interscience, New York, 1983).

ROTH, M., *Turbulent Transfer Characteristics over a Suburban Surface*, Ph. D. Thesis, (University of British Columbia, Vancouver, Canada, 1991).

SORBJAN, Z., *Structure of the Atmospheric Boundary Layer* (Prentice Hall, Englewood Cliffs, New Jersey, 1989).

SORBJAN, Z. (1996.), *Numerical Study of Penetrative and Nonpenetrative Convective Boundary Layers*, J. Atmos. Sci. *53*, 101–112.

STULL, R., *An Introduction to Boundary Layer Meteorology* (Kluwer, Dordrecht, The Netherlands, 1988).

STULL, R. B. (1993), *Review of Transilient Turbulence Theory and Nonlocal Mixing*, Bound.-Layer Meteorol., *62*, 21–96.

ZILITINKEVICH, S. S. (1972), *On the Determination of the Height of the Ekman Boundary Layer*, Boundary-Layer Meteorol. *3*, 141–145.

ZILITINKEVICH, S. S. and DEARDORFF, J. W. (1974), *Similarity Theory for the Planetary Boundary Layer of Time-dependent Height*, J. Atmos. Sci. *31*, 1449–1452.

(Received October 20, 2003, accepted February 9, 2004)
Published Online First June 8, 2005

To access this journal online:
http://www.birkhauser.ch

Pure appl. geophys. 162 (2005) 1747–1778
0033–4553/05/101747–32
DOI 10.1007/s00024-005-2691-x

© Birkhäuser Verlag, Basel, 2005

❘Pure and Applied Geophysics

Characteristics of Stable Flows over Southern Greenland

ANDREW ORR,[1] EDWARD HANNA,[2] JULIAN C. R. HUNT,[1] JOHN CAPPELEN,[3]
KONRAD STEFFEN,[4] and AG STEPHENS[5]

Abstract — The main characteristic features of stable atmospheric flows over a large mountain plateau are summarised and then compared with mesoscale and synoptic scale numerical simulation, meteorological analysis, satellite imagery, and surface observations for the cases of flows over Southern Greenland for four wind directions. The detailed features are identified using the concepts and scaling of stably stratified flow over large mountains with variations in surface roughness, elevation, and heating. For westerly and easterly winds detached jets form at the southern tip, where coastal jets converge, which propagate large distances across the ocean. Near coasts katabatic winds can combine with barrier jets and wake flows generated by synoptic winds. Note how the approach flow rises/falls over southern Greenland for easterly/westerly winds, leading in both cases to more cloud on the western side. Some conclusions are drawn about the large-scale influences of these flows; detached jets in the atmosphere; air-sea interaction; formation of low pressure systems. For accurate simulations of these flows, mesoscale models are necessary with resolutions of order of 20 km or less.

Key words: Greenland, Wind-jets, Air-sea interaction, Synoptic flows.

1. Introduction

Greenland's vast size, remoteness, and inhospitable climate, mean meteorologically it is one of the least studied areas on Earth, especially on the mesoscale (10–100 km) (PUTNINS, 1970; SCORER, 1988). This is despite it being the largest ice-capped structure in the Northern Hemisphere (see Fig. 1). It is known that Greenland greatly influences the climate of northwest Europe as cold, stable air from the Arctic Ocean and the predominantly westerly flow across the North Atlantic is partially blocked (SCORER, 1988). Greenland is critically sensitive to climate change (e.g., HANNA and CAPPELEN, 2003). It is very sensitive to air-sea-ice coupling effects which have a

---

[1] Centre for Polar Observation and Modelling, Dept. of Space and Climate Physics, University College London, UK. Present affiliation: ECMWF, Shinfield Pk, Reading, UK

[2] Department of Geography, University of Sheffield, UK.

[3] Danish Meteorological Institute, Copenhagen, Denmark.

[4] Cooperative Institute for Research in Environmental Sciences, University of Colorado, U.S.A.

[5] British Atmospheric Data Centre, Rutherford Appleton Laboratory, UK.

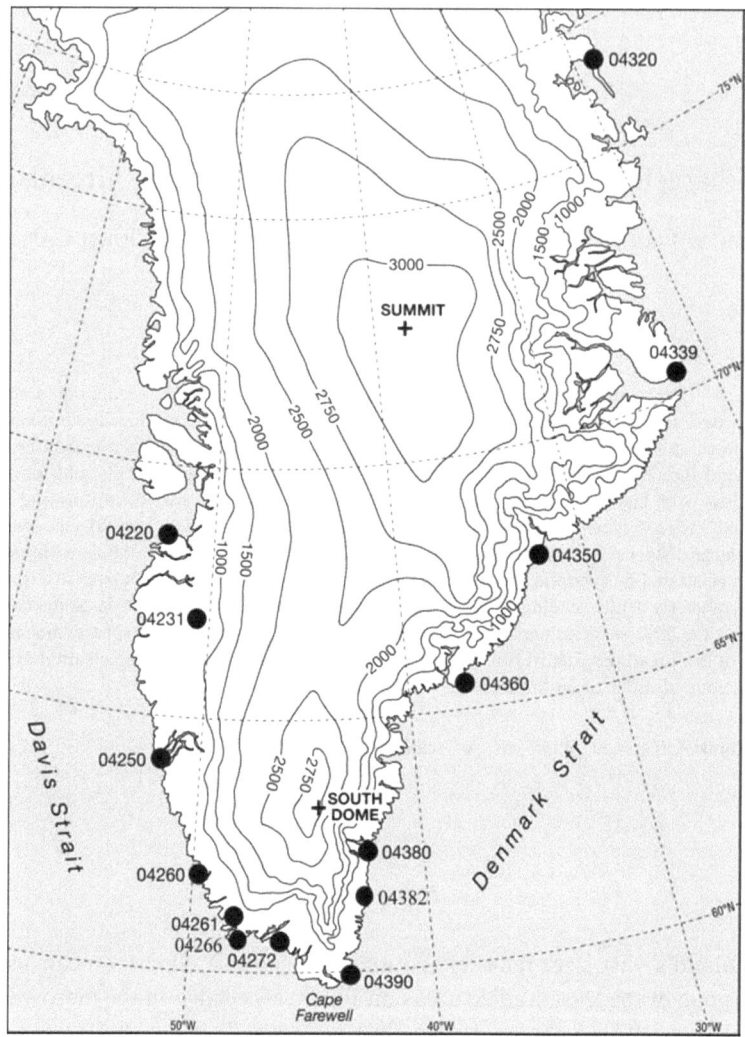

Figure 1
Map of southern Greenland showing orography elevation contours (in meters) and locations of the DMI
and GC-Net stations used in this study.

marked effect on seasonal weather variations in northwest Europe (COLEMAN and
DAVEY, 1999). For example, it acts as a large atmospheric heat sink through ice
albedo feedback. And katabatic winds which flow from the broad gently sloping
interior to the steep coastal margins are important to the energy balance of melting
ice in the ablation zone during summer (MEESTERS, 1994).

The atmospheric flows and processes over Greenland occur on many length
scales, from the microscale, to the mesoscale, to synoptic scales. Flow over and

around Greenland is affected by (i) the large and initially very steep elevation change between the coastal margins and the central plateau (~3 km; see Fig. 1), (ii) the combination of very rough surfaces (roughness length $z_0 \sim 1 - 10$ m) and jagged mountains around the coasts, (iii) the strong katabatic flows from the plateau down to the coasts (~ 100 km) (BROMWICH et al., 1996; KLEIN and HEINEMANN, 2002), (iv) the presence of fjords, such as Kangerlussuaq and Sermilik located in the southwest coastal margins (RASMUSSEN, 1989; KLEIN et al., 2001), (v) the presence of the semi-permanent Icelandic Low (~ 1000 km) (SCORER, 1988), (vi) air-sea-ice interaction processes (STAMMERJOHN et al., 2003), and (vii) associated boundary layer processes (SMEDMAN et al., 2003). Here there are extremely sharp gradients in roughness and elevation. These result in local scale phenomena that have long been observed, for example, coastal wind jets, but as they occur on kilometre length scales, they are only described by numerical mesoscale models when they are run with fine resolution (e.g., CAPON, 2003). A high spatial resolution is also required to adequately resolve the steep orography near the coast (GLOVER, 1999). These local scale phenomena can have large-scale climate effects, e.g. drag, wind waves, downwelling, cyclogenesis, air-sea-ice interaction, etc.

To help define the magnitudes and spatial and temporal scales of these types of stable, mesoscale flows, HUNT et al., (2001) applied a general linearised perturbation model to atmospheric flows over large mountains, in order to predict the overall flow, particularly deflection, waves, and wakes. This model was modified for atmospheric flows where stable inversion layers pass over surface undulations and areas of changing rough resistance (HUNT et al., 2004). Both models showed that there are marked positive and negative perturbations in mean velocity (i.e., wind jets) that decrease gradually over transverse length scales of the order of the Rossby deformation radius L_R (30–300 km). These low-level jets, which do not occur in the absence of Coriolis effects, also exist in wake flows. These jets are associated with rising and falling airstreams which induce changes to cloudiness and precipitation. General concepts arising from these studies are reviewed in section 2.

With regard to Greenland, understanding of these jets is limited, and has mainly focused on the 'Greenland tip jet', a low-level wind jet that forms in the lee of Cape Farewell (see Fig. 1) under westerly or north-westerly synoptic conditions (DOYLE and SHAPIRO, 1999). Recently PICKART et al., (2003) linked deep convection in the Irminger Sea to the southeast of Greenland with forcing by this jet. Other studies have included the wind regime over the Greenland ice sheet, and in particular katabatic flow (e.g. BROMWICH et al., 1996; KLEIN and HEINEMANN, 2002), barrier jets (VAN DEN BROEKE and GALLEE, 1996), and wake flows (PETERSEN et al., 2003).

Weather stations in Greenland are sparse and mostly confined to the coastal margins, predominately the southwest coast. Most of the coastal stations are run by the Danish Meteorological Institute (DMI) and have records going back between several decades and just over a century (CAPPELEN et al., 2001; JORGENSEN and LAURSEN, 2003). Within the last few years an inland network of around 20 automatic weather stations, the Greenland Climate Network (GC-Net), has been established by Konrad Steffen and colleagues (STEFFEN and BOX, 2001). Imagery collected from polar orbiting satellites can give a qualitative analyses of the surface wind patterns. For example RASMUSSEN (1989) showed that strong katabatic winds can result in a dark signature on thermal infrared satellite imagery, arising because of the turbulent mixing bringing warm upper air to the ice-covered surface, allowing the areas covered by such flows to be estimated.

Case studies for four contrasting synoptic conditions over Greenland are studied here using numerical modelling and observations. For westerly, easterly, southerly, and northerly approach winds the flows around Greenland can be approximately related to conceptual models for idealised simulations. Also they provide information which is representative of a wide number of 'in-between' synoptic conditions. The numerical model simulations combine mesoscale simulations with a horizontal resolution of approximately 12 km and ECMWF operational analysis data at $0.5° \times 0.5°$ resolution. The observations combine satellite imagery and the ground-based observations discussed above. The satellite pictures were obtained from a AVHRR (Advanced Very High Resolution Radiometer) onboard a polar orbiting satellite. AVHRR channel 4 (thermal infrared) was employed. Observations are included for the purpose of comparison, to corroborate model results, and to give a more comprehensive overview of meteorological conditions prevailing over southern Greenland during each of the case-study days.

The objective of this study is to further understand the influence of high orography on stable flows, and with regard to Greenland's climate processes, to improve the conceptual understanding of mesoscale meteorology of stable flows over and around Greenland: specifically the existence of low-level wind jets, their implications for atmosphere-ocean coupling, and cloudiness and precipitation. Another aspect of this work is to help improve the representation or parameterisation of mesoscale processes in numerical weather/climate prediction models (WOOD and MASON, 1993; LOTT and MILLER, 1997).

Section 2 will review low-Froude number atmospheric flows over large mountains. Section 3 will analyse the synoptic conditions for each case study. Section 4 describes the numerical mesoscale model used. Section 5 examines the flow over southern Greenland using the numerical modelling results and observations. Section 6 examines air-sea interaction. Section 7 examines cloudiness and precipitation over southern Greenland. Section 8 is a discussion.

2. Stable Flows over Large Mountains

The influence of high orography on stable flows and the local weather have long been recognised by meteorologists (e.g., BLUMEN, 1990). The significance of the additional effects of the Earth's rotation have only been recognised more recently (e.g., HUNT et al., 2001; PIERREHUMBERT and WYMAN, 1985; BOYER and DAVIES, 2000). As the scales of the mountains increase to those of around continental scale plateaus, new phenomena in the characteristic wind patterns arise (see BOYER and CHEN, 1987) which are not currently parameterised or predicted by Numerical Weather Prediction (NWP) models. We focus here on north-south mountains/ plateaus situated in prevailing winds, such as Greenland, the Rockies, the Southern Andes, etc. These flows are also complicated by the fact that the oncoming flow may be far from uniform in magnitude and direction over these vast scales. This causes weather systems to behave somewhat like turbulent eddies passing over or around an obstacle.

Great progress has been made in understanding and quantifying the effects of stable stratification (defined by the Brunt-Väisälä frequency N) on flow of velocity U around isolated mountains of height H, characterised by the Froude number $\mathbb{F}_H = U/NH$ which varies from ∞ to 0. This parameter characterises the strength of the inertial to buoyancy forces. For the cold, stable flows and high mountains and plateaus of interest here ($H \sim 2000 - 3000$ m; and a typical value of $N \sim 0.01$ s^{-1}), $\mathbb{F}_H$ typically lies in the range 0.2 to 0.5.

2.1. Low-Froude Number Flows without Rotation

We first consider low-Froude number flows where the effects of the Earth's rotation have a small influence on the flow near the mountain. (These effects are characterised by the Rossby number Ro ($= U/fD$, where f is the Coriolis parameter and D is the half-width of the mountain), i.e. Ro $\gg$ 1). Note also the Rossby Radius of Deformation L_R ($= HN/f$). For a mountain of mesoscale dimension or smaller $B, D < L_R$ (where B is the half-length of the mountain). In this situation if $\mathbb{F}_H$ is less than about 0.5, the flow splits into the top and middle layers above and below the level z_d of the 'dividing streamline'. This lies at a distance $H_T \sim \alpha \mathbb{F}_H H$ below the top of the mountain (where $\alpha \sim 1$). z_d also varies slightly with the slope of the mountain, shear (i.e. $dU(z)/dz$) in the oncoming flow, and aspect ratio of the mountain (SNYDER et al., 1985; BAINES, 1995; ÓLAFSSON and BOUGEAULT, 1996). See Fig. 2(a).

In the upper layer above z_d the streamlines pass over the mountain, but asymmetrically; ascending on the upwind side where the velocity decreases and descending on the downwind side close to the surface, where the velocity increases and internal waves are generated. In the middle layer below z_d the air flow passes in approximately horizontal planes around the mountain; as it separates on the lee side large recirculating eddies are formed. Lee waves propagate above the mountain and

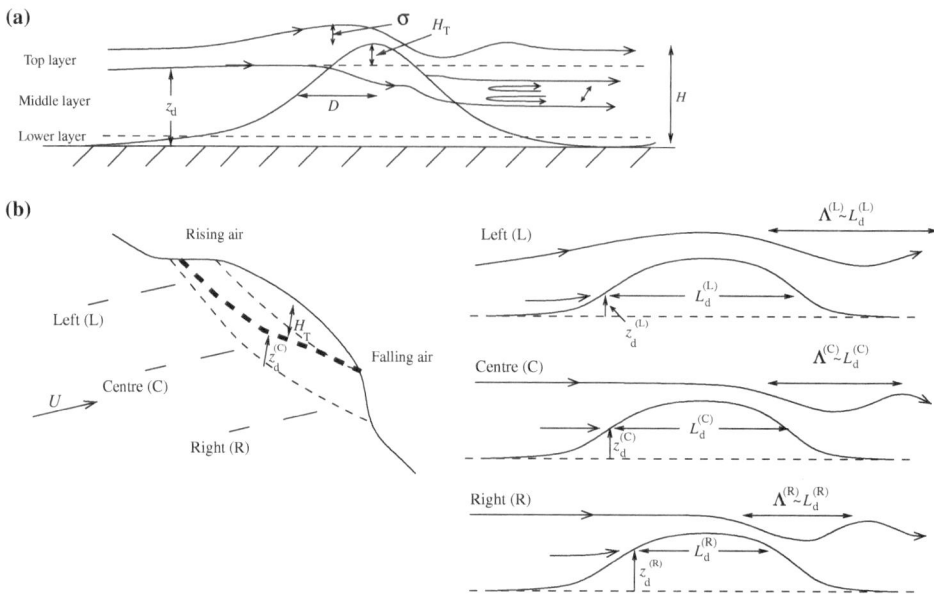

Figure 2

(a) Schematic diagram for low-Froude number flow over a mountain of mesoscale dimension with negligible Coriolis forces (i.e., Ro ≫ 1). D is the half-width of the mountain and H its height. z_d is the level of the 'dividing streamline'. $H_T \sim \alpha \mathbb{F}_H H$ is the distance between the dividing streamline and the mountain top. σ is the vertical streamline deflection. (b) Schematic diagram for low-Froude number flow over a long mountain with significant Coriolis forces (i.e., Ro $\sim 1 - 10$) in the northern hemisphere. Here the air flow rises on the left (looking in the downwind direction) and falls on the right. Consequently z_d is smaller and the wavelength of the lee waves, Λ, is longer on the left than the right, i.e. $z_d^{(R)} > z_d^{(C)} > z_d^{(L)}$ and $\Lambda^{(L)} > \Lambda^{(C)} > \Lambda^{(R)}$. L_d is the length of the hill (in the flow direction) at the dividing streamline height and $L_d^{(L)} > L_d^{(C)} > L_d^{(R)}$.

also down into the middle layer. In the interface between the layers intense shearing motions occur, especially on the downwind side. A third layer (the base or lower layer) is also shown. This lies beneath the middle layer and so here the air must flow around the mountain. See also Fig. 2(a).

No complete analytical model has been derived for the nonlinear flows in the upper and middle layers, though high precision numerical simulations (e.g., SMOLARKIEWICZ and ROTUNNO, 1989; DING and STREET, 2003) and an approximate model of HUNT et al., (1997) have described the main features of the flows observed experimentally in the laboratory and the field (e.g. SNYDER et al., 1985; VOSPER et al., 1999).

2.2. Low-Froude Number Flows with Rotation

For low-Froude number flows where the effects of rotation influences the flow near the mountain (characterised by small values of the Rossby number Ro, i.e., Ro $\sim 1 - 10$), even when Coriolis forces are weak compared to the inertial or buoyancy forces, the flow regime alters substantially and the far-field flow decays slowly on a length scale of order L_R (MERKINE, 1975; NEWLEY, et al., 1991).

For a long mountain when $B \gtrsim L_R$ and $\mathbb{F}_H$ is less than about 0.5 the air flow rises on the left and sinks on the right in the northern hemisphere (looking in the downwind direction) (HUNT et al., 2001). See Fig. 2(b) (based on computations of ÓLAFSSON and BOUGEAULT (1997)). This means that more low level dense air passes over the mountain, causing longer wavelength (Λ) lee waves, and a lower value of z_d, on the left than on the right. A most important feature of this left-right asymmetry and the general vertical deflection of the streamlines over the entire depth of the mountain is that greater drag is produced by the right-hand side of the mountain. Also the drag force per unit width of the mountain is of order $\rho N U H^2 \sim \rho U^2 H / \mathbb{F}_H$ (where ρ is the density of the air), which is considerably greater than the drag force of order $\rho U^2 H$ produced by the top and middle layers of a mountain when $L_R \gg B$ or Ro $\gg 1$ (HUNT et al., 2001; see also LOTT and MILLER, 1997).

When the Coriolis effects become even greater, so that $B/L_R \gtrsim 1/\mathbb{F}_H$, the perturbed lateral pressure gradient ($\partial p / \partial y$) associated with a large stagnant wake (e.g., $B \gtrsim 500$ km) becomes too great to be balanced by the hydrostatic pressure perturbation (Δp) associated with the vertical streamline deflection (σ), i.e., $\Delta p \sim fUB \sim \sigma N^2 H$, and since the deflection downwards cannot exceed H (i.e., $\sigma < H$), this means that $\sigma/H \sim B\mathbb{F}_H/L_R < 1$. Therefore if $B/L_R \gg 1$, to limit the vertical deflection to a physically realistic value (i.e., $\sigma < H$) the flow structure has to change so as to reduce the lateral pressure gradient (again following HUNT et al., 2001).

For a mesoscale mountain or range of mountains in the northern hemisphere where $B \sim L_R$ (e.g. Pyrenees, Alps), if the approach flow is uniform this change must result firstly in flow around the left-hand side of the mountains and secondly in the large wakes or recirculation regions or synoptic eddies becoming detaching from the mountains and moving to a location downwind of the right-hand side of the mountain. This is also consistent with most of the drag being associated with the flow over the right hand part of the mountain. At the same time, because $\mathbb{F}_H < 0.5$, the approach flow streamlines still divide at height z_d between those that pass over (a 'cut-off hill') and those that pass around the mountain. A precise way to define changing flow patterns over obstacles (whether aeronautical or geophysical) is to note the location of the 'singular' points on the surface where the mean velocity is zero, and also whether they are saddle points or node points (HUNT et al., 1978; HUNT and SNYDER, 1980). See Fig. 3(a).

For a synoptic scale range of mountains where $B \gg L_R$ (e.g., Greenland, Rockies) and with uniform approach flow the flow pattern changes to that shown in Fig. 3(b)(i). The saddle point S_2 that was on the downwind side of the dividing streamline in Fig. 3(a) now moves away from the mountain to the left-hand side of the detached synoptic eddies. At the same time the downwind saddle point S_3 moves to the right-hand side of the eddies. The main synoptic feature of this kind of flow is that it deflects the average flow to the right in the northern hemisphere, which leads to large

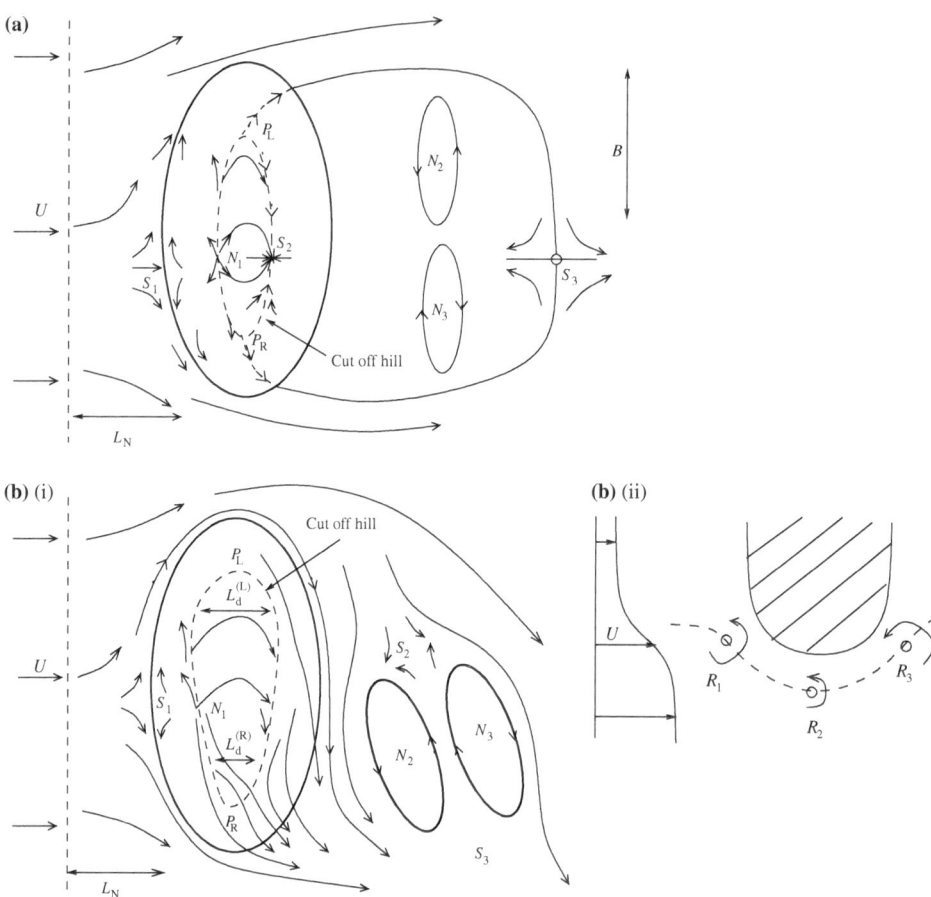

Figure 3

Near surface streamlines for low-Froude number flow in the northern hemisphere over (a) a mesoscale mountain ($B \sim L_R$) and (b) a synoptic scale mountain ($B \gg L_R$). The dividing streamline on the surface (shown as dashed line) is an attachment line upwind of P_L, P_R. S_1, S_2, S_3 are saddle points on the surface. N_1, N_2 are 'nodes'. The area defined as 'cut off hill' (shown as dashed line) is above the dividing streamline. The outline of the mountain is indicated by the thick black line. The distance L_N is the upstream deflection. In (a) the flow is greater around the left-hand side of the mountain and two recirculation regions are situated downwind of the mountain. In (b)(i) the width of the top region is greater on the left than on the right because of Coriolis effects (i.e. $L_d^{(L)} > L_d^{(R)}$). Note the separated detached pair of recirculating cyclones (eddies), which extend up to the height z_d of the dividing streamline. In (b)(ii) the local effect is sketched of north–south shear (with cyclonic vorticity) in the approach flow.

'lift' force on the mountains to the left (HOZUMI and UEDA, 2004). This changed flow pattern was observed in laboratory experiments by BOYER and CHEN (1987) and idealised numerical simulations of ÓLAFSSON (2000).

Fig. 3(b)(ii) sketches the flow pattern if the approach flow is non-uniform. In this case southerly coastal flows are produced on the upwind side of the mountain and northerly coastal flows on the lee side of the mountain. This can lead to concentrations of vorticity at R_1 and R_3, where the velocity decreases, but a decreased concentration of vorticity at R_2. This produces velocity *perturbations* that are in a southerly direction at R_1 and northerly at R_3, which affect the basic flows depicted in Fig. 3(a) and (b)(i).

This changed flow pattern was observed in laboratory experiments by BOYER and CHEN (1987) and idealised numerical simulations of ÓLAFSSON (2000). It is consistent with atmospheric simulations and observations of PETERSEN et al. (2003).

Note that if the atmosphere becomes less stable (increase in Froude number), as occurs in periods of global warming, the flow is changed quite markedly with most of the streamlines passing over the mountains, the recirculating regions becoming smaller, and the overall rightward flow deflection being greatly reduced. This may have considerable implications for the future climate to the east and southeast of Greenland and the Rocky Mountains. When the wind is directed at an acute angle to the mountain and when the approach flow is non-uniform, more complex flow patterns are expected. This is why realistic case studies are necessary.

3. Synoptic Explanation of Case Studies

Below is a short explanation of the synoptic conditions for each case study. Conditions were representative for approximately a 24-hour period. These are based on interpolation and NWP calculations, i.e., 'analysis' of UK Met Office mean sea-level pressure (MSLP). In addition, spot readings are given for representative DMI stations on the southwest coast of Greenland (04260 Paamiut and 04266 Nunarsuit), the east coast (04360 Tasiilaq), the southeastern coast (04380 Timmiarmiut and 04382 Ikermiuarsuk) and the extreme southern coast (04390 Prins Christian Sund). Data are also given for the GC-Net station South Dome (altitude 2901 m) in the southern interior. The data used are temperature at a height of 2 m and the wind velocity and direction at a height of 10 m (note for GC-Net data this height can vary depending on the amount of snow). See Fig. 1 for the locations of the stations.

3.1. Westerly Case Study (7 January, 2002)

Westerly flows are relatively rare. With prevailing easterly and northerly winds over southern Greenland, it was quite difficult to find a clear-cut westerly flow, there just being one or two cases per year.

On the 7th of January, 2002 the MSLP analysis for 00UTC (see Fig. 4(a)) shows low pressures southeast, west and east of Greenland, including in the Labrador Sea, and a weak high pressure (~992 mb) centred over the ice sheet. The infrared satellite picture for approximately 05UTC (see Fig. 4(b)) shows quite a lot of cloud in sea areas west and south of Greenland and wisps of cloud are evident over the ice sheet

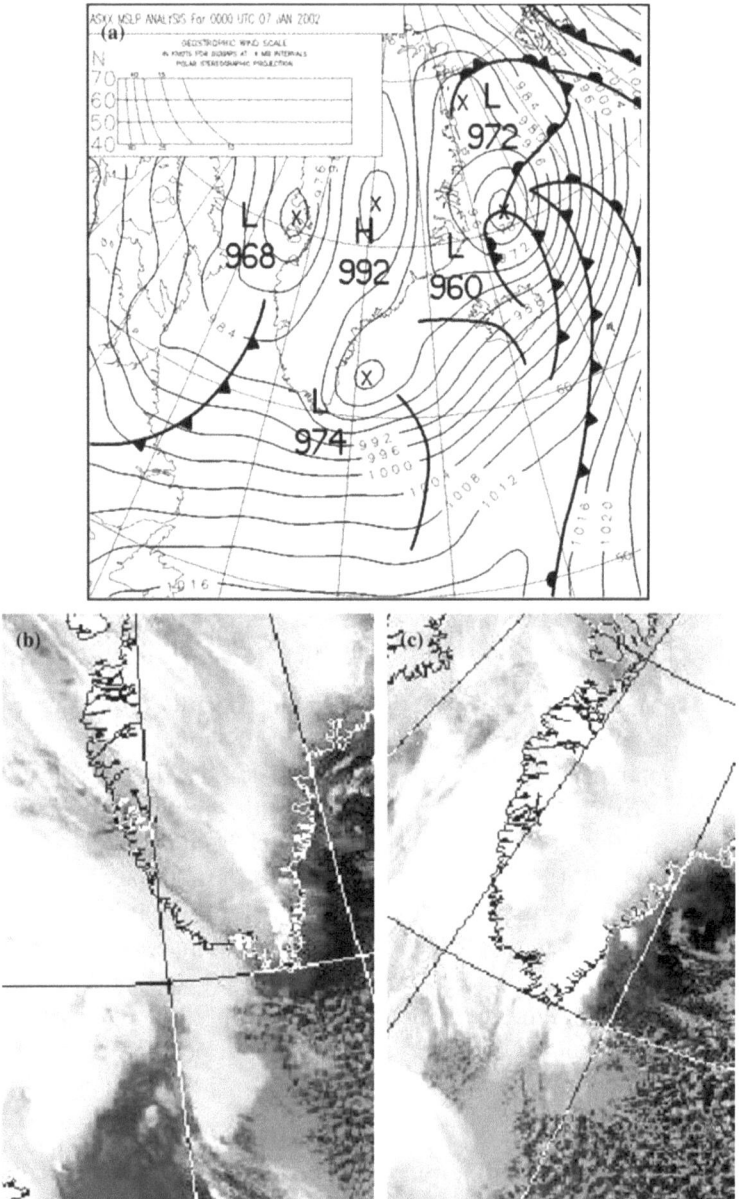

Figure 4
Synoptic analysis for 7 January, 2002 (westerly case). (a) UK Met Office MSLP analysis for 00UTC. (b)
Infrared satellite image at 05UTC. (c) Infrared satellite image at 15UTC.

itself (only partly clear conditions). During the course of the day the cloud travelled across southern Greenland (see Fig. 4(c)). Table 1 shows the station reports for 00UTC indicating a westerly flow incident to the southern tip of Greenland.

Table 1

Station reports for 00UTC 7 January, 2002 (westerly case)

Station	Temperature (°C)	Wind speed (ms^{-1})	Wind direction (°)
04260	−3.4	4.1	280
04360	−3.3	1.5	260
04390	−1.1	8.2	230
South Dome	−28	3.1	299

Conditions were relatively 'mild' for winter under this westerly flow regime. During the course of the day the synoptic airflow changed to more south-westerly as a low developed over the Labrador Sea. Meanwhile, the surface high intensified over the ice sheet, displacing the complex lows to the east and southeast.

3.2. Easterly Case Study (26 January, 2002)

Easterly flows are common in Greenland, especially in the north-west where they predominate ~24–38% of the time (CAPPELEN et al., 2001). They are typically marked by high pressure over northern Greenland.

On the 26th of January, 2002 the MSLP analysis for 00UTC (see Fig. 5(a)) shows a large, deep (964 mb), but not particularly intense, low pressure lying ~500 km south of Greenland, with relatively high pressure over the Canadian Arctic and

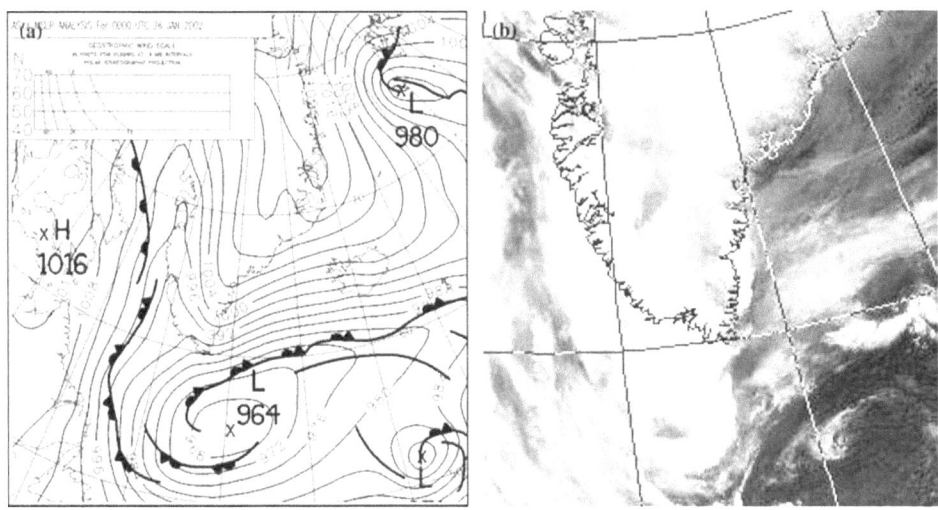

Figure 5
Synoptic analysis for 26 January, 2002 (easterly case). (a) UK Met Office MSLP analysis for 00UTC. (b) Infrared satellite image at 05UTC.

Table 2

Station reports for 12UTC 26 January, 2002 (easterly case)

Station	Temperature (°C)	Wind speed (ms⁻¹)	Wind direction (°)
04260	−8.4	0.5	30
04266	1.2	6.7	60
04360	−12.2	1.0	30
04382	−3.7	22.1	340
04390	−1.9	24.7	20
South Dome	−31	8.8	69

northern Greenland. The depression was fully mature, as marked by its three occluded fronts and broken spiral cloud bands (see Fig. 5(b)), and beginning to fill and decay (969 mb by midnight). The depression was a near-stationary feature throughout the day. Iceland, the Denmark Strait and southern Greenland were engulfed in a strong persistent east/north-easterly flow, causing the development of a foehn trough over southwest Greenland. The infrared satellite picture for approximately 05UTC (see Fig. 5(b)) shows wisps of cloud streaming westwards along the main front across the Norwegian Sea and into southernmost Greenland. However, the bulk of Greenland was under clear skies. Table 2 shows the station reports for 12UTC. Conditions were cold—very cold at South Dome. Most of the station reports show a strong northeast wind incident on Southern Greenland.

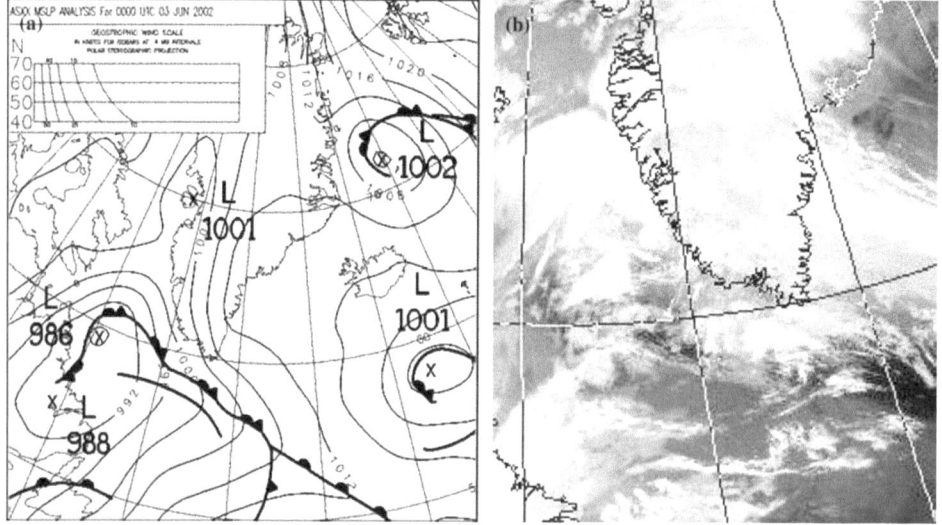

Figure 6
Synoptic analysis for 3 June, 2002 (southerly case). (a) UK Met Office MSLP analysis for 00UTC. (b) Infrared satellite image at 05UTC.

3.3. Southerly Case Study (3 June, 2002)

Southerly flows are by implication mild and moist, emerging as they do from a relatively warmer ocean. Distinct cases are relatively uncommon and they typically occur with a low-pressure system anchored southwest of Greenland and tend to yield precipitation around the southern and south-south-west margins.

On the 3rd of June, 2002 the MSLP analysis for 00UTC (see Fig. 6(a)) shows a complex shallow area of low pressure lying anchored off southwest Greenland, with its occluded frontal system just clipping the southern tip. This fed a relatively mild south-south-easterly airflow over southern Greenland. (This case is unique in being near midsummer; the others are all winter cases.) The infrared satellite picture for approximately 05UTC (see Fig. 6(b)) shows patchy clouds over the southern half of Greenland. There were further low pressure systems situated on the west coast of Greenland and south of Iceland. Table 3 shows the station reports for 12UTC. A couple of the stations (04266 and 04360) have a near southerly wind; the others have something closer to a northerly! Perhaps this resulted from local effects (e.g., sea breeze at 04360) and the relatively slack pressure gradient away from the southernmost tip of Greenland.

3.4. Northerly Case Study (21 February, 2002)

Northerly flows with cold polar air plunging south over Greenland are very typical due to frequent formation and deepening of low pressure systems between Greenland and Iceland. According to a table of 1961–1990 climatological standard normals in CAPPELEN et al., (2001), north is the most frequent wind direction at the capital of Greenland Nuuk on the southwestern coast (21% of cases), and on the northeastern coast and in the extreme southeast (e.g., Prins Christian Sund, 28% of cases). HANNA and CAPPELEN (2003) demonstrated that temperature over southern (particularly southwest coastal) Greenland in the past few decades is strongly

Table 3

Station reports for 12UTC 3 June, 2002 (southerly case)

Station	Temperature (°C)	Wind speed (ms^{-1})	Wind direction (°)
04260	4.9	2.1	320
04266	3.2	14.9	190
04360	2.8	1.0	200
04382	2.6	7.7	330
04390	3.1	9.8	10
South Dome	−11	8.9	138

associated with changes in the North Atlantic Oscillation (NAO), and as the NAO became more positive, overall temperature declined between 1958 and 2001. This is linked in all likelihood with a greater frequency/persistence of strong, cold northerly winds over Greenland.

On the 21st of February, 2002 the MSLP analysis for 00UTC (see Fig. 7(a)) shows a complex and deepening area of low pressure centred in the Denmark Strait (between Greenland and Iceland), with increasingly higher pressure westwards over the Labrador Sea. There was an eastwest pressure gradient of ~40 mb across the Denmark Strait/southern Greenland at 00UTC, which was maintained and even strengthened during the next 24 hours, increasing to ~70 mb by 00UTC on 22 February (not shown). Consequently Greenland was in the grip of a strong northerly air stream. The infrared satellite picture for approximately 15UTC (see Fig. 7(b)) shows a very well-defined frontal cloud band aligned southwest to northeast, over the Atlantic lying well to the south of Greenland, with the centre of the low apparent just west of Iceland. An intricate herringbone structure of partly open cloud cells suggests strong convection in destabilised polar air as it hit open waters around the south of Greenland. The entire Greenland ice sheet appears white and extremely cold in clear skies, leading to large radiative heat loss (~-22 Wm^{-2} at South Dome at 12UTC and -49 Wm^{-2} at 15UTC). The area immediately around Iceland, including the Denmark Strait, was considerably cloudier with very large convective cloud cells. Table 4 illustrates the extremely cold conditions at 12UTC.

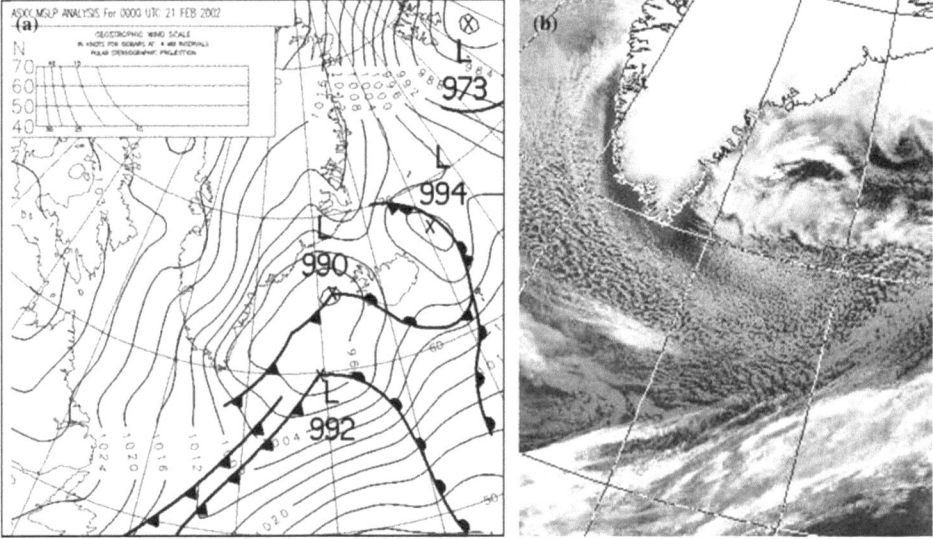

Figure 7
Synoptic analysis for 21 February, 2002 (northerly case). (a) UK Met Office MSLP analysis for 00UTC.
(b) Infrared satellite image at 15UTC.

Table 4

Station reports for 12UTC on 21 February, 2002 (northerly case)

Station	Temperature (°C)	Wind speed (ms⁻¹)	Wind direction (°)
04260	−18.6	5.1	350
04266	−11.4	20.6	320
04360	−16.7	4.1	330
04382	−15.2	11.3	320
04390	−9.5	13.4	270
South Dome	−51	15.2	322

It also shows a north or north-westerly wind travelling parallel to the east and west coasts of Greenland.

4. Numerical Mesoscale Model

The numerical model used is the atmosphere-only UK Met Office Unified Model (UM). The UM is a General Circulation Model (GCM) designed to run either as a climate model or a NWP model. As a climate model it was shown to generally simulate realistically the Antarctic climate, in particular the strong surface winds which are similar to those of Greenland (CONNOLLEY and CATTLE, 1994). Here the model is run in NWP mode as a mesoscale model. Version 4.5 is implemented (UM 4.5).

UM 4.5 employs a non-linear, hydrostatic set of dynamical equations on a horizontal latitude-longitude grid. A Lorenz grid is used in the vertical with a 'hybrid' sigma/pressure coordinate system giving increased resolution near the ground and the troposphere. CULLEN (1993) provides a description of the numerical formulation of the model. The physical parameterisations include schemes to represent boundary layer mixing, convection, precipitation, surface exchange, cloud formation, and shortwave and longwave radiation (CLARK and HOPWOOD, 2001). UK Met Office operational analyses data are used to initialise the model.

A limited-area domain was used with uniform resolution within the domain achieved through a rotated pole, i.e., the coordinate pole is not placed at the centre of the domain. Lateral boundary conditions are generated by nesting the limited-area domain within UM 4.5 running as a global model (CULLEN, 1993). The domain size was a 182 by 146 latitude-longitude grid with 0.11° (~ 12 km) resolution and 38 model levels in the vertical. (The domain configuration for the global model is a 325 by 432 latitude-longitude grid with resolution 0.83° by 0.56° and 30 model levels in the vertical.)

5. Flow over Greenland

In this section we analyse the flow over Greenland for each of the case studies using the combination of modelling and observational data. Note that over southern Greenland the Rossby deformation radius is approximately 230 km (for $f \sim 1.3 \times 10^{-4}$ s^{-1}, $H \sim 3000$ m, and $N \sim 0.01$ s^{-1}).

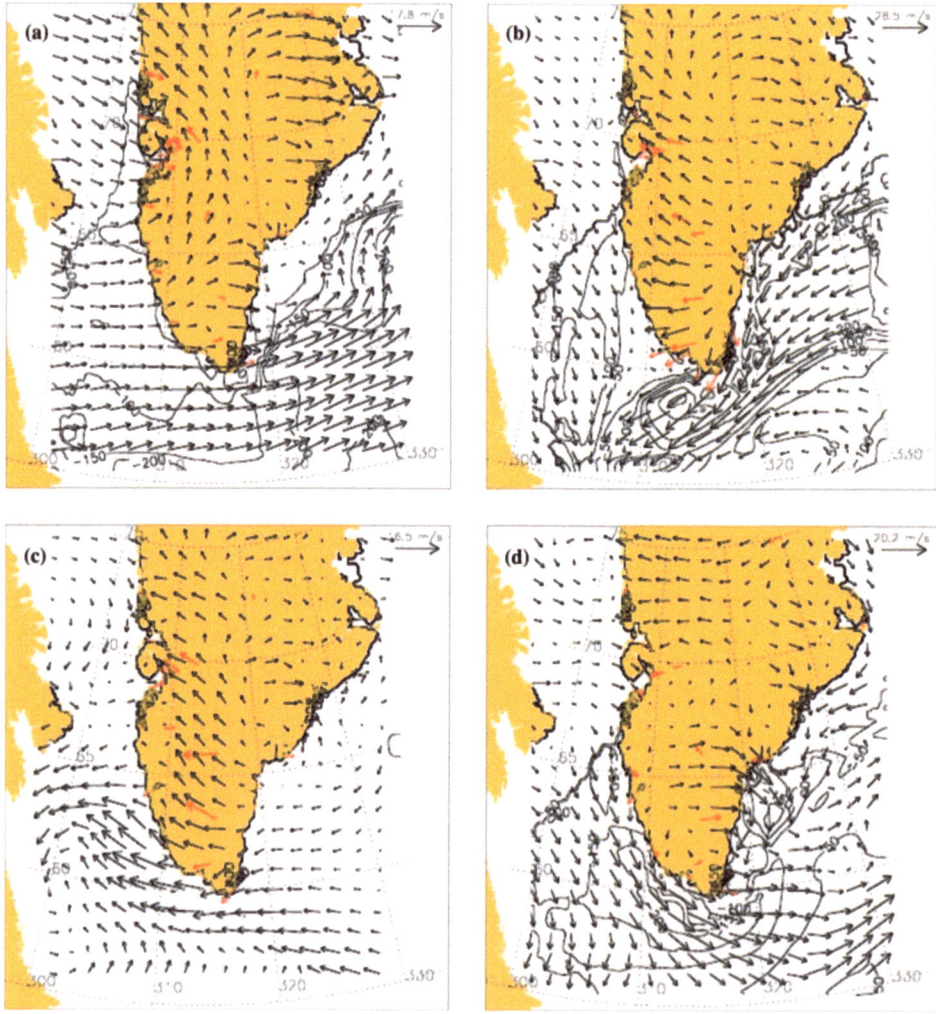

Figure 8

Variation of 10-m wind speed (shown as wind vectors (ms^{-1})) and latent heat flux (shown as contours (Wm^{-2})) over Southern Greenland at (a) 01UTC 7 January, 2002 (westerly case), (b) 06UTC 26 January, 2002 (easterly case), (c) 01UTC 3 June, 2002 (southerly case), and (d) 00UTC 21 February, 2002 (northerly case). Computed using UM 4.5 at a horizontal resolution of 12 km. (Wind speed vectors are averaged over 8 grid points.) DMI and GC-Net observations are shown in red.

5.1. Westerly Approach Flow

Figure 8(a) shows UM 4.5 simulated 10-m wind speeds at 01UTC 7 January, 2002 of a westerly flow of approximately 5–7 ms^{-1} incident to Greenland (giving $\mathbb{F}_H \approx 0.2$) (to the northwest the flow is slightly northerly). Because of the steep coastal terrain the flow slows as it travels into Greenland. Consequently it is deflected to the left due to Coriolis where it combines with the katabatic winds which, due to

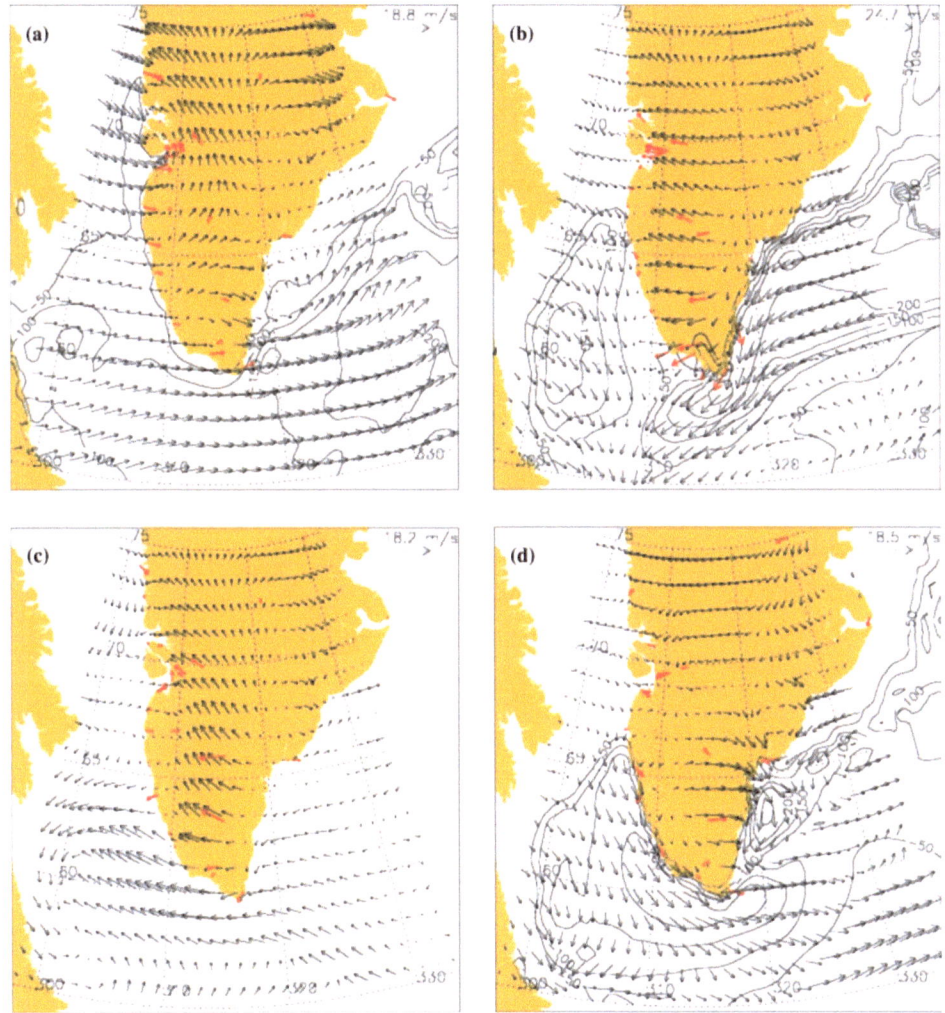

Figure 9

Variation of 10-m wind speed (shown as wind vectors (ms^{-1})) and latent heat flux (shown as contours (Wm^{-2})) over southern Greenland at (a) 00UTC 7 January, 2002 (westerly case), (b) 06UTC 26 January, 2002 (easterly case), (c) 00UTC 3 June, 2002 (southerly case), and (d) 00UTC 21 February, 2002 (northerly case). From ECMWF operational analysis data. DMI and GC-Net observations are shown in red. (The latent heat flux data are averaged over a three-hour period from the times shown.)

the slope of the ice sheet (see Fig. 1), are predominately south-easterly. This is confirmed by the DMI observations shown in red in Fig. 8(a) and 10-m wind ECMWF operational analysis data for 00UTC (see Fig. 9(a)).

At the 700 mb height (which is roughly the same height as the summit of Greenland) the UM 4.5 simulated winds show a strengthening in the upstream velocity, and to the northwest of Greenland the flow is much reduced and

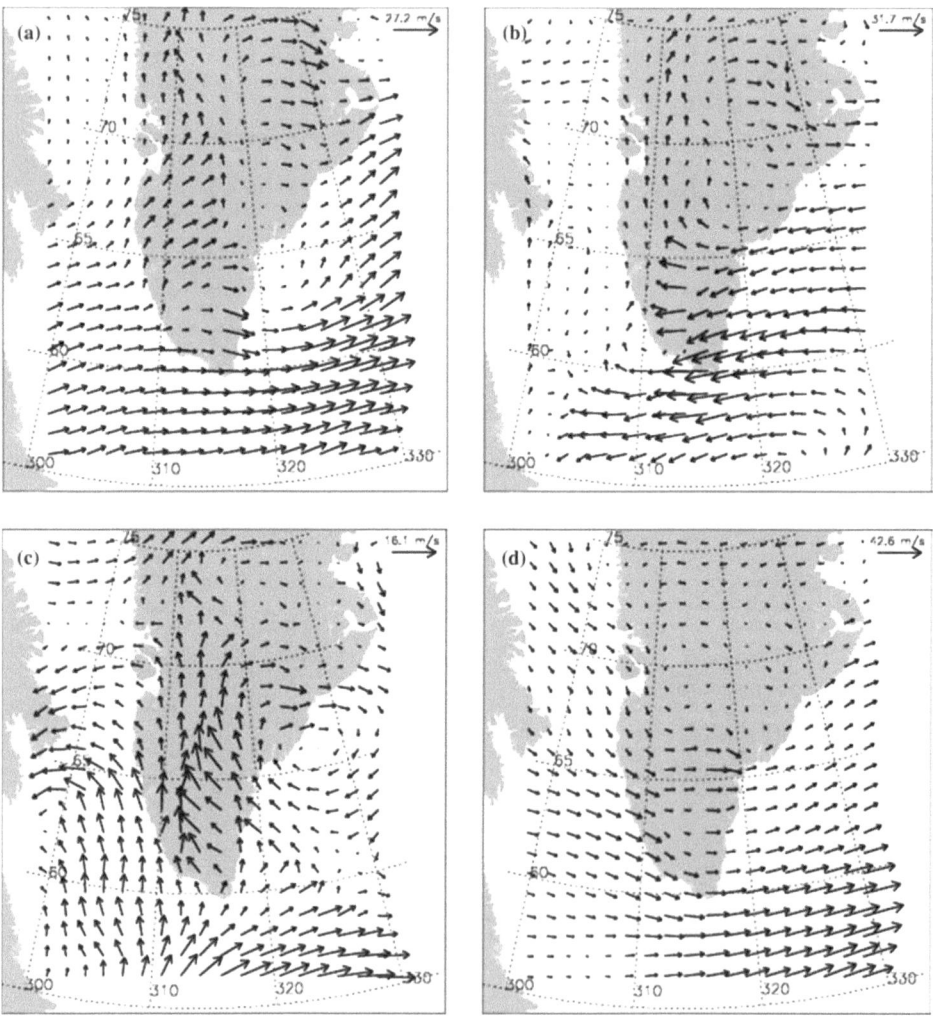

Figure 10

Variation of 700 mb wind speed (shown as wind vectors (ms^{-1})) over southern Greenland at (a) 01UTC 7 January, 2002 (westerly case), (b) 06UTC 26 January, 2002 (easterly case), (c) 01UTC 3 June, 2002 (southerly case), and (d) 00UTC 21 February, 2002 (northerly case). Computed using UM 4.5 at a horizontal resolution of 12 km. (Wind speed vectors are averaged over 8 grid points.)

predominately southerly (see Fig. 10(a)). At this level as the flow travels into Greenland it is only partially slowed and deflected. However, 700 mb flow does not pass over the summit. Rather it is blocked and rapidly deflected to the left where it merges with the katabatic winds over the plateau (c.f. Fig. 3(b)(i)). (Over the plateau the GC-Net observations show good agreement with the UM 4.5 700 mb wind field (not shown).) At 500 mb height (see Fig. 11(a)) the flow passes over Greenland with little deflection (though at this height the upstream winds are predominately

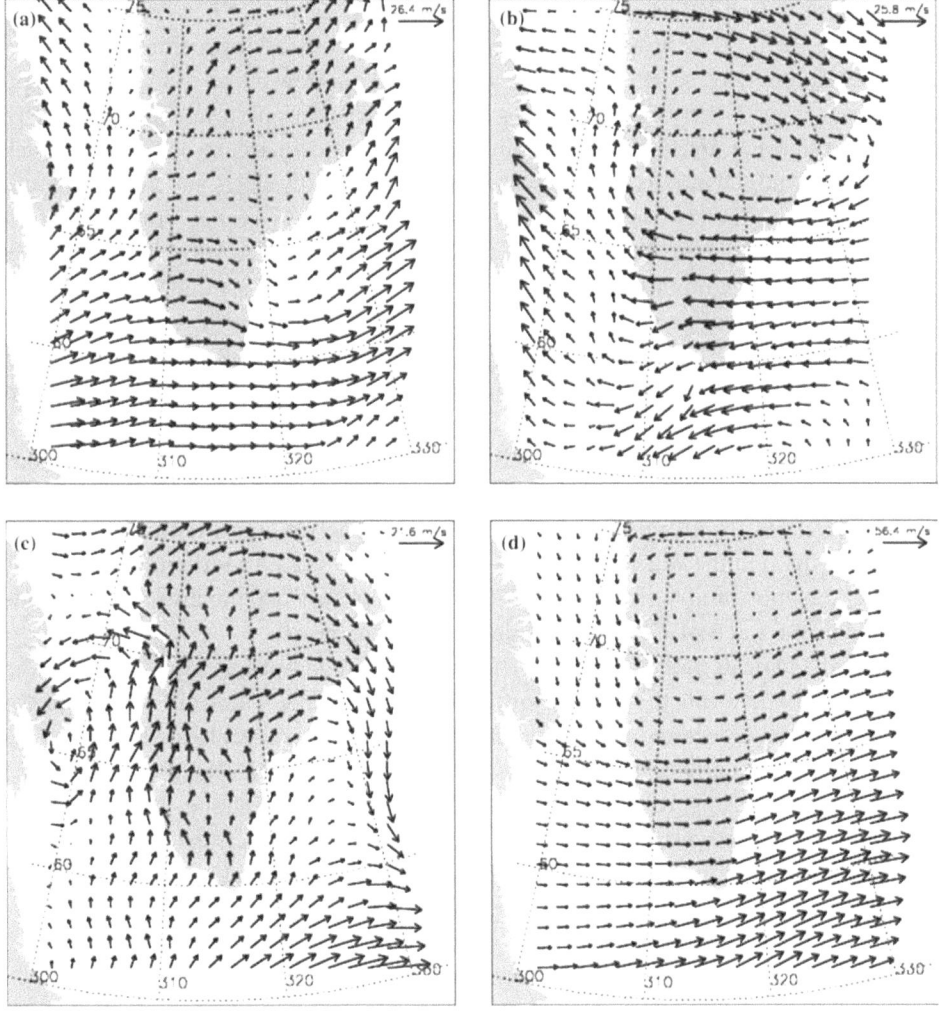

Figure 11
Variation of 500 mb wind speed (shown as wind vectors (ms^{-1})) over Southern Greenland at (a) 01UTC 7 January, 2002 (westerly case), (b) 06UTC 26 January, 2002 (easterly case), (c) 01UTC 3 June, 2002 (southerly case), and (d) 00UTC 21 February, 2002 (northerly case). Computed using UM 4.5 at a horizontal resolution of 12 km. (Wind speed vectors are averaged over 8 grid points.)

southerly due to the low pressure system sitting to the west of Greenland (see Fig. 4(a)).

Over southern Greenland the 10-m wind is well defined upstream and approximately 7–9 ms^{-1} (see Fig. 8(a)). The simulations and observations show the flow passing around the extreme southern tip of Greenland. As the flow to the south of the Greenland approaches the tip it accelerates to around 9–13 ms^{-1}, forming a jet. It continues to accelerate and be deflected to the left as it passes the tip. At station 04390, observations show the wind speed as approximately 8 ms^{-1} and south-westerly (see Table 1 and Fig. 8(a)). The width of the jet here is greater than 500 km (i.e. $> L_R$). The flow reaches its maximum velocity of approximately 17 ms^{-1} in the wake, an increase of over 100%, around 300 km downwind of Cape Farewell. In this wake region the wind vectors in surface (and 700 mb) plots appear to (re)curve slightly to the right up to several hundred kilometres downstream of Cape Farewell (not purely synoptic factors). In the wake the flow gradually slows and becomes more southerly due to the low pressure to the east of Greenland (see Fig. 4(a)). The 10-m wind ECMWF data for 00UTC show the flow being deflected as it passes the tip, though it gives a marginally stronger speed up in the wake of around 18 ms^{-1}. Note that in other synoptic situations over southern Greenland with a more uniform approach flow (e.g., HUNT et al., 2004 – 9 November, 2000), westerly flows lead to southerly winds in the wake of the easterly side of Greenland (as in the idealised and numerical results of Fig. 3(b)(i)).

At the 700 mb height the UM 4.5 simulation shows stronger westerly winds across southern of Greenland. The upstream velocity is approximately 15 ms^{-1}, which accelerates to around 20 ms^{-1} just south of the tip, and eventually reaching around 25 ms^{-1} in the wake. At 500 mb the flow speed across southern Greenland is considerably more uniform, roughly 17–20 ms^{-1}. For both 500 mb and 700 mb winds the flow in the wake of Greenland is deflected to the north.

Over the plateau itself the UM 4.5 and ECMWF data both indicate that near the surface katabatic winds dominate. South-easterly katabatic winds are evident in the western part of Greenland. While on the eastern side of Greenland the winds are predominately westerly reflecting the slope in that region. Both sets of data show the wind speed increasing towards the coastal margins of Greenland and significant interaction with the synoptic flow. This flow pattern is still largely evident at 700 mb, as discussed above. However, winds at 500 mb height are largely dominated by the synoptic flow.

5.2. Easterly Approach Flow

Fig. 8(b) shows UM 4.5 simulated 10-m wind speeds at 06UTC 26 January, 2002 of an upstream region characterized by a relatively strong, low-level east/north-easterly flow of about 13–15 ms^{-1} (giving $\mathbb{F}_H \approx 0.45$). This is blocked by the very steep mountains of the east coast of Greenland, forcing the flow south and parallel to the coast. The flow parallel to the coastline accelerates as it travels south and curves left, forming a barrier jet, and increasing from approximately 8 to 25 ms^{-1} between 70°N and the southern tip, an increase of over 200% (also shown by ECMWF operational analysis in Fig. 9(b)). Observational data show the wind speed at 10 m height at the southeastern coast as approximately 22 ms^{-1} at station 04382 and as approximately 25 ms^{-1} at station 04390 (see Table 2). The width of the jet here is roughly 300–500 km (i.e., of the same order as L_R). Additional UM 4.5 contoured data (not shown) showed very sharp gradients in the velocity perpendicular to the coastline. The velocity of the flow decreases gradually in the wake downstream of Cape Farewell. In this wake region the wind vectors curve slightly to the right due to the deep, low pressure system situated south of Greenland (see Fig. 5(a)). Fig. 8(b) shows easterly katabatic winds from the ice sheet interacting with the synoptic winds at the coastline, increasing these gradients. As the jet passes the tip it curves slightly to the right (looking downwind) and its velocity begins to decrease. Additional contour data show very sharp velocity gradients at the edges of the jet.

At the 700 mb height the upstream region is well defined as a easterly flow with a velocity of around 11–15 ms^{-1}. The flow travels over southern Greenland and is gradually blocked and deflects to the left (looking downwind). An infrared satellite image (see Fig. 5(b)) confirms this as it indicates mid-level frontal cloud being blocked slightly inland of the southeastern coast (c.f. Fig. 3(b)(i)). Over the western side the 700 mb flow is shown accelerating and being deflected to the right. The GC-Net observations over the plateau show good agreement with the flow at this height (not shown). Over the extreme tip of Greenland the flow accelerates, increasing to around 25 ms^{-1}, or around twice the upstream velocity, suggesting that the jet's vertical scale is of the order of ~ 1000 m. A synoptic high is stationed over the main plateau.

The simulated 500 mb flow (see Fig. 11(b)) shows much less deflection due to Greenland itself and a greater emphasis on (mid-tropospheric) synoptic wind fields.

5.3. Southerly Approach Flow

Fig. 8(c) shows UM 4.5 simulated 10-m wind speeds at 01UTC 3 June, 2002 of an upstream region characterised by a low-level southerly flow of 5–7 ms^{-1} (giving $\mathbb{F}_H \approx 0.2$). Part of the upstream flow is south-easterly which deflects the southerly flow to the left and increases its velocity as it approaches Greenland. As the flow travels north parallel to the western coastline its velocity accelerates from around 9 ms^{-1} to 15 ms^{-1} between the tip of Greenland and around 63°N, or by about 80%. The width of the jet here is roughly 300–500 km. Observational data show the wind speed at 10 m height as approximately 15 ms^{-1} at station 04266 (at 12UTC; see Table 3). And also it is deflected to the left due to Coriolis. To the north of 63°N the flow is deflected further to the left by the low pressure system situated off the west coast of Greenland (see Fig. 6(a)). Fig. 8(c) shows predominately south-easterly katabatic winds interacting with the synoptic winds along the western coastal margin. This is confirmed by the DMI observations shown in red in Fig. 8(c) and 10-m wind and ECMWF operational analysis data for 00UTC (see Fig. 9(c)).

The single observation (04390) near the tip of Greenland at 12UTC shows a velocity of around 10 ms^{-1} and in an almost northerly direction. This is also evident 12 hours earlier in the UM 4.5 and ECMWF 10-m wind data (though more easterly). This could represent part of a katabatic (shallow-layer) flow off the ice cap. No equivalent northerly wind vectors are seen in the 700 and 500 mb height data (see Fig. 10(c) and 11(c)), suggesting that the postulated low-level katabatic flow is indeed shallow. The higher level plots also indicate a considerably more linear south-to-north flow over Greenland. The 700 mb plot has flow accelerating quite markedly northwards over southern Greenland as it breaks over the southern ice dome (South Dome in Fig. 1). No such acceleration is evident in the 500 mb plot and this suggests that flow is undisturbed at this level. At both levels the flow to the west and east of Greenland is strongly deflected due to the low pressure systems sitting to the west and east of Greenland (see Fig. 6(a)).

5.4. Northerly Approach Flow

Fig. 8(d) shows UM 4.5 simulated 10-m wind speeds at 00UTC 21 February, 2002 of an upstream region characterised by a relatively weak, low-level westerly flow of $\sim$ 5 ms^{-1} (giving $\mathbb{F}_H \approx 0.1$) between 67°N and 75°N. This is blocked by the mountains of the west coast of Greenland, forcing the flow south. This northerly flow develops into a distinct barrier jet travelling parallel to the coastline and increasing in velocity with distance propagated. Inspection of contoured data show sharp gradients of velocity parallel to the coastline. Between 67°N and the tip of southern Greenland its wind speed increases from about 5 ms^{-1} to 17 ms^{-1}, or by about 240%. Observational data show the wind speed at 10 m height as approximately 5 ms^{-1} at station 04260 and as approximately 21 ms^{-1} at station 04266 (at 12UTC; see Table 4). The width of the jet here is approximately 500 km. As the jet passes around the tip it curves markedly to the left (looking downwind), due to the combination of Coriolis and interaction with the low pressure system centred in the Denmark Strait (see Fig. 7(a)). The wind jet continues as a westerly flow downwind of the tip in its wake, its velocity decreasing gradually. The UM 4.5 10-m wind speeds also show a slight deflection of the low-level flow to the right in the wake of the southern mountains of Greenland (within the region several hundred kilometres southeast of Cape Farewell). This is confirmed by the DMI observations shown in red in Fig. 8(d) and 10-m wind ECMWF operational analysis data for 00UTC (see Fig. 9(d)).

Fig. 8(d) shows strong westerly katabatic winds over southern Greenland being funnelled down to the ocean (perhaps through the Angmagssalik fjord (RASMUSSEN, 1989)) and interacting with the cyclonic flow to the east of Greenland.

Upper air flows at 700 and 500 mb height (see Figs. 10(d) and 11(d), respectively) also show northerly flow west of Greenland, curving westerly and increasing markedly in strength further south around Cape Farewell. However this speed-up could not be termed a jet as it applies to a very large band of the flow (> 1000 km wide) and is linked with the low pressure system centred in the Denmark Strait.

6. Air-sea Interaction around Greenland

Here a negative heat flux is defined as the direction from the surface to the atmosphere (i.e., cooling of the surface).

6.1. Westerly Approach Flow

The strong speed-up jet evident southeast of Greenland and discussed in section 5.1 is commonly called the Greenland tip jet. It is associated with cold temperatures (see Table 1) and corresponds here to regions of large negative surface heat flux. Figs. 8(a) and 9(a) show UM 4.5 and ECMWF modelled latent heat fluxes of up to -200 Wm^{-2} (i.e., strong cooling of the surface due to evaporation). Figs. 12(a) and 13(a) show a slightly smaller sensible heat flux of around -100 to -150 Wm^{-2} in the speed-up jet region. This suggests that for this case study evaporation associated with the jet, more than the jet itself, is cooling the sea-surface southeast of southern Greenland. As the surface water cools it becomes more dense and sinks, resulting in downwelling (PICKART et al., 2003). Additionally, Figs. 8(a) and 9(a) show the jet curving sharply to the left, resulting in a strong, localised wind stress curl. Due to Ekman motion this would create a cyclonic or divergent gyre in the ocean, which will also result in downwelling (GILL, 1982; PICKART et al., 2003).

6.2. Easterly Approach Flow

There are large, negative heat fluxes of up to -200 Wm^{-2} (both sensible and latent) in the Denmark Strait, stretching and narrowing into a region corresponding to the speed-up jet discussed in Section 5.2, which extends southwest of Cape Farewell on the southern tip. The jet here is both colder (see Table 2) and stronger than the previous westerly case, explaining perhaps why both the modelled sensible heat fluxes (see Figs. 12(b) and 13(b)) and latent heat fluxes (see Figs. 8(b) and 9(b)) are slightly larger. We therefore have cooling and evaporation of relatively warm sea water, resulting in downwelling. From Figs. 8(b) and 9(b) there is only a slight curving of the wind vectors and therefore weak wind stress curl. Upstream of Cape Farewell weak downwelling might occur. But downstream of Cape Farewell the rightward turning wind vectors would produce a weak convergent gyre and weak upwelling (GILL, 1982).

6.3. Southerly Approach Flow

There are small, positive sensible heat fluxes of about 50 Wm^{-2} in a narrow region just south and southwest of Greenland, corresponding to the barrier jet travelling north and parallel to the western coastline and discussed in Section 5.3

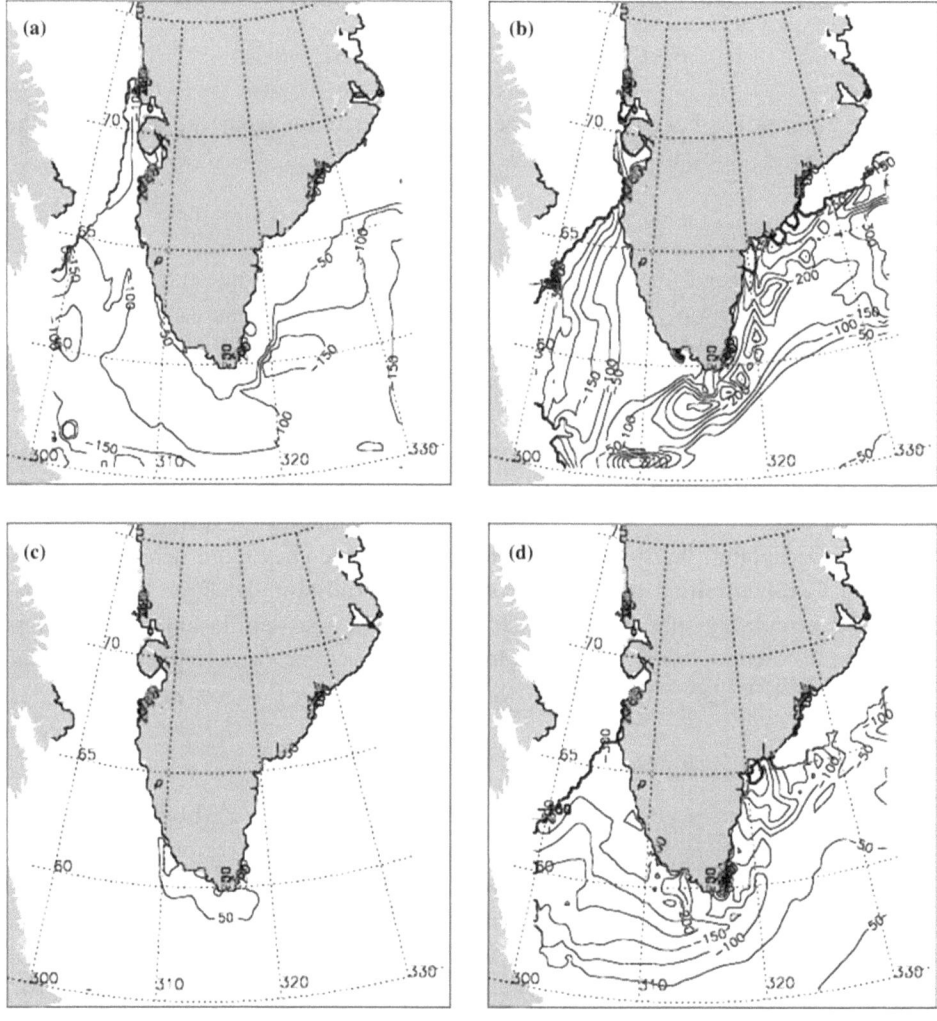

Figure 12

Variations of sensible heat flux (Wm^{-2}) over southern Greenland at (a) 01UTC 7 January, 2002 (westerly case), (b) 06UTC 26 January, 2002 (easterly case), (c) 01UTC 3 June, 2002 (southerly case), and (d) 00UTC 21 February, 2002 (northerly case). Computed using UM 4.5 and a horizontal resolution of 12 km.

(see Figs. 12(c) and 13(c)). There are no significant areas of latent heat fluxes (see Figs. 8(c) and 9(c)). This is because this southerly, near-midsummer flow (air of temperate or even subtropical origin; see Table 3) would not be expected to cool the sea surface or cause evaporation, but it would itself perhaps been modified (cooled) somewhat by passage over the relatively cool ocean (therefore would probably not cause a very substantial warming either). Figs. 8(c) and 9(c) show

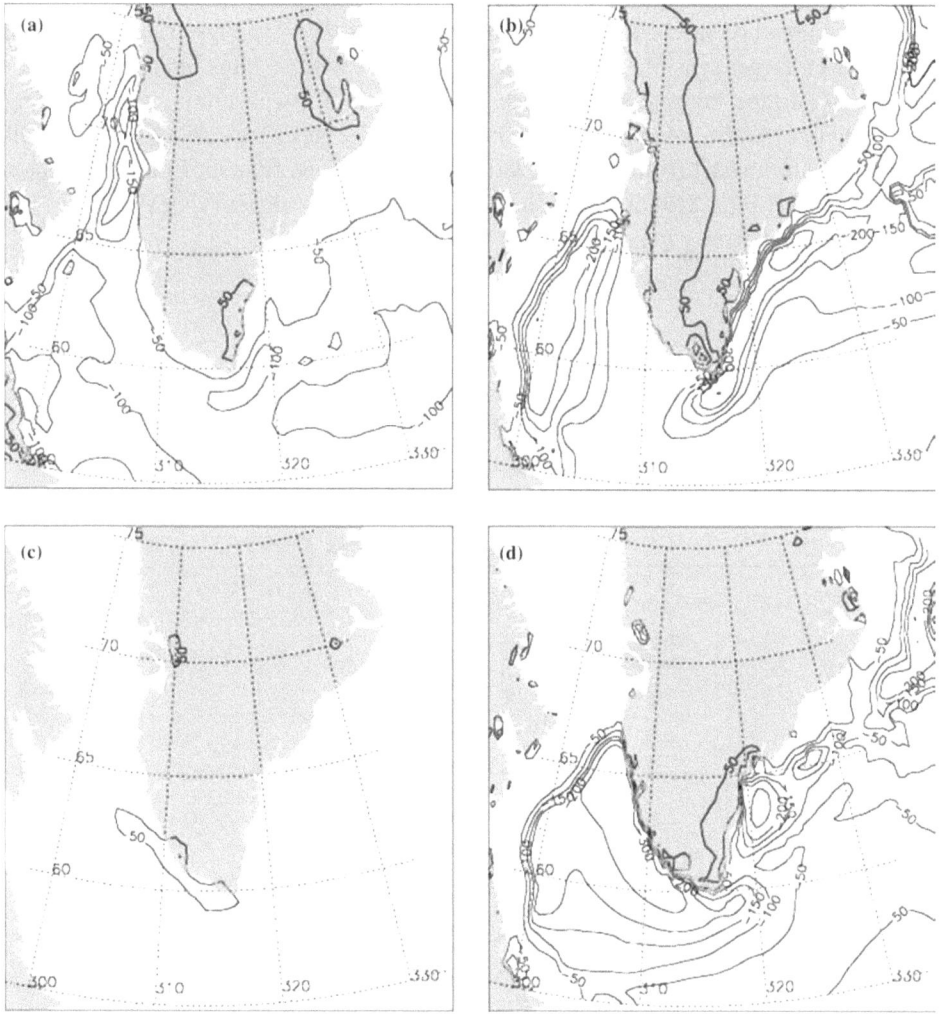

Figure 13
Variations of sensible heat flux (Wm^{-2}) over southern Greenland at (a) 00UTC 7 January, 2002 (westerly case), (b) 06UTC 26 January, 2002 (easterly case), (c) 00UTC 3 June, 2002 (southerly case), and (d) 00UTC 21 February, 2002 (northerly case). From ECMWF operational analysis data. (Data is averaged over a three hour period from the times shown.)

the speed-up jet curving sharply to the left (partly due to low pressure system situated off the west coast of Greenland (see Fig. 6(a))), resulting in a strong, localised wind stress curl and downwelling.

6.4. Northerly Approach Flow

The strong speed-up jet evident around Cape Farewell and accelerating eastward, discussed in Section 5.4, consists of extremely cold, dry polar air and corresponds to regions of large negative surface heat flux. Figs. 8(d) and 9(d) show UM 4.5 and ECMWF modelled latent heat fluxes of up to -200 Wm^{-2}. Figs. 12(d) and 13(d) show a even larger sensible heat flux of up to -300 Wm^{-2}. Thus intense cooling of the ocean surface takes place through advection and evaporation, with resulting downwelling. (Additionally, there is a region of large negative surface heat flux off the southeastern coast from the cold, katabatic winds been funnelled off the ice sheet (see Fig. 8(d)).) Figs. 8(d) and 9(d) show the flow in the wake curving sharply to the left, resulting in strong wind stress curl and downwelling.

Table 5

Cloudiness and precipitation station reports for 12UTC 7 January, 2002 (westerly case)

Station WMO No.	Region of Greenland	Mean cloud cover (oktas)	Precip. (mm)
04220	mid west	8	2
04250	southwest	7.5	42
04260	further down, southwest	8	13.2
04272	extreme south-south-west	8	13.4
04390	extreme south	3.75	3.1
04360	southeast	2.63	7
04339	East	2.25	0.9

Table 6

Cloudiness and precipitation station reports for 12UTC 26 January, 2002 (easterly case)

Station WMO No.	Region of Greenland	Mean cloud cover (oktas)	Precip. (mm)
04220	mid west	4.75	0
04250	southwest	4.5	0
04272	extreme south-south-west	8	0
04390	extreme south	5.9	0
04360	southeast	4	0
04339	east	0.4	0

7. Cloudiness and Precipitation over Southern Greenland

HUNT et al., (2004) confirmed theoretically that for stably stratified flow, in the areas where speed-up jets are shown (see Section 5) there are also changes to the inversion height. Applying their model to Greenland predicts that for westerly flows the air would sink as it crossed the southern tip of Greenland. And hence cloudiness and precipitation would decrease. Similarly, for easterly flows the air would rise as it crossed the southern tip of Greenland. And hence cloudiness and precipitation would increase. Note that their model does not take into account upstream conditions. In this section we examine observational evidence of these predictions.

7.1. Westerly Approach Flow

Table 5 shows observations from DMI coastal stations (averaged over the 24-hour period) for the westerly flow case study (7 January, 2002). The data show that cloud cover is considerably greater and amounts of precipitation higher on the west coast of Greenland than on the east coast. This follows because the synoptic flow was westerly and perhaps introduced moist air from the Labrador Sea. This is clearly seen as areas of bright white cloud around southwest Greenland on the infrared satellite images, especially at 15UTC (see Fig. 4(c)). By comparison the east coast appears comparatively clear on both the 05UTC and 15UTC images (see Figs. 4(b,c)).

7.2. Easterly Approach Flow

Table 6 shows observations from DMI coastal stations (averaged over the 24-hour period) for the easterly flow case study (26 January, 2002). The data show an overcast sky in the south-south-west and that the amount of cloud cover progressively increases southwards down the eastern side of Greenland. This is well seen in the infrared satellite image (see Fig. 5(b)). Especially note the high bright clouds around south and southwest of Greenland.

8. Discussion

8.1. Flow over Greenland

These studies show that typical air movements over Greenland do not generally correspond to the idealised conditions of uniform approach flow shown in Figs. 2 and 3, though such conditions also occasionally exist. Rather the typical upper level flow (at 500 mb) consists of synoptic eddies passing over the plateau. These change the topology of the streamlines, as seen in Fig. 8. But Fig. 3 shows that even if the large scale flow varies significantly over the bulk of the plateau, the characteristic features of the jet formation, with wind speeds of order 2–3 times that of the local winds and a transverse scale of order of the Rossby deformation radius L_R, have an westerly or easterly component (HUNT et al., 2001, 2004). Also evident are sharp gradients in the velocity perpendicular to the coastline and at the edges of the jet. The jets have a vertical scale of order ~ 1000 m and are associated with rising/sinking air and the formation of clouds to the west and clearer air to the east of the tip of Greenland (HUNT et al., 2004). Further theoretical and computational studies are necessary to understand how these synoptically driven surface flows are affected by non-uniform approach flows.

Another general feature of the surface flow is the generation of katabatic winds on the slopes caused by strong radiative longwave cooling over the ice sheet. Over the gently sloping central part of the ice sheet these winds are moderate. Winds accelerate towards the coastal margins where the topography steepens (e.g., see Fig. 8(d)). The strong katabatic flows are important to the energy balance of melting sea ice (e.g., BROMWICH et al., 1996; KLEIN et al., 2001). Over the ice sheet these are shown to be only moderately affected by the synoptic flow. However on the upwind side of Greenland they interact with the barrier jet (VAN DEN BROEKE and GALLÉE, 1996) and on the downside with large-scale wake eddies. But a general conceptual framework for the interaction has yet to be devised.

One of the purposes of this study is to see whether wind speed and direction measured at the coastal stations are representative of the broad-scale synoptic conditions (i.e., wind vectors over the sea shown by the UM 4.5 and ECMWF data). Coastal station winds are sometimes influenced by meso- and micro-scale meteorology (e.g., katabatic winds). However they are quite often, when taken together, broadly representative of the wider picture. On the other hand, it is clear that certain coastal stations are susceptible to particular local and mesoscale effects according to the prevailing synoptic situation.

In general the UM 4.5 modelled 10-m winds, computed using a horizontal resolution of 12 km, compared well both in magnitude and direction with the DMI and GC-Net station observations; as did the ECMWF operational analysis 10-m winds, though typically they were smaller than the observed wind speeds. Comparing the predicted UM 4.5 and ECMWF 10-m winds, the UM 4.5 data typically showed a stronger and more localised jet. Additionally it captured better the sharp gradients in the velocity perpendicular to the coastline and the interaction between the barrier jet and the katabatic winds. For accurate simulations of these flows, mesoscale models are necessary with horizontal resolutions of the order of 20 km or less.

8.2. Air-sea Interaction

The strong winds and cold temperatures of the low-level jets lead to enhanced heat loss in the ocean around southern Greenland. Typically latent heat fluxes were slightly greater than the sensible heat flux, but combined they typically contributed a strong heat flux of between -300 and -400 Wm^{-2}. Additionally the jets often curve markedly to the left due to Coriolis, resulting in strong, localised wind stress curl. Due to Ekman motion this would create a cyclonic or divergent gyre in the ocean. Together the large sea-air heat flux and strong wind curl force oceanic downwelling (GILL, 1982). Recently PICKART *et al.*, (2003) showed that downwelling in the Irminger Sea (to the southeast of Greenland) was associated with the western Greenland tip jet (through strong wind stress curl and large air-sea heat flux). RENFREW *et al.*, (2002) observed deep convection in the western Labrador Sea due to a cooling heat flux comparable with results presented here.

However, the western tip jet is a relatively rare phenomenon (see Section 3.1). Results here suggest that downwelling associated with these jets might be associated at places other than the Irminger Sea and possibly more common than previously thought. Additionally, the wind-jets have implications for the movement of sea-ice, particularly in opening leads and polynyas and warming of the atmosphere (UOTILA, 2001; SMEDMAN, 2003). Note that ice movement is generally southward on the east coast of Greenland near the southern tip, corresponding to the airflow depicted in Fig. 3(b)(i).

In general the predicted UM 4.5 and ECMWF heat fluxes were of the same order of magnitude. However, UM 4.5 fluxes were typically confined to a narrower region, consistent with a more localised jet. UM 4.5 also typically predicted a stronger jet and therefore regions of stronger wind stress curl, suggesting it would predict greater, and possibly more accurate, downwelling (remember the ECMWF operational analysis data are generated using a substantially coarser $0.5° \times 0.5°$ grid). This indicates that perhaps improved parameterisation of air-sea coupling is required in numerical weather/climate prediction models.

8.3. Cloudiness and Precipitation

Are cloudiness and precipitation greater on the west coast of Greenland than on the east coast, in accordance with the model of HUNT *et al.*, (2004)? Data from the westerly and easterly case studies tentatively suggested this.

Measurements of cloud cover are prone to uncertainties as it is based on visual estimates by observers and partly missing records. Nevertheless from climatological records (CAPPELEN *et al.*, 2001) a pattern emerges of the east coast being relatively cloud-free compared with the west. The two cloudiest stations during 1961–1990 were 04250 (70% mean cloud cover (MCC)) and 04260 (71% MCC) in southwest Greenland, whereas 04320 in the northeast had only 50% MCC. 04390 near the southern tip of Greenland had an intermediate value of 63% MCC, and 04272 (also in the south) had 65% MCC. However, 04360 in the east had a relatively high value of 68% MCC. CAPPELEN *et al.*, (2001) suggest the southwest coast is about $\sim 10\%$ cloudier, on average, than the southeast coast but there are large local variations.

Regarding precipitation, this is also notoriously difficult to measure in the polar regions due to problems with blizzards and wind-blown snow. CAPPELEN *et al.*,

(2001) gives yearly precipitation amounts averaged over the period 1961–1990, showing that precipitation is highest at the southern stations, especially 04390 (extreme south; average of 2474 mm), followed by 04380 (southeast; 1535 mm), 04261 (extreme south-south-west; 1040 mm), 04360 (southeast; 984 mm), 04260 (south-south-west; 874 mm), and 04272 (extreme south-south-west; 858 mm). This suggests that precipitation is in fact greater in the southeast than the southwest, perhaps due to prevailing easterlies hitting the south-east coast of Greenland and being orographically forced. Thus further theoretical and computational studies are necessary to understand this aspect of Greenland's mesoscale climatology.

Acknowledgments

It is a pleasure to contribute to this volume in honour of Professor M. R. Singh, who stimulated much important research on geophysical fluid dynamics; Julian Hunt particularly benefited from his support and encouragement. We are grateful to the University of Dundee Satellite Receiving Station and NOAA for the satellite images. The MSLP analysis charts are copyright of the UK Met Office and were obtained from the Uni. Wetterzentrale Topkarten website. We are grateful to the British Atmospheric Data Centre for providing access to data from the European Centre for Medium-Range Weather Forecasts. Thanks to Paul Coles for the map of Greenland. Andrew Orr would like to thank Andy Shepard and Paul Taylor. This work was supported by grants from NERC to the Depts. of Space & Climate Physics and Mathematics at University College London. Julian Hunt is grateful to the Mechanical and Aerospace Engineering Department, Cornell University where he is the Mary B. Upson visiting Professor.

REFERENCES

BAINES, P. G., *Topographic Effects in Stratified Flows* (Cambridge University Press, UK, 1995).

BLUMEN, W. (Editor), *Atmospheric Processes over Complex Terrain* (Meteorological Monographs, Ame. Meteorol. Soc., *23*, 1990).

BOYER, D. L. and CHEN, R. (1987), *Laboratory Simulation of Mountain Effects on Large-scale Atmospheric Motion Systems: The Rocky Mountains*, J. Atmos. Sci. *44*, 100–123.

BOYER, D. L. and DAVIES, P. A. (2000), *Laboratory Studies of Orographic Effects in Rotating and Stratified Flows*, Annu. Rev. Fluid Mech. *32*, 165–202.

BROMWICH, D. H., Du, Y., and HINES, K. M. (1996), *Wintertime Surface Winds over the Greenland Ice Sheet*, Mon. Wea. Rev. *124*, 1941–1947.

CAPON, R. (2003), *Wind Speed-up in the Dover Straits with the Met Office New Dynamics Model*, Meteorol. Appl. *10*, 229–237.

CAPPELEN, J., JORGENSEN, B. V., LAURSEN, E. V., STANNIUS, L. S., and THOMSEN, R. S., *The Observed Climate of Greenland, 1958–99, with Climatological Standard Normals, 1961–90.* (Tech. Rpt. 00-18, Danish Meteorol. Institute, Copenhagen, Denmark, 2001).

CLARK, P. A. and HOPWOOD, W. P. (2001), *One-dimensional Site-specific Forecasting of Radiation Fog. Part 1: Model Formulation and Idealised Sensitivity Studies*, Meteorol. Appl. *8*, 279–286.

COLEMAN, A. and DAVEY, M. (1999), *Prediction of Summer Temperature, Rainfall and Pressure in Europe from Proceeding Winter North Atlantic Ocean Temperature*, Int. J. Climatol. *19*, 513–536.

CONNOLLEY, W. M. and CATTLE, H. (1994), *The Antarctic Climate of the UKMO Unified Model*, Antarctic Science *6*, 115–122.

CULLEN, M. J. P. (1993), *The Unified Forecast/Climate Model*, The Meteorol. Mag. *122*, 81–94.

DING, L. and STREET, R. L. (2003), *Numerical Study of the Wake Structure behind a Three-dimensional Hill*, J. Atmos. Sci. *60*, 1678–1690.

DOYLE, J. D. and SHAPIRO, M. A. (1999), *Flow Response to Large-scale Topography: The Greenland Tip Jet*, Tellus *51A*, 728–748.

GILL, A. E., *Atmosphere-Ocean Dynamics* (Academic Press, London, UK, 1982).

GLOVER, R. (1999), *Influence of Spatial Resolution and Treatment of Orography on GCM Estimates of the Surface Mass Balance of the Greenland Ice Sheet*, J. Climate *12*, 551–563.

HANNA, E. and CAPPELEN, J. (2003), *Recent Cooling in Coastal Southern Greenland and Relation with the North Atlantic Oscillation*, Geophys. Res. Lett. *30*, 1132.

HOZUMI, Y. and UEDA, H. (2005), *Numerical Study on Vortices in the Middle Layer of Flow around a Large Mountain under Rotating Stratified Conditions*, Pure. Appl Geophys. *162* (10) this issue.

HUNT, J. C. R., ABELL, C. J., PETERKA, J. A., and WOO, H. G. C. (1978), *Kinematic Studies of the Flow around Free or Surface-mounted Obstacles: Applying Topology to Flow Visualisation*, J. Fluid Mech. *86*, 179–200.

HUNT, J. C. R. and SNYDER, W. H. (1980), *Experiments on Stably Stratified Flow over a Model Three-dimensional Hill*, J. Fluid Mech. *96*, 671–704.

HUNT, J. C. R., FENG, Y., LINDEN, P. F., GREENSLADE, M. D., and MOBBS, S. D. (1997), *Low Froude Number Stable Flows past Mountains*, Il Nuovo Climento *20C*, 261–272.

HUNT, J. C. R., ÓLAFSSON, H., and BOUGEAULT, P. (2001), *Coriolis Effects on Orographic and Mesoscale Flows*, Q. J. Roy. Met. Soc. *127*, 601–633.

HUNT, J. C. R., ORR, A., ROTTMAN, J. W., and CAPON, R. (2004), *Coriolis effects in mesoscale flows with sharp changes in surface conditions*, Q. J. Roy. Met. Soc., *130*, 2703–2731.

JORGENSEN, P.V. and LAURSEN, E. V., *DMI Monthly Climate Data Collection 1860–2002, Denmark, Faroe Island and Greenland* (An Update of: NACD, REWARD, NORDKLIM and NARP datasets Version 1. Tech. Report 03-26, Danish Meteorological Institute, Copenhagen, Denmark, 2003).

KLEIN, T., HEINEMANN, G., and GROSS, P. (2001), *Simulation of the Katabatic Flow near the Greenland Ice Margin using a High-resolution Nonhydrostatic Model*, Meteorologische Zeitschrift *10*, 221–339.

KLEIN, T. and HEINEMANN, G. (2002), *Interaction of Katabatic Winds and Mesoscyclones near the Eastern Coast of Greenland*, Meteorol. Appl. *9*, 407–422.

LOTT, F., and MILLER, M. J. (1997), *A New Subgrid-scale Orographic Drag Parameterization: Its Formulation and Testing*, Q. J. Roy. Met. Soc. *123*, 101–127.

MEESTERS, A. (1994), *Dependence of the Energy Balance of the Greenland Ice Sheet on Climate Change: Influence of Katabatic Wind and Tundra*, Q. J. Roy. Met. Soc. *120*, 491–517.

MERKINE, L. O. (1975), *Steady Finite-amplitude Baroclinic Flow over Long Topography in a Rotating, Stratified Atmosphere*, J. Atmos. Sci. *32*, 1881.

NEWLEY, T. M. J., PEARSON, H. J., and HUNT, J. C. R. (1991), *Stably Stratified Rotating Flow through a Group of Obstacles*, Geophys. Astrophys. Fluid Dyn. *58*, 147–171.

ÓLAFSSON, H., and BOUGEAULT, P. (1996), *Non-linear Flow past an Elliptic Mountain Ridge*, J. Atmos. Sci. *53*, 2465–2489.

ÓLAFSSON, H. and BOUGEAULT, P. (1997), *The Effect of Rotation and Surface Friction on Orographic Drag*, J. Atmos. Sci. *54*, 193–210.

ÓLAFSSON, H. (2000), *The Impact of Flow Regimes on Asymmetry of Orographic Drag at Moderate and Low Rossby Numbers*, Tellus *52A*, 365–379.

PETERSEN, G. N., ÓLAFSSON, H., and KRISTJÁNSSON, J. E. (2003), *Flow in the Lee of Idealized Mountains and Greenland*, J. Atmos. Sci. *60*, 2183–2195.

PIERREHUMBERT, R. T. and WYMAN, B. (1985), *Upstream Effects of Mesoscale Mountains*, J. Atmos. Sci. *42*, 977–1003.

PICKART, R. S., SPALL, M. A., RIBERGAARD, M. H., MOORE, G. W. K., and MILLIFF, R. F. (2003), *Deep convection in the Irminger Sea Forced by the Greenland Tip Jet*, Nature *424*, 152–156.

PUTNINS, P., *The climate of Greenland*. In *World Survey of Climatology Volume 14: Climates of the Polar Regions* (eds. Orvig, S.; Landsberg, H. E.) (Elsevier Publishing Company, New York, 1970) pp. 3–128.

RASMUSSEN, L. (1989), *Greenland Winds and Satellite Imagery*, VEJRET (in English) (ed. Nilsen, N. W.), Danish Meteorol. Soc., pp. 32–37.

RENFREW, I. A., MOORE, G. W. K., GUEST, P. S., BUMPKE, K. A. (2002), *A Comparison of Surface-layer Heat Flux and Surface Momentum Flux Observations over the Labrador Sea with ECMWF and NCEP Reanalyses*, J. Phys. Oceanogr. *32*, 383–400.

SCORER, R. S. (1988), *Sunny Greenland*, Q. J. Roy. Met. Soc. *114*, 3–29.

SMEDMAN, A. S., HÖGSTRÖM, U., and HUNT, J. C. R. (2003), *Effects of Shear Sheltering in a Stable Atmospheric Boundary Layer with Strong Shear*, Q. J. Roy. Met. Soc. *130*, 31–50.

SMOLARKIEWICZ, P. K. and ROTUNNO, R. (1989), *Low Froude Number Flow past Three-dimensional Obstacles. Part I: Baroclinically Generated Lee Vortices*, J. Atmos. Sci. *46*, 1154–1164.

SNYDER, W. H., THOMPSON, R. S., ESKRIDGE, R. E., LAWSON, R. E., CASTRO, I. P., LEE, J. T., HUNT, J. C. R., and OGAWA, Y. (1985), *The Structure of Strongly Stratified Flow over Hills: Dividing-streamline Concept*, J. Fluid Mech. *152*, 249–288.

STAMMERJOHN, S. E., DRINKWATER, M. R., SMITH, R. C., and LIU, X. (2003), *Ice-atmosphere Interactions during Sea-ice Advance and Retreat in the Western Antarctic Peninsula Region*, J. Geophys. Res. *108*, 3329.

STEFFEN, K. and BOX, J. (2001), *Surface Climatology of the Greenland Ice Sheet: Greenland Climate Network 1995–1999*, J. Geophys. Res. *106*, 33,951–33,964.

UOTILA, J. (2001), *Observed and Modelled Sea-ice Drift Response to Wind Forcing in the Northern Baltic Sea*, Tellus *53A*, 112–128.

VAN DEN BROEKE, M. R. and GALLÉE, H. (1996), *Observation and Simulation of Barrier Winds at the Western Margin of the Greenland Ice Sheet*, Q. J. Roy. Met. Soc. *122*, 1365–1383.

VOSPER, S. B., CASTRO, I. P., SNYDER, W. H., and MOBBS, S. D. (1999), *Experimental Studies of Strongly Stratified Flow past Three-dimensional Orography*, J. Fluid Mech. *390*, 223–249.

WOOD, N., and MASON, P. (1993), *The Pressure Force Induced by Neutral, Turbulent Flow over Hills*, Q. J. Roy. Met. Soc. *119*, 1233–1267.

(Received January 20, 2004, accepted April 12, 2004)

To access this journal online:
http://www.birkhauser.ch

Pure appl. geophys. 162 (2005) 1779–1793
0033–4553/05/101779–15
DOI 10.1007/s00024-005-2692-9

© Birkhäuser Verlag, Basel, 2005

❚ **Pure and Applied Geophysics**

Numerical Study on Vortices in the Middle Layer of Flow around a Large Mountain under Rotating Stratified Conditions

Yu Hozumi and Hiromasa Ueda

Abstract—Generation of cyclonic vortices in the middle layer of flow around a large mountain like Tibet and Rocky was investigated by means of a 3-D nonhydrostatic meteorological prognostic model. Special attention was paid to the effects of the earth's rotation and stratification on the vortices detached successively from the slope of a high and large horizontal scale mountain. It was found the successive formation and detachment of such 'von Karman-like vortices' occurred in the flow regime at high Rossby numbers Ro and low Froude numbers Fr. It was successfully divided by the criterion of baroclinic instability. This means that if the condition is unstable baroclinically, a lee vortex is destabilized into a three-dimensional one, while under baroclinically stable conditions the lee vortex with vertical axis retains its standing structure and remains long lasting in the middle layer.

Key words: Lee vortex, Lee wave, flow around mountain, stratification effect, Coriolis effect, rotating stratified flow, numerical model.

1. Introduction

Flow around a great mountain such as Tibet and Rocky has drawn increasingly more attention relating to the global circulation and regional- and meso-scale climate. Such a mountain supplies an enormous amount of heat and water to the middle and upper tropospheric atmosphere and so influences the global circulation (YANAI, *et al.* 1992; YANAI and LI, 1994). On the regional- and meso-scales it frequently causes severe storms. They are not only strong winds and draught weather but also at times hail and heavy rain. Lee vortex generated behind a great mountain is sometimes coupled with a baroclinic cyclone in the lower troposphere after traveling for long distances and then intensifies rapidly to produce heavy rain (HOZUMI and UEDA, 2003).

Such lee vortex formation and vortex shedding at the lee side of a mountain slope proceed under the interaction between surface friction and adverse pressure

[1]Disaster Prevention Research Institute, Kyoto University, Uji, 611-0011, Kyoto, Japan.
E-mail: hozumi@storm.dpri.kyoto-u.ac.jp

gradient, which cause the flow to separate from the surface. If the oncoming flow is stratified stably, the formation of the lee vortex is influenced by the internal gravity waves and if the mountain has a large horizontal scale, the vortex shedding is also affected by Coriolis force. An excellent schematic model for the flow around a mountain at low Froude numbers Fr was presented by HUNT et al. (1997) and HUNT et at. (2001). They distinguished the stratified flow field into two layers. They were called the top layer and the middle layer. The top layer was characterized by the flow passing over the mountain crest and by the lee wave and the lee vortex with horizontal axis which sometimes resulted from the internal gravity wave. While in the middle layer the flow detoured around the mountain side slope and lee vortices with vertical axis were created. This implies that both of the vortices with vertical axis and those with horizontal axis coexist in the wake region of the mountain at low Fr numbers.

Numerous studies have been performed on the effects of stratification on the flow around a mountain. These studies include laboratory experiments by LIN et al. (1992) and HUNT and SNYDER (1980) and numerical models by HANAZAKI (1988) and BAINES and SMITH (1993). An excellent review was also presented by SMITH (1979). However, in the case of a great mountain with large horizontal scale effects of the earth's rotation becomes dominant on the flow around the mountain, together with the stratification effects. At low Rossby numbers Ro the earth's rotation brings about a difference of the flow velocities on the left and right hand sides of the mountain looking leeward. This horizontal wind shear has the potential to change the regime of vortex generation in the wake (e.g., HUPPERT, 1975; HUPPERT and BRYAN, 1976; CHAPMAN and HAIDVOGEL, 1993; BOYER and CHEN, 1987). Moreover the geostrophic balance due to Coriolis force compels creation of different vertical wind shears and horizontal temperature gradients on the left and right hand sides of the mountain side slope. Thus the rotating effect produces the change of the dynamical state of vortex formation in addition to the change of flow patterns in the middle layer.

In this paper rotating and stratified flows around a great mountain were simulated by a 3-D non-hydrostatic meteorological prognostic model (SHA et al., 1996, 1998), and the vortex formation, vortex shedding and the effects of rotation and stratification on their dynamical stability were investigated. Special attention was paid to the vertical vortices in the middle layer. Here it should be noted that the range of Froude number Fr is limited to Fr < 1. In that condition the middle layer is created where the oncoming flow detours around the mountain side. In section 2, the numerical model and numerical experiments are described. In section 3 a detection method of the vertical vortex in the near-wake region is devised. Characteristics of lee vortices in the middle layer and its dependence on Fr and Ro are discussed in section 4, and the conclusions are summarized in section 5.

2. Model Description and Outline of Numerical Experiments

The numerical model used is a 3-D non-hydrostatic meteorological prognostic model (SHA et al., 1996, 1998). It consists of the primitive equations with anelastic and Boussinesq approximations. A terrain following the σ coordinate system is adopted for its vertical ordinate. It is gridded nonequally ($\Delta\sigma$: 100 m $\sim$ 2000 m). Horizontal grids are equally spaced ($\Delta x = \Delta y = 50$ km). The calculation domain is x: 6000 km $\times$ y: 6000 km $\times$ z: 25 km divided by the total grids 120 $\times$ 120 $\times$ 44.

As the initial condition a uniform westerly wind U and uniform thermal stability (constant at Brunt Väisälä frequency N) are assumed. The vertical eddy diffusivity of momentum is solved by the scheme of BLACKADAR (1962), and that of heat is calculated by the turbulent Prandtl number Pr_t depending on the Richardson number. The Coriolis effect is considered for the constant Coriolis parameter f plane.

For the side boundary conditions the uniform westerly wind and thermal stability are specified at the west side boundary, while the other three lateral boundaries are set to a radiation condition (SHA et al., 1991). The upper boundary condition is fixed to the initial constant value for wind and temperature with a dumping layer from 15 km to 25 km, while the lower boundary is constrained by a no-slip condition. As for the dynamical pressure, a 3-D Poisson equation is solved with the Neumann boundary condition (SCHUMANN and VOLKERT, 1984). An "impulsive start" is adopted, and the time integral is executed until a nondimensional time τ (= Ut/L) = 25.0.

Numerical simulations were performed for mountains of various sizes, however the mountain shape was assumed to be fixed as

$$h(x,y) = h_m \cdot \exp\left\{-\frac{(x-x_0)^2 + (y-y_0)^2}{L^2}\right\}, \tag{1}$$

where h_m is the height of the crest, L is the horizontal scale of the mountain, and (x_0, y_0) is the center position of the mountain. Typically, $h_m = 4$ km and $L = 500$ km so that in the present work a great massif such as Tibet in the mid-latitudes is assumed. In this study, the effect of local heating or cooling of air flow on the mountain slope is absent.

A number of simulations were made for various flow parameters and mountain sizes. They are the approach wind speed U, Brunt- Väisälä frequency N, Coriolis parameter f, mountain height h_m and mountain horizontal scale L. The U and N are fixed uniformly. Two series of simulations were performed. One is that U and N are variable but the remainders are fixed, i.e., f (= f_0)=8.338 $\times$ 10^{-5} s^{-1}, $h_m = 4$ km and $L = 500$ km. Another series is that f , h_m and L are variable instead of the fixed $U = 10$ ms^{-1} and $N = 0.02$ s^{-1}. They are summarized in Tables 1 and 2.

Table 1

Flow parameters for simulation series I. (U and N are varied, but $f(=f_0)=8.338 \times 10\text{-}5 \text{ s}^{-1}$, $h_m=4$ km and L = 500 km)

U (m s^{-1})	3.0, 5.0, 6.0, 8.0, 10.0, 12.0, 14.0, 15.0, 16.0, 18.0, 20.0, 22.0, 24.0, 26.0, 28.0, 30.0, 32.0, 36.0, 40.0
N (s^{-1})	0.0050, 0.0100, 0.0150, 0.0200

Table 2

Flow parameters for simulation series □ (f, h_m and L are varied, but $U=10$ m s^{-1}, $N=0.02$ s^{-1} and $f_0 =8.338 \times 10\text{-}5 \text{ s}^{-1}$)

Coriolis factor f / f_0	0.10, 0.20, 0.40, 0.60, 0.80, 1.20, 1.40, 1.60
Mountain height h_m (km)	2.0, 3.0, 5.0, 6.0, 7.0, 8.0
Mountain horizontal scale L (km)	100, 200, 250, 400, 750

3. Detection Method of Vortex Center in Near-wake Region

To see the vortex formation and its dependence on Ro, Fr and mountain parameters, the following criterion was defined for the existence of the vortex, that is

$$\mathbf{u}(x, y) = \mathbf{0},$$

$$D_k(x, y) = \frac{\partial(k \times u(x, y)) \cdot i_k}{\partial(r \cdot i_k)} > 0 \quad \text{for all k or } D_k(x, y) < 0 \quad \text{for all } k, \qquad (2)$$

where *u* (x,y) denotes the horizontal wind vector and *k* is the vertical unit vector. In the detector function $D_k(x,y)$, *r* is the horizontal vector of examination and i_k is the horizontal unit vector along the vector of examination *r* in θ_k direction. Those are defined as

$$r = x\mathbf{i} + y\mathbf{j},$$

$$i_k = \cos \theta_k \mathbf{i} + \sin \theta_k \mathbf{j}, \quad \theta_k = 2\pi k/n (k = 0, 1, \ldots, n - 1), \qquad (3)$$

where *i* and *j* denote the unit vectors in streamwise(x) and spanwise(y) directions, respectively and *n* is the total count of examination. For this criterion $\mathbf{u}(x, y) = \mathbf{0}$ is added for the detection of vortex in the near-wake region.

▶

Figure 1(a)
Wind vectors and vertical vorticity (isopleth for every 20×10^{-6} s^{-1}) in the horizontal plane of h = 1500 m at $\tau(= tU/L) = 4.8$ for RUN 1. •: cyclonic vortex center, □: anticyclonic vortex center. Open circle is the cross section of the mountain.
Figure 1(b)
Same as Figure 1(a), but $\tau = 6.6$.
Figure 1(c)
Same as Figure 1(a), but $\tau = 8.4$.

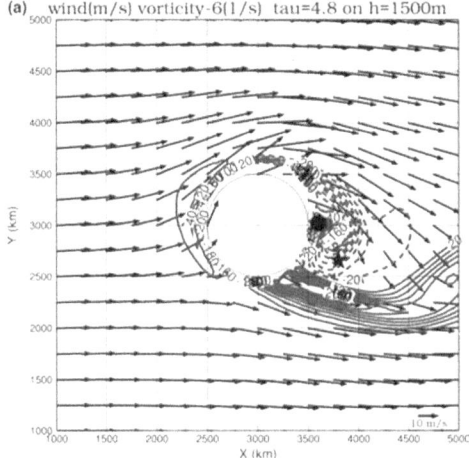

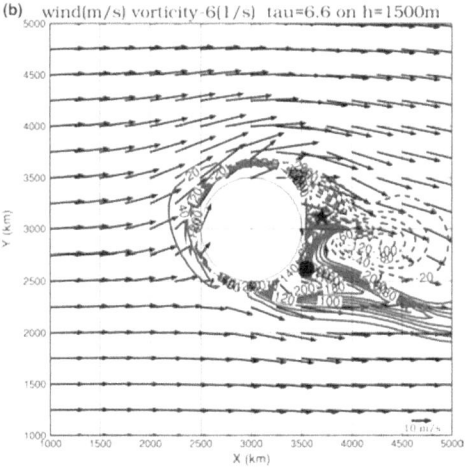

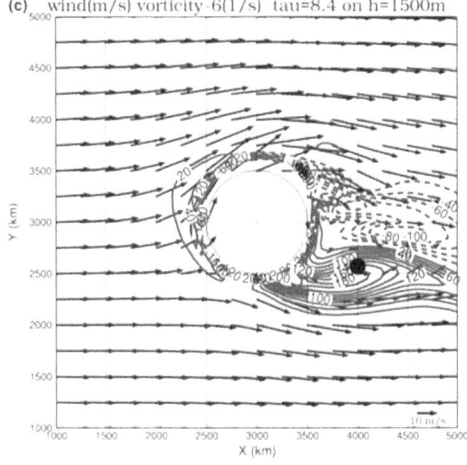

Adopting $n = 8$, this criterion was applied to the flow fields after $\tau = 3.6$ in each run. If every $D_k(x, y)$ is larger than zero at a certain point (x, y), a cyclonic vortex is assumed to exist at that point, and if $D_k(x, y)$ is less than zero for all k, an anticyclonic vortex is formed at that point. On the other hand, when $D_k(x, y)$ is larger than zero for some of k and $D_k(x, y)$ is less than zero for other k' at a certain point (x, y), it means no vortex exists. Examples of the location of the vortex center in the near-wake region detected by this criterion are shown in Figures 1(a)–(c) and Figure 3.

4. Results and Discussion

4.1 Flow Configuration and Vortex Generation in the Middle Layer under Rotating Stratified Conditions

A typical flow configuration associated with vortex formation and vortex shedding in the middle layer is presented in Figure 1, 2 and 3. Here the flow parameters are $U =$

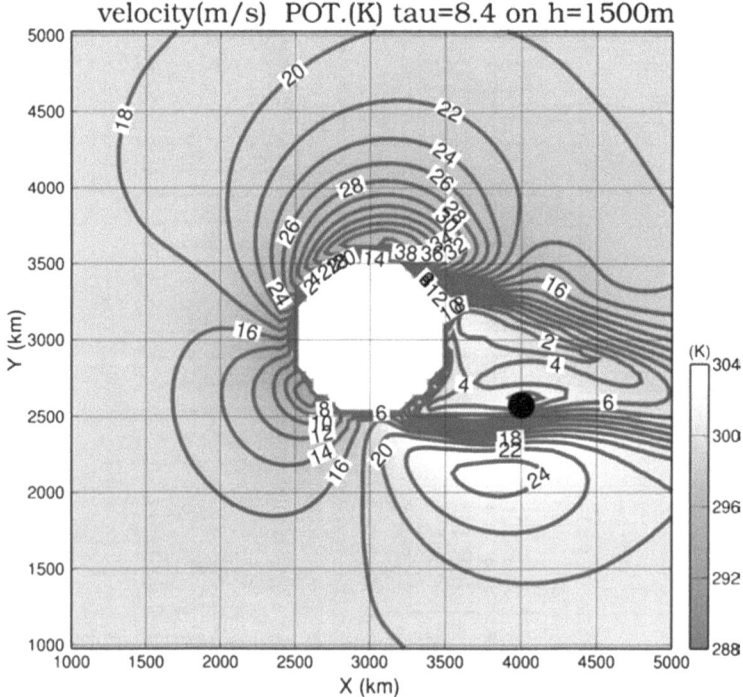

Figure 2

Horizontal wind speed (isopleth for every 2 ms^{-1}) and temperature (shaded) in the $h = 1500$ m plane at τ = 8.4. •: cyclonic vortex center. Open circle is the cross section of the mountain.

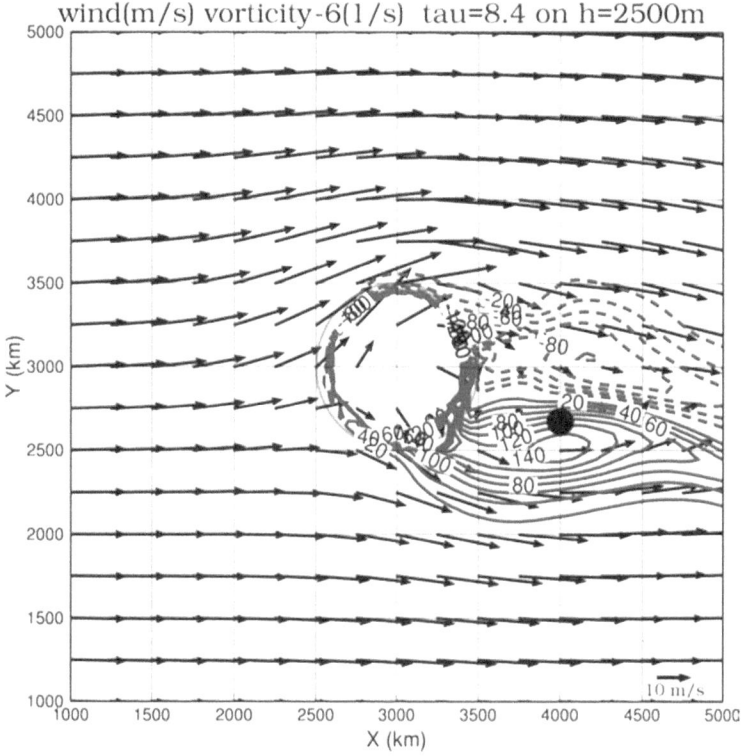

Figure 3
Same as Figure 1(c), but in the $h = 2500$ m plane.

18 m s^{-1}, $N = 0.015$ s^{-1}, Fr $= 0.3$ and Ro $= 0.43$. It is called hereafter the RUN 1. As seen in Figure 1(a)–(c), the flow pattern is not symmetric due to the Coriolis effect. On the upstream side of the mountain the stagnation point shifts significantly to the south. Because of the stable stratification (Fr $= 0.3$) the oncoming flow to the stagnation point cannot rise to the mountain top but rises for a short distance. It creates the minimum potential temperature region in the upstream splitting region, as seen in Figure 2. Because of the stagnation point shift to the south, the split flow to the north is accelerated more than that to the south and the local peak velocity on the side slope attain 38 m s^{-1} that is two times larger than the oncoming flow $U = 18$ m s^{-1}, while the peak velocity is 24 m s^{-1} in the split flow to the south. This is typically seen in the flow around a great massif in the Northern Hemisphere. This asymmetric flow pattern and the accelerated wind peaks are well-explained qualitatively by the quasi-geostrophic potential vorticity conservation theorem of HUPPERT and BRYANW (1976).

On the lee side the flow pattern changes with time. Corresponding to the vortex formation and detachment, the structure of the mountain wake changes drastically. The distribution of the vertical vorticity, together with the horizontal wind vectors, is illustrated in Figures 1(a)–(c). In those diagrams, two minimum velocity areas with almost zero velocity are clearly seen in the wake region behind the mountain. Cyclonic and anticyclonic vortices are generated in this wake region.

They are shed successively and move downstream like 'von Karman vortex pair'. Nondimensional frequency of the vortex shedding, i.e., Strouhal number St was around 0.15 for most simulations, when the average horizontal size of the mountain in the middle layer is adopted as the reference length scale. This value St = 0.15 approximates the value of 0.2 for two-dimensional flow around a circular cylinder.

In order to observe the vertical structure of these vortices, the flow pattern at the height of $h = 2500$ m is depicted in Figure 3. When compared with Figure 1(c) for $h = 1500$ m, the flow pattern and the vorticity distribution are similar. In particular, the cyclonic vorticity anomaly and the positions of vortex centers are located at a same x-y position in spite of different heights. This suggests that the axis of this vortex tube stands vertically.

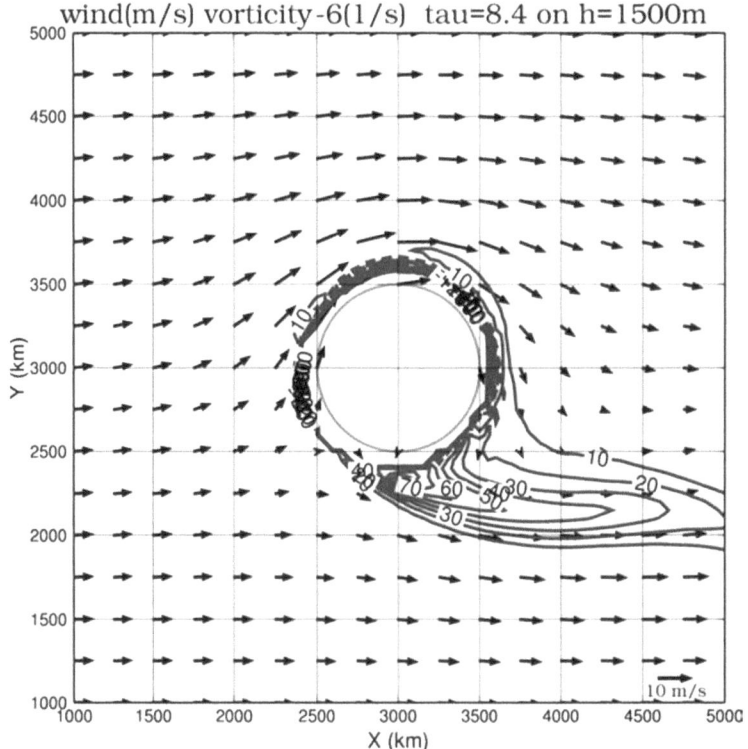

Figure 4

Wind vectors and vertical vorticity (isopleth for every 10×10^{-6} s^{-1}) in the $h = 1500$ m plane at $\tau = 4.8$ for RUN 2.

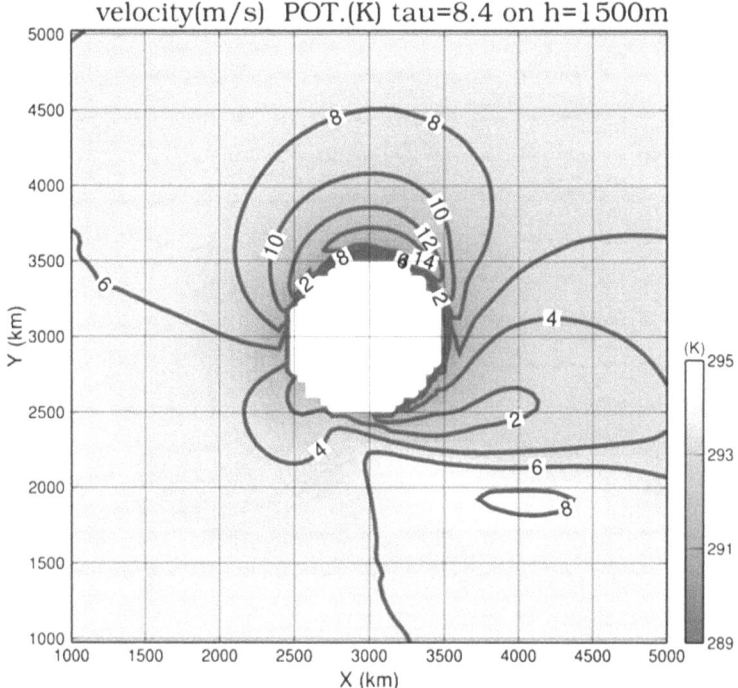

Figure 5
Horizontal wind speed (isopleth for every 2 m s^{-1}) and temperature (shaded) in the $h = 1500$ m plane at
$\tau = 8.4$.

As a typical case in which vortex shedding does not occur, results of RUN 2 for U = 6 ms^{-1}, N = 0.005 s^{-1}, Fr = 0.3 and Ro = 0.14 are presented in Figure 4, 5 and 6. In this case Froude number Fr is the same but Rossby number Ro is less than that in RUN 1, and so the earth's rotation and/or low wind speed work to suppress the vortex formation. In those diagrams any moving cyclone-anticyclone pair does not appear. The oncoming flow pattern (Fig. 4) is similar to that of Figure 1(c) except for a much larger shift of the upstream stagnation point. However, the wake flow patterns are also different. In this case the split flow on the northern mountain side detours to the south on the back side of the mountain and deters the flow separation. As a result, the separated area is restricted in the southern part of the mountain back side and only cyclonic vorticity dominates in that area. While the anticyclonic vorticity is distributed on the front and northern mountain sides, as seen clearly at height $h = 2500$ m in Figure 6. This is a major contrast to RUN 1 in which both the cyclone and anticyclone are generated in the wake region on the back side of the mountain. In RUN 2, there is no longer any vortex detected by the criterion (2). When Figures 4 and 6 are compared, the horizontal shear line on the southern boundary of the wake seems to tilt to $+y$ direction with height.

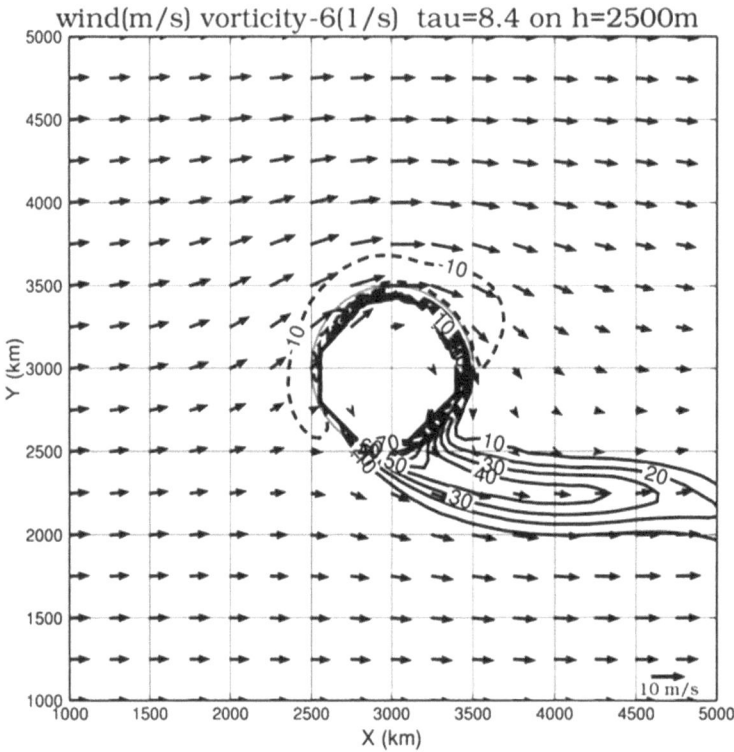

Figure 6
Same as Figure 4, but in the $h = 2500$ m plane.

4.2 Flow Classification on the Basis of Vortex Shedding

There appear many vortices with various sizes and lifetimes under different rotating and stratified conditions. Based on the vertical vortices detected by the criterion (2) in the near-wake region and on whether, the vortex shedding occurs or not, classification of the rotating and stratified flows around the mountain was made. The result was illustrated in Figure 7. In the diagram the symbol "○" indicates the existence of a vortex which has a dimensionless lifetime τ_{lf} longer than 4.0 and moves for a distance d_{tr} more than the horizontal mountain size L before it disappeared, while the symbol "□" denotes a vortex with shorter lifetime but longer than 1.0. The symbol "×" indicates that no vortex or a vortex with its lifetime less than 1.0 exists. In this diagram it is clearly seen that flows associated with the vortex shedding are classified on the basis of Fr and Ro, and vortex formation occurs in the range of lower Fr and higher Ro and no vortex is formed in the range of high Fr and low Ro.

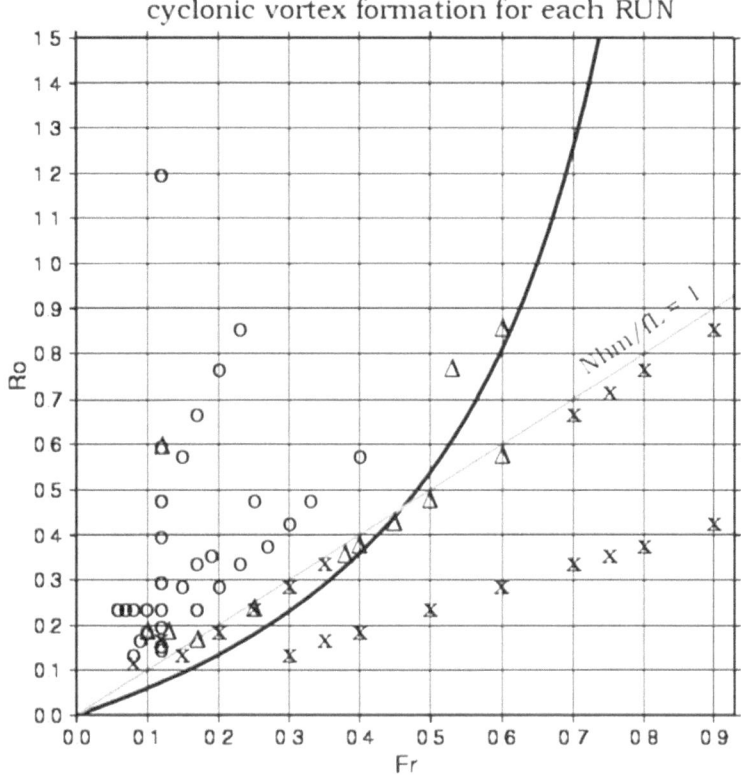

Figure 7

Diagram of flow regime associated with vortex formation in the middle layer. ○: vortex with $\tau_{lf} > 4.0$ and $dt_r > L$, □ : $1.0 < \tau_{lf} < 4.0$, × : .$\tau_{lf} < 1.0$ or no vortex formation. Thick line is the critical one for baroclinic instability (right-hand side: unstable, left: stable).

4.3 Dynamic Stability of Vertical Vortices in the Middle Layer of Near-wake Region

As seen in Figure 7, vortex formation occurs in the range of lower Fr and higher Ro. In the middle layer stably stratified flow around the mountain slope is quasi-two-dimensional and behaves similarly to the flow around a circular cylinder. It separates from the surface of the mountain slope and tends to create a vortex pair in the wake region. Since buoyancy works to tilt the vortex axis to the vertical direction under the stably stratified conditions, vertical vortices become dominant. On the other hand, the flow in the top layer is assumed to be the same as the flow around the obstacle at Fr number equal to one. It is because the lower boundary of the top layer is assumed to be located at the so-called dividing streamline and therefore the Fr number based on the height from the dividing streamline to the mountain top is equal to one. Thus, on the lee side of the mountain a strong downslope wind is created. It induces a strong internal gravity wave and at times a transverse horizontal vortex in the top layer.

As the Fr number decreases, the thickness of the middle layer becomes larger and that of the top layer decreases, and thus the horizontal vortex in the top layer will become relatively weak and hardly interact with the middle layer. As a result, the vertical vortex in the middle layer is thought to remain as it is.

On the other hand, as the Ro number is increased, the vortex formation and shedding tend to be activated. It is easily understood from the definition of Ro = U/fL. Here, the increase of U and/or decrease of L tend to generate stronger vortical motion along the side slope of the mountain. Coriolis effect is somewhat complicated to explain. As the Coriolis parameter f is increased, a westerly oncoming wind, for example, tends to flow to the north in front of the mountain and the northern split flow detours down to the south on the back side of the mountain. At a critical state the upstream streamline and the lee side streamline merge on the southern side of the mountain and form a closed loop. This Coriolis effect was analyzed extensively by a linear model by HUPPERT and BRYAN (1976) and was certified by a numerical model in the previous work (HOZUMI and UEDA, 2003). This detouring flow caused by the Coriolis effect is considered to work to suppress the vortex formation or to decrease the vorticity on the lee side in the middle layer. This is in contrast to the flow for $f = 0$ in that strong vortex pairs are created successively in the separated region. Thus it can be concluded that an increase in Ro number causes activation of the vertical vortices in the middle layer.

This qualitative explanation of the effects of stratification and the earth's rotation on the vortex formation and vortex shedding from the mountain slope was examined by the baroclinic stability criterion in the light of whether the disturbance in the form of a vertical vortex can be retained as the two-dimensional one or be destabilized to a three-dimensional one under a baroclinic instability condition.

4.4 Stability Criterion of Vertical Vortices in the Middle Layer

Referring to PEDLOSKY (1987), at first we consider the condition of the baroclinic instability for the disturbance of streamwise wavelength k and spanwise one l located in the rotating stratified fluid layer at constant f_o and N. Considering the disturbance in the linearized set of quasi-geostrophic potential equations and temperature equations which are affected by stratification and baroclinicity, the necessary condition for it to grow is derived as

$$\frac{\mu}{2} < \coth \frac{\mu}{2}, \tag{4}$$

where μ is defined as

$$\mu^2 = \frac{N_s^2(k^2 + l^2)(p_b - p_t)^2}{f_0^2}, \tag{5}$$

and

$$N_s = \alpha_0 N / g, \tag{6}$$

a_0 is the specific volume of the air at reference state, g is the acceleration due to gravity, and p_t and p_b are the pressures at the top and bottom of the fluid layer. Thus the necessary condition (4) for the disturbance to grow is satisfied under the condition of

$$0 < \mu/2 < \mu_c/2 = 1.1997. \tag{7}$$

Now we examine this instability condition for the disturbance in the middle layer of the rotating stratified flow around the mountain. Assuming that the upper boundary of the middle layer can be represented by the level of dividing streamline and the lower boundary by the zero ground level, their heights can be written as

$$h_t = h_m \left(1 - \frac{U}{Nh_m} \right) = h_m (1 - \text{Fr}), \tag{8}$$

$$h_b = 0. \tag{9}$$

Thus their pressure difference is approximated by

$$p_b - p_t = (g/\alpha_0) h_m (1 - \text{Fr}). \tag{10}$$

Assuming that the streamwise and spanwise wave lengths are represented by $k \sim l \sim 2\pi/2L$, the baroclinic instability criterion (7) is written as

$$\frac{Nh_m}{f_0 L} < \frac{2.4}{2\sqrt{2\pi}} \frac{1}{1 - \text{Fr}}, \tag{11}$$

$$\text{Ro} < \frac{2.4}{2\sqrt{2\pi}} \frac{\text{Fr}}{1 - \text{Fr}}. \tag{12}$$

This criterion of the baroclinic instability is plotted by a thick curve in Figure 7. The right hand side area of this curve represents the baroclinically unstable regime for the disturbance and the left represents the regime stable. The two regimes with and without detectable lee vortices or 'von Karman-like vortices' can be distinguished clearly by this baroclinic instability curve. It suggests that if the rotating and stratified flow on the lee side of the mountain is unstable baroclinically, a lee vortex is destabilized into a three-dimensional one, while under baroclinically stable conditions the lee vortex with a vertical axis retains its standing structure and remains long-lasting as a 'von Karman-like vortex' in the middle layer.

5. Conclusions

In order to investigate the effects of the earth's rotation and stratification on the flow around a great mountain with a large horizontal scale and height, numerical experiments were performed by means of a 3-D nonhydrostatic meteorological prognostic model, changing the oncoming flow velocity, Coriolis parameter, Brunt-Väisälä frequency and horizontal and vertical scales of the mountain. Special attention was paid to the formation of vortices with vertical axis and the vortex shedding in the middle layer around the mountain slope.

It was ascertained that the vortex formation and vortex shedding occurred in the flow regime at high Rossby numbers Ro and low Froude numbers Fr. Vertical vortices were shed successively and moved downstream like a von Karman vortex pair. Nondimensional frequency of the vortex shedding, i.e., Strouhal number St was around 0.15 for most simulations, when the average horizontal size of the mountain in the middle layer is adopted as the reference length scale. This value St = 0.15 approximates the value of 0.2 for the two-dimensional flow around a circular cylinder.

The two regimes with and without detectable lee vortices or 'von Karman-like vortices' could be distinguished clearly by the baroclinic instability curve. This suggests that if the condition is unstable baroclinically, a lee vortex is destabilized into a three-dimensional one, while under baroclinically stable conditions the lee vortex with a vertical axis retains its standing structure and persists as a 'von Karman-like vortices' for a long time in the middle layer.

Acknowledgements

Delightful appreciation to Professor. M.P. Singh for useful discussions and suggestions. This work was supported by the Ministry of Education, Culture, Sports, Science and Technology, Japan under the contract of Research Revolution 2002 (RR2002). Numerical computation was made possible by the team of the Computer Center in the National Institute for Environmental Studies whom we thank for their kind help and enthusiasm.

REFERENCES

BAINES, P.G. and SMITH, R.B. (1993), *Upstream Stagnation Points in Stratified Flow Past Obstacles*, Dyn. Atmos. Ocean *18*, 105–113.

BLACKADAR, A.K. (1962), *The Vertical Distribution of Wind and Turbulent Exchange in a Neutral Atmosphere*, J. Geophys. Res. *67*, 3095–3102.

BOYER, D.L., CHEN, R. (1987), *Laboratory Simulation of Mountain Effects on Large-scale Atmospheric Motion Systems: The Rocky Mountains*, J. Atmos. Sci. *44*, 100–123.

BOYER, D.L., ZHANG, X. (1990), *The Interaction of Time-dependent Rotating and Stratified Flow with Isolated Topography*, Dyn. Atmos. Ocean *14*, 543–575.

CHAPMAN, D.C., HAIDVOGEL, D.B. (1993), *Generation of Internal Lee Waves Trapped over a Tall Isolated seamount*, Geophys. Astrophys. Fluid Dyn. *69*, 33–54.

DING, L., STREET, R.L. (2003) *Numerical Study of the Wake Structure behind a Three-dimensional Hill*, J. Atmos. Phys. *60*, 1678–1690.

FERRERO, E., LONGLISCI, N., LONGHETTO, A. (2002), *Numerical Experiments of Barotropic Flow Interaction with a 3-D Obstacle*, J. Atmos. Sci., *59*, 3239–3253.

HANAZAKI, H. (1988), *A Numerical Study of Three-dimensional Stratified Flow Past a Sphere*, J. Fluid Mech. *192*, 393–419.

HOLLAND, D.M. (2001), *Explaining the Weddell Polynya: A Large Ocean Eddy Shed at Maud Rise*, Science *292*, 1697–1700.

HOZUMI, Y., UEDA, H. (2003), *Numerical Studies on the Vortices Generated in the Flow around a Great Mountain under Rotating Stratified Conditions*, J. Atmos. Sci. (submitted).

HUNT, J.C.R., SNYDER, W.H. (1980), *Experiments on Stably and Neutrally Stratified Flow over a Model Three-dimensional Hill*, J. Fluid Mech. *96*, 671–704.

HUNT, J.C.R., FENG, Y., LINDEN, P.F., GREENSLADE, M.D., MOBBS, S.D. (1997), *Low Froude Number Stable Flows Past Mountains*, Il Nuovo Cimento. *20C*, 261–272.

HUNT, J.C.R., OLAFSSON, H., BOUGEAULT, P. (2001), *Coriolis Effects on Orographic and Mesoscale Flows*, Quart. J. Roy. Met. Soc. *127*, 601–633.

HUPPERT, H.E. (1975), *Some Remarks on the Initiation of Inertial Taylor Columns*, J. Fluid Mech 67, 397–412.

HUPPERT, H.E., BRYAN, K. (1976), *Topographical Generated Eddies*, Deep-Sea Res. *23*, 655–679.

LIN, Q., LINDBERG, W.R., BOYER, D.L., FERNANDO, H.J.S. (1992), *Stratified Flow Past a Sphere*, J. Fluid Mech. *240*, 315–354.

PEDLOSKY, J., *Geophysical Fluid Dynamics* (Springer-Verlag, New York, 1987).

SCHAR, C., DURRAN, D.R. (1997), *Vortex Formation and Vortex Shedding in Continuously Stratified Fows Past Isolated Topography*, J. Atmos. Sci. *54*, 534–554.

SCHUMANN, U., VOLKERT, H. (1984), *Three-dimensional mass- and momentum-consistent Helmholtz equation in terrain-following coordinates*. In *Efficient Solutions of Elliptic Systems*, (Ed. Hackbusch, W) (Friedrick Vieweg and Son, Kiel 1984), pp. 109–131.

SHA, W., KAWAMURA, K., UEDA, H. (1991), *A Numerical Study on Sea/Land Breezes as a Gravity Current: Kelvin-Helmholtz Billows and Inland Penetration of the Sea-breeze Front*, J. Atmos. Sci. 48, 1649–1665.

SHA, W., GRACE, W., PHYSICK, W. (1996), *A Numerical Experiment on the Adelaide Gully Wind of South Australia*, Aust. Meteor. Mag. *45*, 19–40.

SHA, W., NAKABAYASHI, K., UEDA, H. (1998), *Numerical Study on Flow over a Three-dimensional Obstacle under a Strong Stratification Condition*, J. Appl. Meteorol. 37, 1047–1054.

SHEPPARD, P.A. (1956), *Airflow over Mountains*, Quart. J. Roy. Met. Soc. *82*, 528–529.

SMITH, R.B. (1979) *The Influence of Mountains on the Atmosphere, Advances in Geophysics*, vol. 21, (Academic Press 1979), pp. 87–230.

SUN, W.-Y., CHERN, J.-D. (1994), *Numerical Experiments of Vortices in the Wake of Large Idealized Mountains*, J. Atmos. Sci. *51*, 191–209.

YANAI, M.H., LI, C.F., SONG, Z.S. (1992), *Seasonal Heating of the Tibetan Plateau and its Effects on the Evolution of the Asian Summer Monsoon*, J. Meteor. Soc. Japan *70*, 319–351

YANAI, H., LI, C.F. (1994), *Mechanism of Heating and the Boundary-layer over the Tibetan Plateau*, Mon. Wea. Rev. *122*, 305–323.

(Received December 15, 2003; accepted March 25, 2004)
Published Online First

To access this journal online:
http://www.birkhauser.ch

Pure appl. geophys. 162 (2005) 1795–1809
0033–4553/05/101795–15
DOI 10.1007/s00024-005-2693-8

© Birkhäuser Verlag, Basel, 2005

| Pure and Applied Geophysics

A Model Study of the Strong and Weak Wind, Stably Stratified Nocturnal Boundary Layer: Influence of Gentle Slopes

S. G. Gopalakrishnan[1], Frank Freedman[2], Maithili Sharan[3],
T. V. B. P. S. Rama Krishna[3]

Abstract—With the exception of intermittency and waves, a brief review of the observed and modeled mean structure of the nocturnal boundary layer (NBL) is presented. The effect of gentle slopes on strong and weak wind NBL was investigated here using a one-dimensional model, with a simple correction term to account for the slope effects, identical to the one used by Brost and Wyngaard (1978). The study indicates that the wind profiles, temperature profiles and surface layer turbulence characteristics are extremely sensitive to the imposed geostrophic wind when small slopes are present especially for light winds. This is due to the complex interaction between the buoyancy driven slope flow and the imposed geostrophic wind that in turn influence the shear generation of turbulence. Finally, the current issues in the modeling of weak wind boundary layer are discussed.

Key words: Turbulence, weak wind NBL, inertial oscillations, inversion depth, nocturnal jet, nocturnal boundary layer.

1. Introduction

The structure of the nocturnal boundary layer (NBL) is significantly affected by the synoptic and mesoscale forcings. Past studies, including the Wangara (Clarke et al., 1971) and Cabauw (Garratt, 1982 and Niewstadt, 1984) field experiments, theoretical studies of Niewstadt (1985) and Derbyshire (1990), and the numerical studies by Yamada and Mellor (1975), Andre et al. (1978), Garratt and Brost (1981), Stull and Driedonks (1987), Moeng and Sullivan (1994) have all led to a good understanding of the structure and evolution of the NBL over fairly flat and homogeneous domains, and under windy conditions. Andre and Mahrt (1982), for instance, examined the fair-weather, NBL data from the Wangara and Voves experiments and inferred that the lower part of the nocturnal inversion layer

---

[1]SAIC and EMC/NCEP/NOAA, 5200 Auth Road, Camp Springs, MD, 20746, U.S.A.
[2]EMC/NCEP/NOAA, 5200 Auth Road, Camp Springs, MD, 20746, U.S.A.
[3]Centre for Atmospheric Sciences, Indian Institute of Technology, Hauz Khas, New Delhi, 110 016, India.

normally appears to be turbulent but strongly stratified. The thicker part of the inversion layer is characterized by weaker stratification which appears to be mostly generated by clear-air radiative cooling. The radiatively cooled, nocturnal inversion layer thickens significantly as night proceeds. However, the thickness of the turbulent layer normally varies slowly during the night, but differs significantly from night to night.

Often, well within the inversion layer, as thermal stratification progresses, a low-level wind maxima or a low-level nocturnal jet (see MAHRT et al.,1979; LENSCHOW et al.,1988 and KURZEJA et al., 1991, for instance) generally appears. The wind maxima in the NBL is formed in response to an undamped inertial oscillation that is set-up in the lower atmosphere, at sunset, when the turbulent fluxes start to decay rapidly and represents one of the earliest manifestations of a NBL phenomenon amenable to simple analytical treatment (BUAJITTI and BLACKADAR, 1957; BLACK-ADAR, 1957). For several years Professor M.P. Singh and his group have been pioneers in developing more complex analytical models and have significantly contributed to the understanding of the dynamics of these oscillations, and the subsequent development of the nocturnal jet over flat and homogeneous domains. SINGH et al. (1993), for instance, developed an analytical model for studying variations in the diurnal wind structure in the planetary boundary layer and the evolution of the low-level nocturnal jet. A time-dependent eddy-diffusivity coefficient corresponding to solar input is used. This analytical framework was used to examine the role of boundary-layer shear in the dispersion process (MCNIDER et al.,1993; SINGH et al., 1997). The model assumed constant eddy diffusivity and a linear variation of the geostrophic wind with height. Further, SINGH et al. (1999) have obtained an analytical solution by accounting for time-dependence of geostrophic wind in their earlier model (SINGH et al., 1993). The wind maxima that occurs at the top of the layer of significant turbulence is usually referred to as the top of the momentum layer (MAHRT et al., 1979). It is significantly influenced by several factors, such as, baroclinicity, surface cooling rate, radiative cooling of air above the land-surface, conditions at sunset and the terrain. Most of the numerical studies (ANDRE et al., 1978; ZEMAN, 1979; MCNIDER and PIELKE, 1981; 1984; STULL and DRIEDONKS, 1987; MCNIDER et al.,1988) have been more or less successful in modeling this phenomena over flat domains and mountainous terrain, as well. However, in most cases the ambient forcing was strong.

Weak wind conditions prevail all over the globe for a considerable period of time and assume special importance in a stable atmosphere due to their high air pollution potential (see, for instance, SHARAN et al., 1995, 1996, 2000; GOPALAKRISHNAN and SHARAN, 1997 for the impact of weak wind on the dispersion of Methyl Isocyanate during the infamous Bhopal gas accident). Theoretical studies of ESTOURNEL and GUEDALIA (1985, 1987) and GOPALAKRISHNAN et al. (1998) indicate that, over a fairly flat and homogeneous domain, the radiative and turbulent processes that control the evolution of the NBL differ considerably between strong and weak wind conditions. While ESTOURNEL and GUEDALIA (1985) illustrated that the stable inversion layer evolves in a weak wind NBL, whereas, the depth of the layer varies little under strong wind conditions, GOPALAKRISHNAN et al. (1998) numerically showed that when shear-driven turbulence is weak, clear air radiative cooling plays a dominant role in the integrated cooling budget within the NBL. More recently, RAMAKRISHNA et al. (2003) studied the effects on the NBL of geostrophic wind. The major objective of their study was to determine if the mean structure and evolution of the weak wind NBL is very different from those under windy conditions. Some meteorological data collected during the plume validation experiment conducted by Electric Power Research Institute (EPRI) over a nearly flat and homogeneous terrain at Kincaid, USA (39°35N and 89°25W) and an improved version of one-dimensional meteorological boundary layer model originally developed by PIELKE (1974) and further modified by SHARAN and GOPALAKRISHNAN (1997) with TKE mixing length closure and a layer-by-layer emissivity based radiation scheme (MAHRER and PIELKE, 1977; GOPALAKRISHNAN et al., 1998) was used for that purpose. The study shows that, over the flat and homogeneous domain, while larger shear resulting from stronger wind produced NBL with an average depth of over 300 m in which the shear production was larger than the buoyancy consumption at least within the lower 100 m (layer cooled by turbulence), a weak geostrophic wind produced shallow NBL with an average depth of about 100 m in which shear production was weak and comparable to buoyancy consumption even within the turbulence layer. As part of our continuous effort to understand the structure of weak wind stable boundary layer here we perform sensitivity experiments to an existing column model framework by including a simple correction term for explaining mild terrain slope effects.

The study of slope flows in the NBL is not of recent origin. It is well known that the air near surface cools due to the rapid radiative cooling on a clear night, and a flow develops over an inclined surface when the cooled air layer is heavy enough to overcome friction. Commencing with studies in 1951 by Defant (see MCNIDER and PIELKE, 1984 for an excellent review), simplified models supported by several observations of nocturnal slope flows have illustrated several important features of the NBL that develop over such an heterogeneity (MANINS and SAWFORD, 1979; RAO and SNODGRASS, 1981; GARRATT, 1983; MAHRT, 1982; DORAN and HORST, 1983). BROST and WYNGAARD (1978) used a second-order turbulence model and showed that even slight terrain slopes on the order of 0.002 and a cooling rate of 2 K/h have very strong effects on the NBL. Here we illustrate the influence of geostrophic forcing on their results. In the next section we briefly describe the numerical model and the experimental set-up. Section 3 presents the results, and the implications of this analysis are discussed in the conclusions (Section 4).

2. Numerical Model and Experimental Set-up

A detailed description of the model used in this study is given in FREEDMAN and JACOBSON (2003). Thus, for brevity, only a summary of its major features relevant to the present study is provided here. A one-dimensional, turbulence closure model, with a simple correction term to account for the slope effects identical to the one used by BROST and WYNGAARD (1978) is used. The correction term is derived assuming a hydrostatic, Boussinesq atmosphere, and therefore valid only for "weak" slopes (MAHRT, 1982). The model was originally developed by one of the authors (Freedman) and is undergoing continuous development. The following is the set of governing equations for mean stream-wise flow, u, oriented along the direction of the geostrophic wind and the cross-stream flow, v, perpendicular to it:

$$\partial_t u = fv - (g/\theta)\theta'|\beta|\cos\gamma + \partial_z(K_M\partial_z u), \tag{1}$$

$$\partial_t v = f(u - G) - (g/\theta)\theta'|\beta|\sin\gamma + \partial_z(K_M\partial_z v), \tag{2}$$

$$\partial_t \theta = \partial_z(K_H\partial_z\theta) \tag{3}$$

$$\partial_t E = K_M\left[(\partial_z u)^2 + (\partial_z v)^2\right] - (g/\theta)(K_H\partial_z\theta) + C\partial_z(K_M\partial_z E) - [0.2E]^{3/2}/\ell \tag{4}$$

where ∂_t is used here to represent the partial derivative with respect to time, t, and ∂_z is the partial derivative with respect to z; G is the geostrophic wind speed; θ is the potential temperature; f is the Coriolis parameter; g is the acceleration due to gravity; K_M and K_H are, respectively, the vertical exchange coefficients of momentum and heat, related to the turbulence kinetic energy, E, by dimensional argument (i.e., $K_M = [C_M/C_\xi]\ell E^{1/2}$ and $K_H = [C_H/C_\xi]\ell E^{1/2}$); C_M, C_H and C_ξ are evaluated applying level-2–5 simplification provided in FREEDMAN and JACOBSON (2003); C is a constant; The mixing length, ℓ, is computed following MELLOR and YAMADA (1982). θ' is the deviation from the adiabatic base state which is taken to be the value existing throughout the boundary layer at transition. Slope gradient, β, was set to 0.002 (i.e., 0.002 km rise for every 1 km in the horizontal). γ is the angle, measured counterclockwise, from the "fall-line" vector pointing down the slope to the x-axis. This model version also contains calculation of long- and shortwave radiative fluxes that reach the ground surface for the determination of the surface energy budget and, consequently, a predictive equation for the surface temperature. For the present case, atmospheric longwave flux is switched off, however, its surface values are computed and prescribed for surface energy budget very similar to the procedure followed in BROST and WYNGAARD (1978; see their equation 4).

In the current numerical framework, the atmosphere, which is assumed to extend to 6 km from the surface, was segmented into 124 vertical levels. sixty levels were below 1.3 km. The lowest model level is located at 10 m above ground surface. Surface roughness length was chosen to be 0.01 m and the Coriolis parameter was taken to represent a typical mid-latitude location (i.e., 45° N). The model was integrated through a diurnal cycle starting at 6 LST. A time step of 10 s was used. Initially, a background static stability of 5 K/km was set above 1.3 km. Below this height, dry-adiabatic potential temperature corresponding to a well mixed state was assumed. Although these profiles are highly idealized, the results may not be critically dependent on the base state temperature. Geostrophic wind was prescribed initially and the wind within the boundary layer was barotropically initialized using an Ekman type of balance (FREEDMAN and JACOBSON, 2003; RAMAKRISHNA et al., 2003).

Two control cases simulating flat, homogeneous domains with two different geostrophic winds of 10 m/s and 3 m/s, respectively, were considered here. While the former case is representative of strong wind conditions, the latter is taken to represent a weak wind NBL. In order to study the influence of a flat but slightly inclined surface, four different geostrophic wind directions[1] (i.e., γ changes) were considered for the two geostrophic forcings. While the surface slopes downward from west to east, westerly flow is 0°, wind from the east is 180°, southerly flow is 90° and wind from the north is 270°. However for the diurnal evolution of the boundary layer that was considered in the current NBL study, all the numerical experiments corresponding to the strong wind cases are consistent with BROST and WYNGAARD (1978). We discuss the results of these ten numerical experiments.

3. Results and Discussions

One of the major objectives of this study was to evaluate the slope effects in the NBL under weak wind conditions, however we first describe the results of the comparison of our model simulations with those by BROST and WYNGAARD (1978) for the strong wind cases. Although these comparisons were initially intended to examine the consistency of our results and mostly as a reference case, as explained later in this section, we found interesting differences between the two studies.

Figure 1 depicts the evolution of the surface frictional velocity for different slope orientations in strong (Fig. 1a) and weak wind conditions (Fig. 1b). The control cases simulating, respectively, these conditions are also shown for comparison. Small terrain slopes have a significant effect on surface-driven turbulent mixing within the NBL. For a terrain-slope orientation in a west-east direction, the NBL is least stable either for a southerly (90°) or, to a lesser extent, an easterly (180°) geostrophic flow. It should be noted that since the geostrophic wind in either of these directions aids upslope flow during the daytime, stronger winds, stronger shear production and large

---

[1] Directional changes in geostrophic wind were provided in terms of changes to γ in the governing equations (1) and (2) and, consequently, the direction of the slope is after all apparent, and is relative to the geostrophic wind always pointing along the x axis. This implies that the component equations presented here are couched along streamwise, geostrophic coordinate and cross-stream component.

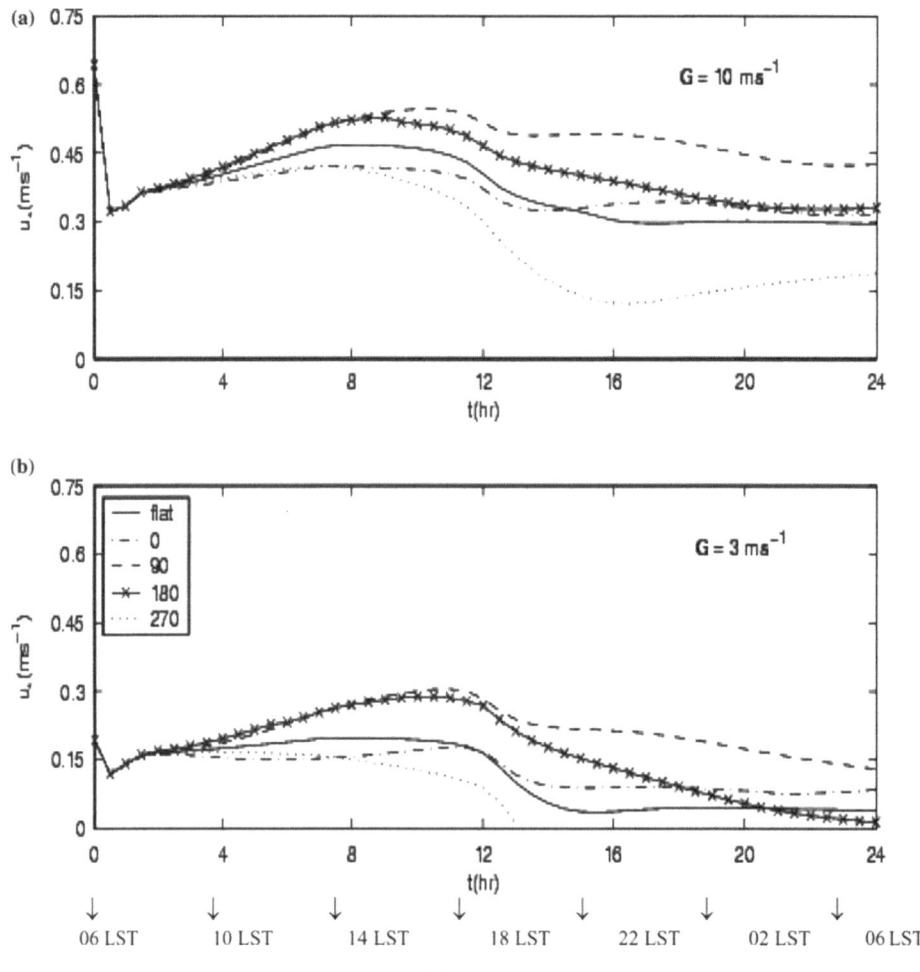

Figure 1
Diurnal evolution of the surface frictional velocity for different slope orientations in (a) strong and (b) weak wind conditions. While the surface slopes downward from west to east at 0.002 km rise for every 1 km, westerly flow is 0°, geostrophic wind from the east is 180°, southerly flow is 90° and wind from the north is 270°.

turbulence production during the day and, subsequently, during early evening may not be off-set by radiative cooling and the consequent development of down-slope flow in the NBL. All this leads to a more turbulent boundary layer during most of the night (Figs. 1a and b). However, as expected, when the easterly, upper-level flow become weak, the opposing down-slope terrain-driven flow within the NBL begins to dominate towards the end of the night (Fig. 1b). Interestingly, it appears that inclusion of even a gentle terrain slope effect, where need be so, in numerical models may improve simulations leading to the study of the weak-wind NBL where most models underpredict turbulence.

For the same west-east slope orientation, the case of the northerly flow (270°) and, to a lesser extent, a westerly geostrophic flow (0°) are more stable (Figs. 1a and b). A northerly flow at the upper level tends to oppose near-surface upslope flow resulting in minimum turbulence during the daytime. Strong cooling near the surface further annihilates turbulence as well as shear production leading to a near collapse of the strong wind NBL (Fig. 1a) and a more adversely effected NBL under weak wind conditions (Fig. 1b). Nevertheless, although a westerly geostrophic wind (0°) directly opposes near-surface upslope flow during the daytime, strong shear production in the NBL leads to enhanced turbulence in the stable layer. Interestingly, a comparison between values of u_* over a flat domain and those over an inclined surface for the westerly case shows that the latter exceeds the former during the later part of night, perhaps due to the development of the shallow down slope, terrain-

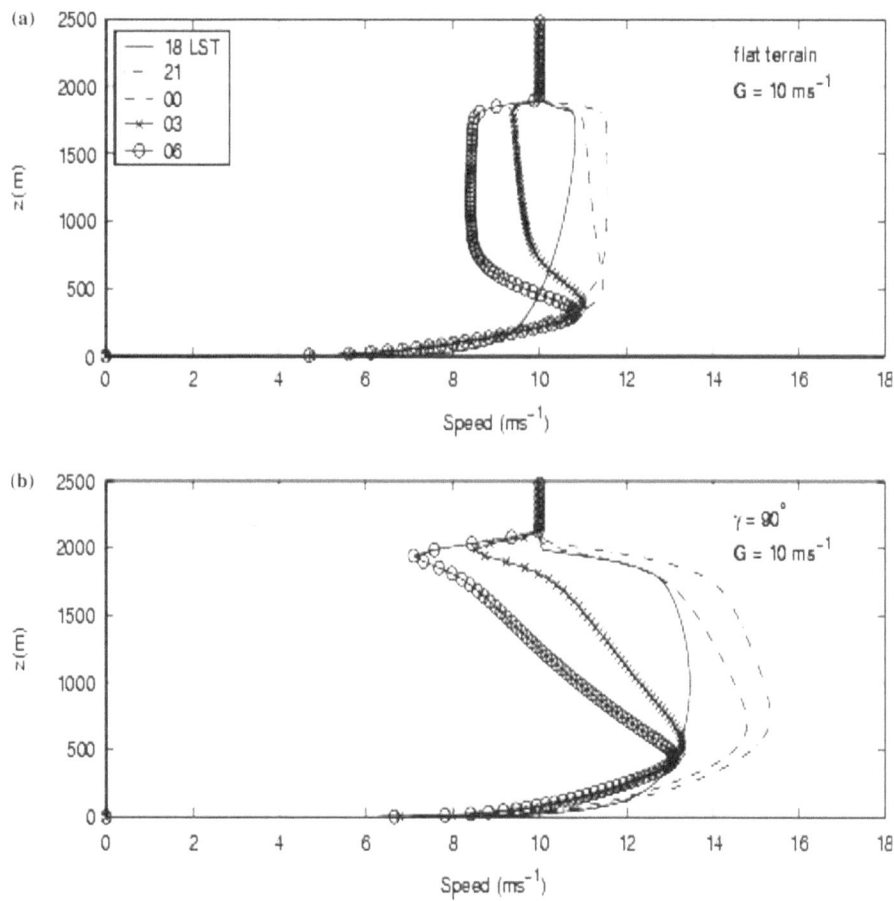

Figure 2
Evolution of the mean wind structure in a strong wind nocturnal boundary layer over (a) a flat terrain and (b) west to east sloping terrain with a southerly geostrophic wind (90°).

driven flow in the NBL that aids the upper level mean wind. Although all these results are consistent with BROST and WYNGAARD (1978), one should note the swapping of results between various cases. It appears that the inclusion of anabaric flows during the day in evolving the NBL through a diurnal cycle results in these differences.

Figure 2 presents profiles of mean wind speed over a flat domain and those over an inclined surface for the southerly geostrophic flow of 10 m/s. For brevity of presentation we describe here only the results pertaining to the nocturnal evolution of wind. Clearly, a nocturnal jet develops in both these cases and, as expected, the jet is stronger over the sloping terrain (Fig. 2b). It can be easily shown that for a time invariant geostrophic wind, the super geostrophic component developing later at night due to inertial oscillations is largely influenced by ageostrophic wind at sunset

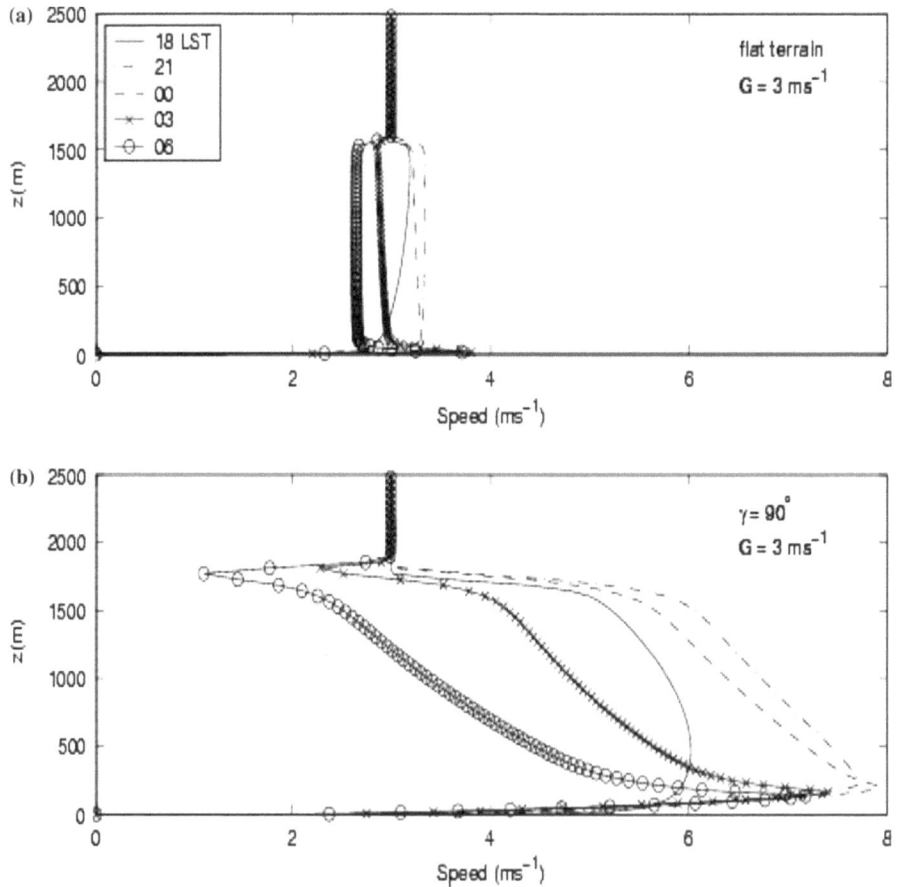

Figure 3

Evolution of the mean wind structure in a weak wind nocturnal boundary layer over (a) a flat terrain and (b) west to east sloping terrain with a southerly geostrophic wind (90°).

(i.e., $\partial[\mathbf{V} - \mathbf{V_g}]/\partial t = [\mathbf{V} - \mathbf{V_g}]_{\text{sunset}}\exp\{\text{-ift}\}$; where $\mathbf{V}$ is the wind vector, f is the Coriolis parameter and t is the time). Since a southerly geostrophic flow over the gently sloping terrain aids upslope, anabaric flow during the daytime and at sunset, the ageostrophic component is larger than those over a flat surface (Fig. 2a) and, consequently, a stronger nocturnal jet is produced in this case (Fig. 2a).

Figure 3 illustrates the influence of gentle slopes in the weak wind NBL. Note that over a flat terrain, while a strong geostrophic forcing (Fig. 2a) produced a nocturnal jet with a maxima well above 200 m altitude, the wind maxima is located at an altitude of less than 100 m under weak wind conditions (Fig. 3a). These are consistent with some observed and modeled features. RAMAKRISHNA et al. (2003) attributed these differences to weaker and shallower diffusion in the weak wind boundary layer during transition. However, even a very small slope appears to have a more dominant influence on the evolution of the nocturnal jet under weak wind conditions (Fig. 3b). Although the behavior of the inertial oscillations and, consequently, the nocturnal jet, consistent with what was explained earlier, may be attributed to stronger ageostrophic forcing at sunset, the effects are, yet, nonlinear. For instance, while a strong geostrophic forcing of 10 m/s produced a nocturnal jet with a maximum of less than 14 m/s at around 03 LST (Fig. 2b), weaker geostrophic flow resulted in a wind maxima of over 7 m/s at approximately the same time (Fig. 3b).

Figures 4 and 5 display the thermodynamic structure in a strong and weak wind NBL and the influence of small terrain slopes in either case. Clearly, for a west to east sloping terrain, a southerly geostrophic wind (90°) aids turbulence production and consequently, the profiles show a more well mixed behavior than those depicted over the flat terrain (Figs. 4a and 5a). As expected, the influence of gentle terrain slope is more visible in a weak wind NBL (Fig. 5b). For instance, one can notice a well-defined evolution of the nocturnal inversion layer in response to surface cooling in this case (Fig. 5b). It should be noted, that the effects of longwave radiation in the atmosphere were neglected in this study. Nevertheless, we expect that our conclusions may not be significantly altered by inclusion of these effects. However, what remains to be seen in observations is the "stronger than normal" cooling produced over a flat terrain in this model (Fig. 5a) as well as other models under weak wind conditions. Although this is clearly not the focus of our study, GOPALAKRISHNAN (1996) ascertained that although turbulence closure scheme could produce the mean fields quite well under weak wind conditions, the schemes underestimated the depth of frictional layer as well as wind maxima. It appears that the turbulence diffusion in numerical models decreases with the geostrophic wind at a much faster rate in comparison with the real atmosphere, over a flat and homogeneous domain. Perhaps surface layer parameterization schemes used in models, in a weak wind NBL that develops over flat and homogeneous domains need to be studied and better validated.

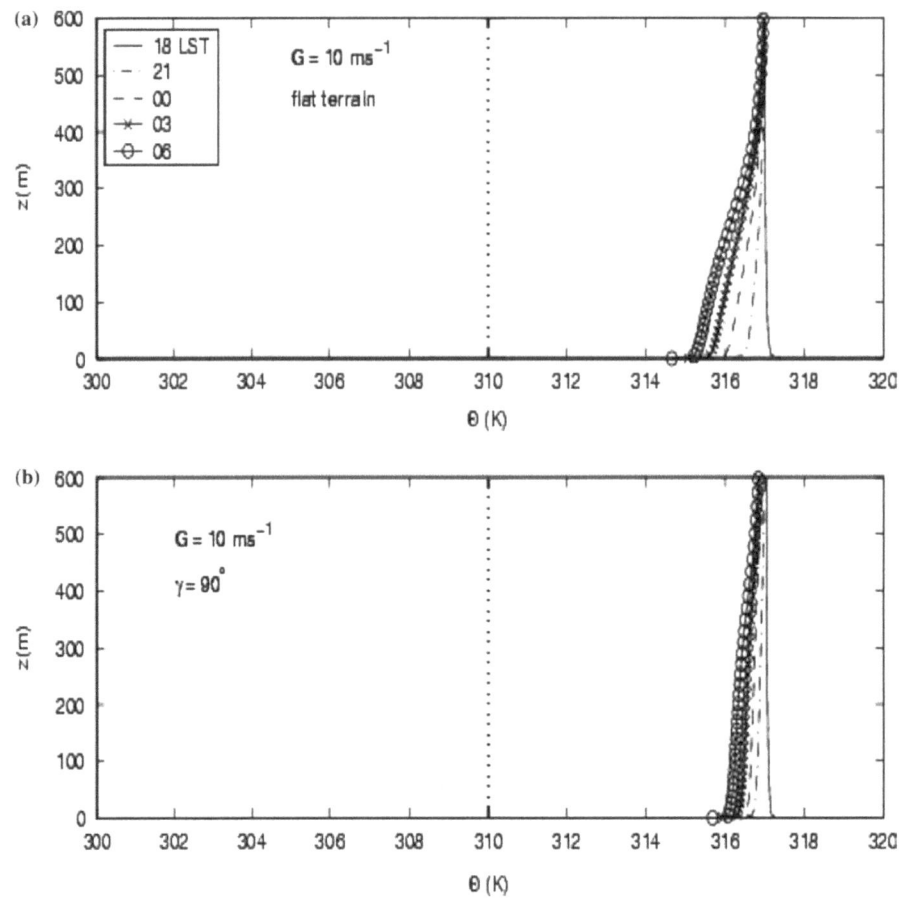

Figure 4
Evolution of the potential temperature structure in a strong wind nocturnal boundary layer over (a) a flat terrain and (b) west to east sloping terrain with a southerly geostrophic wind (90°).

4. Conclusions and Recommendations

Although terrain may sometimes look flat to the eye, nevertheless, slopes may be significant enough to produce "drainage" accelerations in a nocturnal boundary layer (NBL). The effect of gentle slopes on a weak and strong-wind NBL was investigated here using a one-dimensional model, with a simple correction term to account for the terrain effects, very similar to the one used by BROST and WYNGAARD (1978). While real world application needs a fully three-dimensional model to describe the flow and structure of the NBL and we have idealized our modeling efforts considerably, we expect that these results may yet be useful in developing a more complete picture of the NBL over complex terrains. For strong wind and weak

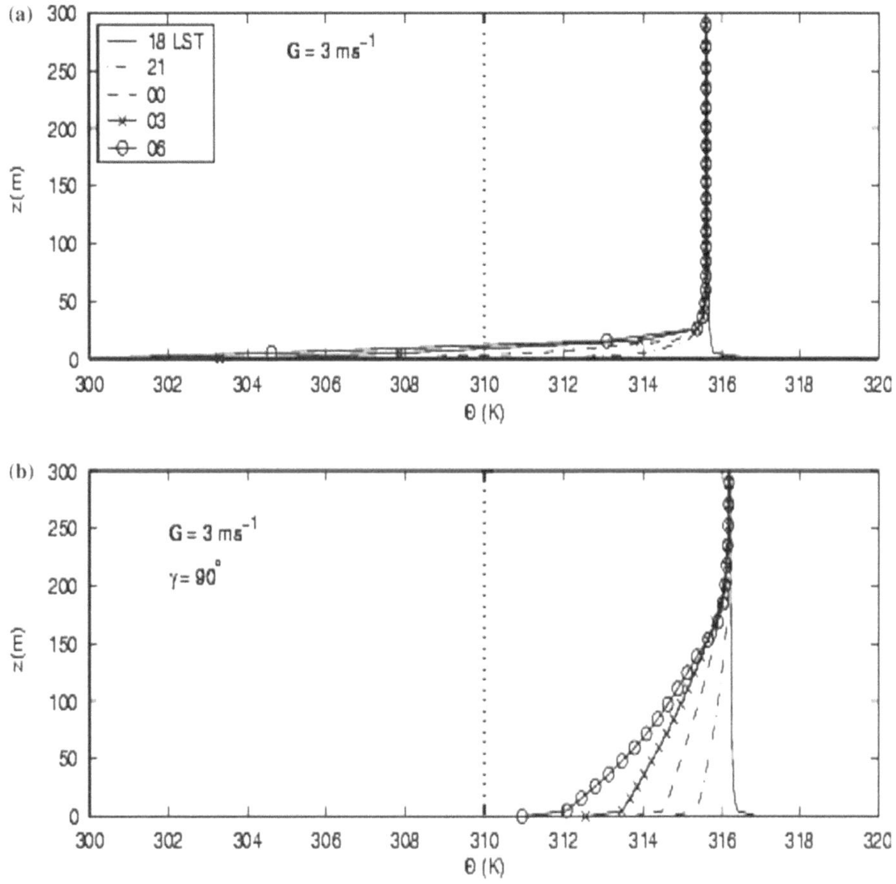

Figure 5

Evolution of the potential temperature structure in a weak wind nocturnal boundary layer over (a) a flat terrain and (b) west to east sloping terrain with a southerly geostrophic wind (90°).

wind NBL, the model produced some of the salient features observed over flat and homogeneous domain (RAMAKRISHNA et al., 2003). It appears that depending on the direction of the terrain slope with respect to the geostrophic wind, the slope may have a profound influence on the evolution of the nocturnal jet under weak wind conditions. Interestingly, for a slope oriented from west to east, geostrophic wind from the east and southerly directions, respectively, was strongly affected, in terms of turbulence and shear, by the orientation of the terrain. Whereas BROST and WYNGAARD (1978) found that for the above orientation of slope, the geostrophic wind from the west and north had the strongest influence. The inclusion of anabaric flows during the day in evolving the NBL through a diurnal cycle results in these differences.

Observations in the NBL, in general, are limited. Further, very few observations are available in stable, weak wind conditions. It is expected that the CASES99 data set may offer better in sight on these issues. Further, even gentle slopes may have significant influence on the structure of the NBL. For instance, Wangara field site in Australia has slopes on the order of 0.007 and Great Plains in USA slopes at about 0.001 (STULL, 1988). Several studies, including the current one, have found significant influence of these slopes on the mean profile in the NBL (BROST and WYNGAARD, 1978; MAHRT, 1982). Figure 6, for instance, shows a comparison of the observed profile of the components of wind profiles at 05 LST on May 27, 1981, at Kincaid, Illinois, USA in a weak wind NBL with a column model (RAMAKRISHNA et al., 2003). A comparison between Figures 3 and 6 illustrates some effects of mild slopes, especially in the upper part of the surface inversion layer. It appears that inclusion of a simple correction term to account for the terrain effects to column models may improve the results of such intercomparisons.

In the absence of measurements, while high-resolution LESs have offered several important insights in understanding turbulence and dispersion in the convective boundary layer (CBL) that develops over homogeneous (DEARDORFF, 1974; LAMB, 1984) and heterogeneous domains (GOPALAKRISHNAN and AVISSAR, 2000; GOPALA-KRISHNAN et al., 2000), such models had very limited success in the NBL. Apart from large computational requirements (very high resolution on the order of a few meters is required for these simulations), co-existence of small-scale gravity waves

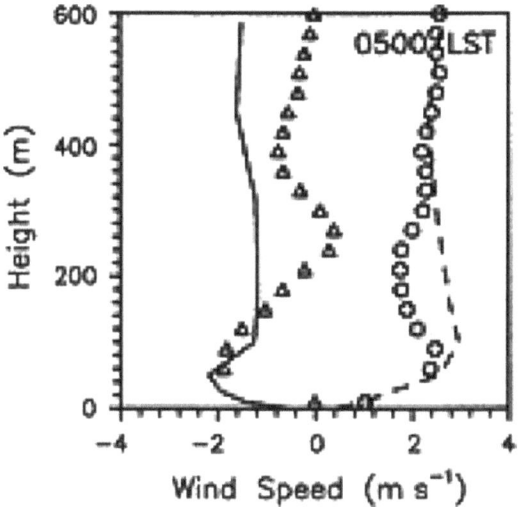

Figure 6
Comparison of *u* and *v* components of wind profiles computed from a column model with those observed in the plume-validation experiment over a flat and homogeneous terrain in Kincaid at 0500 LST under weak wind conditions. Computed *u* — —; Computed *v* - - - - - -; Observed *u* O O O O; Observed v □ □ □ □ (adopted from RAMAKRISHNA et al., 2003, J. Appl. Meteorol. 42, p 958).

(FINNIGAN and EINAUDI, 1981), discontinuous/intermittent turbulence (STULL, 1988), pivotal effects of radiation (GOPALAKRISHNAN et al., 1998) and sensitivity of the flow to surface heterogeneities (BROST and WYNGAARD, 1978) have all made such a modeling effort complex. Furthermore, NBL seldom reaches steady-state and, consequently, it may not be possible to attain a steady-state turbulence structure especially in a continuously evolving weak wind condition. Until recently, most studies focused on the development of an LES subgrid scale parameterization scheme appropriate for NBL (MASON and DERBYSHIRE, 1990 and SULLIVAN et al., 1994). However, more recently, SAIKI et al. (2000) examined a highly cooled yet fairly windy NBL, using LES which reached a quasi-steady state, and found small-scale structure in the NBL and a well-defined nocturnal jet above it. In an earlier simulation, ANDREN (1995) demonstrated that with finite stratification aloft, indeed, eddies in the turbulent layer trigger wave motion aloft. However, no discernible effects of these waves on turbulence statistics was found. The future challenge in LES lies in simulating strongly stable, weak-wind NBL.

Acknowledgements

The views expressed here are those of the authors and do not necessarily represent the official policy or position of the organization. We thank the reviewers for offering useful suggestions in elevating the quality of this manuscript.

REFERENCES

ANDRE, J. C., DE MOOR, G., LACARRERE, P., THERRY, G., and DU VACHAT, R. (1978), *Modelling the 24-hour Evolution of the Mean and Turbulent Structure of the Planetary Boundary Layer*, J. Atmos. Sci. *35*, 1861–1883.

ANDRE, J. C., and MAHRT, L. (1982), *The Nocturnal Surface Inversion and Influence of Clear-air Radiational Cooling*, J. Atmos. Sci. *39*, 864–878.

ANDREN, A. (1995), *Stably Stratified Atmospheric Boundary Layers*, Q. J. R. Meteorol. Soc. *121*, 961–985.

BROST, R. A. and WYNGAARD, J. C. (1978), *A Model Study of the Stably Stratified Planetary Boundary Layer*, J.Atmos. Sci. *35*, 1427–1440.

BLACKADAR, A. K. (1957), *Boundary Layer Wind Maxima and their Significance for the Growth of Nocturnal Inversion*, Bull. Am. Meteor. Soc. *38*, 283–290.

BUAJITTI, K. and BLACKADAR, A. K. (1957), *Theoretical Studies of Diurnal Wind Variations*, Quart. J. Roy. Meteorol. Soc. *83*, 486–500.

CLARKE, R. H., DYER, A. J., BROOK, R. R., REID, D. G., and TROUP, A. G. (1971), *The Wangara Experiment: Boundary Layer Data*, Tech. Paper 19, Div. Meteor. Pys., (CSIRO, Australia).

DEARDORFF, J. W. (1974), *Three-dimensional Study of the Height and Mean Structure of a Heated Planetary Boundary Layer*, Boundary-Layer Meteorol. 7, 81–106.

DERBYSHIRE, S. H. (1990), *Nieuwstadt's Stable Boundary Layer Revisited*, Quart. J. Roy. Meteorol. Soc. *116*, 27–158.

DORAN, J. C. and HORST, T. W. (1983), *Observations and Models of Simple Nocturnal Slope Flows*, J. Atmos. Sci. *40*, 708–717.

ESTOURNEL, C., and GUEDALIA, D. (1985), *Influence of Geostrophic Wind on Atmospheric Nocturnal Cooling*, J. Atmos. Sci. *42*, 2695–2698.

ESTOURNEL, C., and GUEDALIA, D. (1987), *A New Parameterization of Eddy Diffusivities for Nocturnal Boundary-layer Modelling*, Bound.- Layer Meteor. *39*, 191–203.

FREEDMAN, F. R. and JACOBSON, M. J. (2003), *Modification of the Standard ε-Equation for the Stable ABL through Enforced Consistency with Monin-Obukhov Similarity Theory*, Bound.- Layer Meteor. *106*, 383–410.

FINNIGAN, J. J. and EINAUDI, F. (1981), *The Interaction between an Internal Gravity Wave and the Planetary Boundary Layer, Part II: The Effect of the Wave on the Turbulence Structure*, Quart. J. Meteorol. Soc. *107*, 807–832.

GARRATT, J. R. and BROST, R. A. (1981), *Radiative Cooling Effects within and above the Nocturnal Boundary Layer*, J. Atmos. Sci. *38*, 2730–2746.

GARRATT, J. R. (1982), *Observations in the Nocturnal Boundary Layer*, Bound.-Layer Meteor. *22*, 21–48.

GARRETT, A.J. (1983), *Drainage Flow Prediction with a One-dimensional Model Including Canopy, Soil and Radiation Parameterization*, J. Clim and Appl. Meteor. *22*, 79–91.

GOPALAKRISHNAN, S. G., *Mesoscale Dispersion Modelling in a Weak Wind Stable Boundary Layer with a Special Reference to the Bhopal Gas Episode* (Ph.D. Thesis, Indian Institute of Technology, Delhi, India, 1996) 177 pp.

GOPALAKRISHNAN, S. G. and SHARAN, MAITHILI, (1997), *A Lagrangian Particle Dispersion Model for Marginally Heavy Gas Dispersion*, Atmos. Environ. *31*, 3369–3392.

GOPALAKRISHNAN, S. G., SHARAN, MAITHILI, MCNIDER, R. T., and SINGH, M. P. (1998), *A Study of Turbulent and Radiative Processes in the Stable Boundary Layer under Weak Wind Conditions*, J. Atmos. Sci. *55*, 954–960.

GOPALAKRISHNAN, S. G. and AVISSAR, R. (2000), *An LES study of the Impacts of Land Surface Heterogeneity on Dispersion in the Convective Boundary Layer*, J. Atmos. Sci. *57*, 352–371.

GOPALAKRISHNAN, S. G., ROY, S. B., and AVISSAR, R.(2000), *An Evaluation of the Scale at which Topographical Features Affect the Convective Boundary Layer Using Large Eddy Simulations*, J. Atmos. Sci. *57*, 334–351.

KURZEJA, R. J., BERMAN, S., WEBER, A. H. (1991), *A Climatological Study of the Nocturnal Planetary Boundary Layer*, Boundary-Layer Meteorol. *54*, 105–128.

LAMB, R.G. (1984), *Diffusion in the Convective Boundary Layer. Atmospheric Turbulence and Air Pollution Modelling* (Nieuwstadt, F.T.M. and Van Dop, H., eds.) (D. Reidel Publishing Company, Holland, 1982) pp. 69–106.

LENSCHOW, D. H., LI, X. S., ZHU, C. J., and STANKOV, B. B. (1988), *The Stably Stratified Boundary Layer over the Great Plains. I. Mean and Turbulence Structure*, Bound.-Layer Meteor. *42*, 95–121.

MAHRER, Y. and PIELKE, R. A. (1977), *A Numerical Study of Airflow over Irregular Terrain*, Beiträge zur Physik der Atmosphäre *50*, Band seite 98–113.

MAHRT, L., HERALD, R. C., LENSCHOW, D. H., and STANKOV, B. B. (1979), *An Observational Study of the Structure of the Nocturnal Boundary Layer*, Bound. Layer Meteorol. *17*, 247–264.

MAHRT, L. (1982), *Momentum Balance of Gravity Flows*, J. Atmos. Sci. *39*, 2701–2711.

MANINS, P. C. and SAWFORD, B.L.(1979), *A Model of Katabatic Winds*, J. Atmos. Sci. *36*, 619–630.

MASON, P. J. and DERBYSHIRE, S. H. (1990), *Large-eddy Simulation of the Stably–stratified Atmospheric Boundary Layer*, Bound. Layer Meteorol. *90*, 375–396.

MELLOR, G. L. and YAMADA, T. (1982), *Development of a Turbulence Closure Model for Geophysical Fluid Problems*, Rev. Geophys. Space Phys. *20*, 851–875.

MCNIDER, R. T., and PIELKE, R. A. (1981), *Diurnal Boundary-layer Development over Sloping Terrain*, J. Atmos. Sci. *38*, 2198–2212.

MCNIDER, R. T. and PIELKE, R. A. (1984), *Numerical Simulation of Slope and Mountain Flows*, J. Clim. and Appl. Meteorol. *23*, 1441–1453.

MCNIDER, R. T., MORAN, M. D., PIELKE, R. A. (1988), *Influence of Diurnal and Intertial Boundary Layer Oscillations on Long-Range Dispersion*, Atmos. Environ. *22*, 2445–2462.

MCNIDER, R. T., SINGH, M. P., and LIN, J. T. (1993), *Diurnal Wind Structure Variations and Dispersion of Pollutants in the Boundary Layer*, Atmos. Environ. *27* A (14), 2199–2214.

MOENG, C. H. and SULLIVAN, P. J. (1994), *A Comparison of Shear and Buoyancy Driven Planetary Boundary Layer Flows*, J. Atmos. Sci. *51*, 999–1022.

NIEUWSTADT, F. T. M. (1984), *The Turbulent Structure of the Stable Nocturnal Boundary Layer*, J. Atmos. Sci. *41*, 2202–2216.

NIEUWSTADT, F. T. M. (1985), *Some Aspects of the Turbulent Stable Boundary Layer*, Bound. Layer Meteorol. *30*, 31–56.

PIELKE, R. A. (1974), *A Three-dimensional Numerical Model of Sea Breeze over South Florida*, Mon. Wea. Rev. *102*, 115–139.

RAMAKRISHNA, T. V. B. P. S, SHARAN MAITHILI, GOPALAKRISHNAN, S. G., and ADITI (2003), *Mean Structure of the Nocturnal Boundary Layer under Strong and Weak Wind Conditions: EPRI Case Study*, J. Appl. Meteorol. *42*, 952–969.

RAO, K. S. and SNODGRASS, H.F. (1981), *A Nonstationary Nocturnal Drainage Flow Model*, Bound. Layer Meteorol. *20*, 309–320.

SAIKI, E. M., MOENG, C. H., and SULLIVAN, P. P. (2000), *Large Eddy Simulation of the Stably Stratified Planetary Boundary Layer*, Bound. Layer Meteorol. *95*, 1–30.

SHARAN, MAITHILI, MCNIDER, R. T., GOPALAKRISHNAN, S. G., and SINGH, M. P. (1995), *Bhopal Gas Leak: A Numerical Simulation of Episodic Dispersion*, Atmos. Environ. *29*, 2061–2070.

SHARAN, MAITHILI, GOPALAKRISHNAN, S. G., MCNIDER, R. T. and SINGH, M. P. (1996), *Bhopal Gas Leak: A Numerical Investigation of Prevailing Meteorological Conditions*, J. Appl. Meteorol. *3*, 1637–1656.

SHARAN, MAITHILI and GOPALAKRISHNAN, S. G. (1997), *Comparative Evaluation of Eddy Exchange Coefficients for Strong and Weak Wind Stable Boundary Layer Modelling*, J. Appl. Meteorol. *36*, 545–559.

SHARAN, MAITHILI, GOPALAKRISHNAN, S. G., MCNIDER, R. T., and SINGH, M.P. (2000), *Bhopal Gas Leak: A Numerical Investigation on the Possible Influence of Urban Effects on the Prevailing Meteorological Conditions*, Atmos. Environ. *34*, 539–552.

SINGH, M. P., MCNIDER, R. T., and LIN, J. T. (1993), *An Analytical Study of Diurnal Wind-structure Variation in the Boundary Layer and the Low-level Nocturnal Jet*, Bound. Layer Meteorol. *63*, 397–423.

SINGH, M. P., MCNIDER, R. T., MEYERS, R. and GUPTA, S. (1997), *Nocturnal Wind Structure over Land and Dispersion of Pollutants: An Analytical Study*, Atmos. Environ. *31*, 1, 105–115.

SINGH, M. P., YADAV, A. K., MCNIDER, R. T., MEYERS, R., SHARAN, MAITHILI, and LATIF, AZAHAR (1999), *Presented in International Conference cum Workshop on Air Quality Management* 1999.

STULL, R. B. and DRIEDONKS, A. G. M. (1987), *Applications of the Transilient Turbulence Parameterization to Atmospheric Boundary-layer Simulations*, Bound. Layer Meteorol. *40*, 209–239.

STULL, R., *An Introduction to Boundary Layer Meteorology* (Kluwer Academic Publishers, Netherlands 1988).

SULLIVAN, P. P., MCWILLIAMS, J. C., and MOENG, C. H. (1994), *A Subgrid-scale Model for Large Eddy Simulation of Planetary Boundary Layer Flows*, Bound. Layer Meteorol. *71*, 247–276.

TJEMKES, S. A. and DUYNKERKE, P. G. (1989), *The Nocturnal Boundary Layer: Model Calculations Compared with Observations*, J. Appl. Meteorol. *28*, 161–175.

YAMADA, T. and MELLOR, G. (1975), *A Simulation of the Wangara Atmospheric Boundary Layer Data*, J. Atmos. Sci. *32*, 2309–2329.

ZEMAN, O. (1979), *Parameterization of the Dynamics of Stable Boundary Layer and Nocturnal Jets*, J. Atmos. Sci. *36*, 792–804.

(Received November 30, 2003, accepted April 5, 2004)
Published Online First June 8, 2005

To access this journal online:
http://www.birkhauser.ch

Pure appl. geophys. 162 (2005) 1811–1829
0033–4553/05/101811–19
DOI 10.1007/s00024-005-2694-7

© Birkhäuser Verlag, Basel, 2005

| Pure and Applied Geophysics

On the Behavior of the Stable Boundary Layer and the Role of Initial Conditions

Xingzhong Shi,[1] Richard T. McNider,[2] M. P. Singh,[3] David E. England,[4] Mark J. Friedman,[4] William M. Lapenta,[5] and William B. Norris[2]

Abstract—Previous studies of the stable atmospheric boundary layer using techniques of nonlinear dynamical systems (McNider *et al.*, 1995) have shown that the equations support multiple solutions in certain parameter spaces. When geostrophic speed is used as a bifurcation parameter, two stable equilibria are found—a warm solution corresponding to the high-wind regime where the surface layer of the atmosphere stays coupled to the outer layer, and a cold solution corresponding to the low-wind, decoupled case. Between the stable equilibria is an unstable region where multiple solutions exist. The bifurcation diagram is a classic S shape with the foldback region showing the multiple solutions. These studies were carried out using a simple two-layer model of the atmosphere with a fairly complete surface energy budget. This allowed the dynamical analysis to be carried out on a coupled set of four ordinary differential equations. The present paper extends this work by examining additional bifurcation parameters and, more importantly, analyzing a set of partial differential equations with full vertical dependence. Simple mathematical representations of classical problems in dynamical analysis often exhibit interesting behavior, such as multiple solutions, that is not retained in the behavior of more complete representations. In the present case the S-shaped bifurcation diagram remains with only slight variations from the two-layer model. For the parameter space in the foldback region, the evolution of the boundary layer may be dramatically affected by the initial conditions at sunset. An eigenvalue analysis carried out to determine whether the system might support pure limit-cycle behavior showed that purely complex eigenvalues are not found. Thus, any cyclic behavior is likely to be transient.

Key words: Stable boundary layer, predictability, bifurcation analysis, dynamical system.

1. Introduction

Understanding the behavior of the stable boundary layer is critical for weather forecasting and air-quality modeling. In recent years there has been increased interest in the proper formulation of stable boundary-layer parameterizations in

[1]Alabama A & M University, Huntsville, Alabama, U.S.A.
[2]University of Alabama in Huntsville, Huntsville, Alabama, U.S.A.
[3]Ansal Institute of Technology, Gurgaon, India.
[4]Florence, Alabama, U.S.A.
[5]NASA Marshall Space Flight Center, Huntsville, Alabama, U.S.A.

models of all scales. Evidently, application of traditional Monin-Obukhov local closure with stability functions having relatively sharp cutoffs at theoretical critical Richardson numbers (BLACKADAR, 1979) can produce decoupled solutions that are too cold, leading to deterioration of model performance (DERBYSHIRE, 1999; VITERBO *et al.*, 1999). The failure to incorporate the intermittency of turbulence of the real atmosphere into model parameterizations may be a key factor in the poor performance problem (VAN DER WIEL *et al.*, 2002; MAHRT, 1999). Another potential factor may be the failure to handle non-local effects in the application of the parameterizations (McNIDER and PIELKE, 1981).

Unlike the daytime boundary layer, which is dominated by incoming solar heat flux, the stable boundary layer is affected by many physical parameters and processes (MAHRT, 1999). These include surface roughness, surface heat capacity, geostrophic wind, low-level jets, and latent heat in the soil. The dynamic complexity enters as nonlinearities in the diffusion terms. Thus, the system may support exotic behavior such as stable and unstable attracting solutions, limit cycles, and hysteresis (loss of predictability).

The purpose of this paper is to use techniques of nonlinear analysis (SEYDEL, 1988; DOEDEL and KERNEVÉZ, 1986; DOEDEL *et al.*, 1991) to explore the influence of parameters on the behavior of the equations used in forecast and air-quality models. Nonlinear analysis allows an understanding of the equations as functions of imposed parameters such as surface roughness, pressure gradient, downward longwave radiation, and heat capacity. A previous paper (McNIDER *et al.*, 1995) truncated the boundary-layer equations to a two-layer system that allowed construction of bifurcation diagrams, definition of bifurcation points (turning points), and compu- tation of eigenvalues in order to define the local stability of the steady solutions. That study (referred to in this paper as MCN95) established that the truncated system did indeed support hysteresis so that, for a range of geostrophic winds and surface- roughness values, the system had multi-valued solutions that reduce the predictabil- ity of the system and make it sensitive to initial conditions using a BLACKADAR (1979) closure.

Although MCN95 showed this exotic behavior in the truncated system, a question remained as to whether the behavior might be an artifact of the two-layer system and not a property of the full PDE system. The present study analyzes an *n*-layer system to determine whether the essential characteristics of the two-layer equations are retained. Also, it more fully explores the role of the initial conditions in the final solutions and carries out an eigenvalue analysis to determine whether pure limit cycles exist. We employ new tools in nonlinear analysis, particularly the numerical analysis system AUTO developed by DOEDEL and KERNEVÉZ (1986) and DOEDEL *et al.* (1991).

2. Statement of the Problem

We consider a simple, one-dimensional (single column) model $u = u(z,t)$, $v = v(z,t)$, and $\theta = \theta(z,t)$ (see STULL,1988) that closely represents the type columns embedded within weather forecast and air pollution models. Symbols used in the equations below that are not explained following the equations are defined in Table 1.

$$\frac{\partial u}{\partial t} = f_{co}\left(v - v_G\right) + \frac{\partial}{\partial z}\left(K_m(z, u_z, v_z, \theta_z)\frac{\partial u}{\partial z}\right), \quad (2.1)$$

$$\frac{\partial v}{\partial t} = f_{co}\left(u_G - u\right) + \frac{\partial}{\partial z}\left(K_m(z, u_z, v_z, \theta_z)\frac{\partial v}{\partial z}\right), \quad (2.2)$$

$$\frac{\partial \theta}{\partial t} = R_c(\theta) + \frac{\partial}{\partial z}\left(K_h(z, u_z, v_z, \theta_z)\frac{\partial \theta}{\partial z}\right), \quad (2.3)$$

with the boundary conditions (for $t \geq 0$),

$$u(z_0, t) = 0, \qquad u(z,t) = \frac{z - z_0}{z - Z}u_G$$

Table 1

Variable definitions and parameter values

f_{co}	Coriolis parameter	10^{-4} s^{-1}
u_G	x component of geostrophic wind	8 m s^{-1}
v_G	y component of geostrophic wind	0 m s^{-1}
C_a	heat capacity per unit area	1.46×10^4 J K^{-1} m^{-1}
σ	Stefan-Boltzmann constant	5.669×10^{-8}
κ_m	Soil transfer coefficient	2.78×10^{-3} s^{-1}
θ_m	Mean temperature for the substrate	275 K
Q_a	Specific humidity at the height z_0	3×10^{-3} g kg^{-1}
ρ	Air density	1.225 kg m^{-3}
C_p	Specific heat at constant pressure	1005 kg m^{-3}
κ	von Karman constant	0.4
G	Gravitational acceleration	9.81 m s^{-1}
θ_A	Ambient air potential temperature	300 K
θ_Z	Potential temperature at the top	300 K
a	Constant in the surface radiative cooling	0.0 (or 0.2)
b	Constant in the surface turbulent flux	
Ri$_c$	Critical Richardson number	0.25
z_0	Roughness height	1 m
z_1	First level height in the two-layer mode	5 m
z_2	Second level height in the two-layer mode	50 m
Z	Upper level height for the boundary layer	50 m
K_m^0	Constant exchange coefficient for momentum	1 m^2 s^{-1}

$$v(z_0, t) = 0, \qquad v(z, t) = \frac{z - z_0}{z - Z} v_G$$

$$\theta(z_0, t) = \theta_g(t), \qquad \theta(z, t) = \theta_A = \theta_g(0),$$

coupled with the surface energy budget equation for the ground potential temperature $\theta_g = \theta_g(t)$ for $z = 0$, $t > 0$,

$$\frac{d\theta_g(t)}{dt} = \frac{1}{C_b} \left(I_1(\theta)|_{z=z_0} - \sigma \theta_g^4(t) - H_0(z, u_z, v_z, \theta_z)|_{z=z_0} \right) - \kappa_m (\theta_g(t) - \theta_m), \qquad (2.4)$$

and the initial conditions

$$u(z, 0) = u_G, \quad v(z, 0) = v_G, \quad \theta(z, 0) = \theta_A = \theta_g(0).$$

In equation. (2.3), $R_c(\theta)$ is the clear-air radiative cooling rate for the air temperature and is taken to be zero. In equation (2.4), I_1 is the longwave back radiation from the atmosphere, given by STALEY and JURICA (1972):

$$I_1 \equiv I_1(\theta)|_{z=z_0} = 0.67\sigma (1670 Q_a)^{0.08} \theta^4|_{z=z_0},$$

and H_0 is the heat flux carried away from the surface by turbulence, given by BLACKADAR (1979)

$$H_0 \equiv H_0(z, u_z, v_z, \theta_z)|_{z=z_0} = -\rho\, C_p (K_h(z, u_z, v_z, \theta_z) \partial\theta_z/\partial z)|_{z=z_0}.$$

Here,

$$K_m = K_m(z, u_z, v_z, \theta_z) \text{ and } K_h = K_h(z, u_z, v_z, \theta_z)$$

are the exchange coefficients for momentum and heat, which, in the absence of geostrophic wind shear, are functions of the space variable z and the partial derivatives:

$$u_z = \partial u(z, t)/\partial z, \qquad v_z = \partial v(z, t)/\partial z, \qquad \theta_z = \partial\theta(z, t)/\partial z.$$

The nonlinearity in the equations enters through these terms, originally defined as

$$K_m = f_m(\text{Ri})l^2 s, \qquad K_h = f_h(\text{Ri})l^2 s.$$

The stability functions f_m and f_h (namely, the Richardson-number formulations) are given by ENGLAND and MCNIDER (1995):

$$f_m(\text{Ri}) = f_h(\text{Ri}) = \begin{cases} (1 - \text{Ri}/\text{Ri}_c)^2, & \text{if } 0 \leq \text{Ri} \leq \text{Ri}_c, \\ 0, & \text{if } \text{Ri} \geq \text{Ri}_c. \end{cases}$$

The length scale l and vertical wind shear s are defined as

$$l = \kappa z, \quad s \equiv s(u_z, v_z) = \sqrt{u_z^2 + v_z^2}.$$

2.1 A Two-layer Model

To simplify the analysis of the time-dependent and steady-state behavior of the solutions while hoping to capture the essential physical and mathematical properties of the PDE system (2.1)–(2.4), we follow MCN95 in developing a simple two-layer model by employing a finite-difference representation (PIELKE, 1984) and the Monin-Obukhov similarity theory. The space interval $z \in [z_0, Z]$ is divided into two layers: $z \in [z_0, z_1]$ and $z \in [z_1, z_2]$, with $z_2 = Z$. The wind speed is denoted by $u_i = u(z_i)$, $v_i = v(z_i)$ and the potential temperature by $\theta_i = \theta(z_i)$ at the first and second levels with $z = z_i$, $i = 1, 2$, and the ground potential temperature θ_g is assumed to be constant at the roughness level $z = z_0$. With $u_2 = u_G$, $v_2 = v_G$, $\theta_2 = \theta_Z$ given and fixed, we are to solve for $u = u_1$, $v = v_1$ and $\theta = \theta_1$ at the first level $z = z_1$, and $\theta = \theta_g$ at the roughness level $z = z_0$. See Figure 1 for the schematic of the coordinate system for this discretization.

Throughout the "ground" layer $z \in [0, z_0]$, the potential temperature is assumed to be independent of the height z, and is a function of the time t only: $\theta_g = \theta_g(t)$. The wind field is identically zero: $u \equiv v \equiv 0$.

Using centered finite differences with respect to z for the level $z = z_1$ yields

$$\frac{\partial}{\partial z}\left(K_i \frac{\partial \phi}{\partial z}\right)_1 = \frac{\left(K_i \frac{\partial \phi}{\partial z}\right)_{3/2} - \left(K_i \frac{\partial \phi}{\partial z}\right)_{1/2}}{z_{3/2} - z_{1/2}}, \quad \phi = u, v, \theta, \quad i = m, h$$

and for the level $z = (z_1 + z_2)/2$,

$$\left(K_i \frac{\partial \phi}{\partial z}\right)_{3/2} = \frac{K_{i,3/2}(\phi_2 - \phi_1)}{z_2 - z_1}, \quad \phi = u, v, \theta, \quad i = m, h.$$

On the other hand, applying the Monin-Obukhov similarity theory for the lowest level $z = z_1/2$ yields

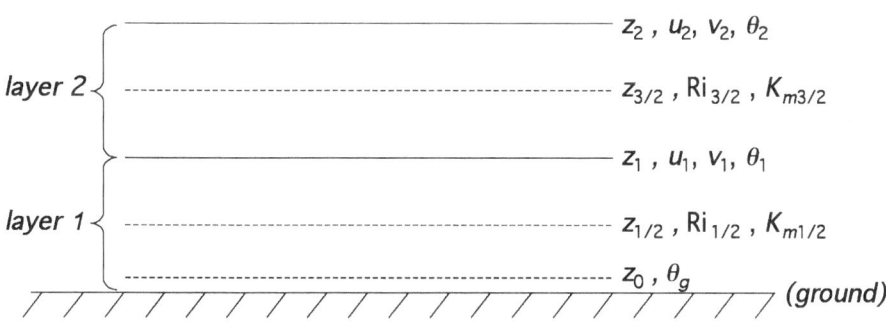

Figure 1
Schematic of two-layer model.

$$\left(K_m\frac{\partial u}{\partial z}\right)_{1/2} = u_*^2\cos(\psi), \quad \left(K_m\frac{\partial v}{\partial z}\right)_{1/2} = u_*^2\sin(\psi), \quad \left(K_m\frac{\partial \phi}{\partial z}\right)_{1/2} = u_*\phi_*.$$

Therefore, we reach a system of four first-order ODEs for the two-layer model:

$$\frac{du_1}{dt} = f_{co}(v_1 - v_G) + \frac{1}{z_{3/2} - z_{1/2}}\left(\frac{K_{m,3/2}(u_2 - u_1)}{z_2 - z_1} - u_*^2\cos(\psi)\right), \tag{2.5}$$

$$\frac{dv_1}{dt} = f_{co}(u_G - u_1) + \frac{1}{z_{3/2} - z_{1/2}}\left(\frac{K_{m,3/2}(v_2 - v_1)}{z_2 - z_1} - u_*^2\sin(\psi)\right), \tag{2.6}$$

$$\frac{d\theta_1}{dt} = \frac{1}{z_{3/2} - z_{1/2}}\left(\frac{K_{h,3/2}(\theta_2 - \theta_1)}{z_2 - z_1} - u_*\theta_*\right), \tag{2.7}$$

$$\frac{d\theta_g}{dt} = \frac{1}{C_b}\left(I_1 - \sigma\theta_g^4 - H_0\right) - \kappa_m(\theta_g - \theta_m), \tag{2.8}$$

with initial conditions $u_1(0) = u_G, v_1(0) = v_G$, and $\theta_1(0) = \theta_A$, where u_* and θ_* are the friction velocity and friction temperature, given by

$$u_* = \frac{\kappa\sqrt{f_m(Ri_{1/2})}\sqrt{u_1^2 + v_1^2}}{\ln(z_1/z_0)},$$

$$\theta_* = \frac{\kappa(\theta_1 - \theta_g)f_h(Ri_{1/2})}{\sqrt{f_m(Ri_{1/2})}\ln(z_1/z_0)},$$

respectively, and

$$\psi = \arctan(v_1/u_1),$$

$$z_{1/2} = z_1/2,$$

$$z_{3/2} = (z_1 + z_2)/2,$$

$$K_{m,3/2} = f_m(Ri_{3/2})\, l_{3/2}^2 s_{3/2},$$

$$K_{h,3/2} = f_h(Ri_{3/2})\, l_{3/2}^2 s_{3/2},$$

$$Ri_{3/2} = \frac{g}{\theta_A}\frac{(z_2 - z_1)(\theta_2 - \theta_1)}{(u_2 - u_1)^2 + (v_2 - v_1)^2},$$

$$Ri_{1/2} = \frac{g}{\theta_A}\sqrt{z_0 z_1}\ln\left(\frac{z_1}{z_0}\right)\frac{\theta_1 - \theta_g}{u_1^2 + v_1^2},$$

$$l_{3/2} = \kappa z_{3/2},$$

$$s_{3/2} = \frac{\sqrt{(u_2 - u_1)^2 + (v_2 - v_1)^2}}{z_2 - z_1}.$$

The system of ODEs (2.5)–(2.8) is an initial value problem. We can specify typical observation data at sunset as the initial values and compute the time evolution of the wind field u_1, v_1 and the temperature field θ_1, θ_g during the whole night, using any accurate numerical schemes to solve ODEs, e.g., the fourth-order Runge-Kutta scheme.

2.2 A Multi-layer Model

The two-layer model (2.5)–(2.8) is obviously very crude from a physical point of view, nonetheless it is relatively simple and easy to use and provides physically meaningful results that can be used to explain the predictability of some physical phenomena. Now we use a highly accurate orthogonal collocation method to discretize the ODE system into a "multi-layer" model. Our goal is to study the problem (2.1)–(2.4) more carefully and, in particular, to assess the validity of steady-state solutions for the two-layer model by performing similar numerical analyses.

To solve for the steady-state solution is to solve for the equilibria of (2.1)–(2.4), i.e., from the system of second-order non-autonomous ODEs (for $z \in (z_0, Z)$)

$$\frac{d}{dz}\left(K_m(z, \bar{u}', \bar{v}', \bar{\theta}')\frac{d\bar{u}}{dz}\right) = f_{co}(v_G - \bar{v}), \tag{2.9}$$

$$\frac{d}{dz}\left(K_m(z, \bar{u}', \bar{v}', \bar{\theta}')\frac{d\bar{v}}{dz}\right) = f_{co}(\bar{u} - u_G) \tag{2.10}$$

$$\frac{d}{dz}\left(K_h(z, \bar{u}', \bar{v}', \bar{\theta}')\frac{d\bar{\theta}}{dz}\right) = 0, \tag{2.11}$$

we solve for the vectors $\bar{u} = \bar{u}(z)$, $\bar{v} = \bar{v}(z)$, $\bar{\theta} = \bar{\theta}(z)$, $\bar{\theta}_g =$ scalar variable. The boundary conditions are

$$\bar{u}(z_0) = 0, \quad \bar{u}(Z) = u_G,$$

$$\bar{v}(z_0) = 0, \quad \bar{v}(Z) = v_G,$$

$$\bar{\theta}(z_0) = \bar{\theta}_g - \frac{bH_0(z, \bar{u}', \bar{v}', \bar{\theta}')\,|_{z=z_0}}{a\rho\,C_p z_0}, \quad \bar{\theta}(Z) = \theta(Z),$$

coupled with the algebraic equation θ_g:

$$I_1\Big|_{z=z_0} - \sigma\bar{\theta}_g^4 - H_0(z, \bar{u}', \bar{v}', \bar{\theta}')\Big|_{z=z_0} - C_b k_m(\theta_g - \theta_m) = 0.$$

Here, $K_m = K_m(z, \bar{u}', \bar{v}', \bar{\theta}')$ and $K_h = K_h(z, \bar{u}', \bar{v}', \bar{\theta}')$ are functions of z, and the derivatives $\bar{u}' = d\bar{u}(z)/dz$, $\bar{v}' = d\bar{v}(z)/dz$, $\bar{\theta}' = d\bar{\theta}(z)/dz$ with u_z, v_z, θ_z replaced by $\bar{u}'$, $\bar{v}'$, $\bar{\theta}'$, respectively, and $\bar{I}_1 = I_1(\bar{\theta}(z_0))$, $\bar{H}_0 = H_0(z_0, \bar{u}'(z_0), \bar{v}'(z_0), \bar{\theta}'(z_0))$ with θ, u_z, v_z, θ_z replaced by $\bar{\theta}$, u', $\bar{v}'$, $\bar{\theta}'$, respectively.

Note that the boundary conditions for θ at $z = z_0$ are replaced by a uniform expression for the cases $b = 0$ and $b \neq 0$, respectively. Hence forth, K_h will no longer appear, since it is taken to be identically equal to K_m. Also, the arguments of K_m, H_0 and I_1 will usually be omitted to simplify the notation.

We introduce the notation $u_1(z) = \bar{u}(z)$, $u_2(z) = d\bar{u}(z)/dz$, $u_3(z) = \bar{v}(z)$, $u_4(z) = d\bar{v}(z)/dz$, $u_5(z) = \bar{\theta}(z)$, $u_6(z) = d\bar{\theta}(z)/dz$, and $u_7(z) = z$ and rewrite the steady-state problem (2.9)–(2.11) as an equivalent system of seven first-order autonomous ODEs as required by AUTO (where $z \in (z_0, Z)$):

$$u_1' = u_2, \quad K_m u_2' + K_m' u_2 = f_{co}(v_G - u_3), \quad u_3' = u_4, \quad K_m u_4' + K_m' u_4 = f_{co}(u_1 - u_G),$$
$$u_5' = u_6, \quad K_m u_6' + K_m' u_6 = 0, \quad u_7' = 1.$$

The boundary conditions are

$$u_1(z_0) = 0, \quad u_1(Z) = u_G,$$
$$u_3(z_0) = 0, \quad u_3(Z) = v_G,$$
$$u_5(z_0) = \theta_g - \frac{bH_0}{a\rho C_p z_0}, \quad u_5(Z) = \theta(Z),$$

and

$$u_7(z_0) = z_0,$$

coupled with the algebraic equation for θ_g:

$$I_1 - \sigma \theta_g^4 - H_0 - C_b k_m(\theta_g - \theta_m) = 0,$$

where $u_i' = du_i(z)/dz$, $i = 1, \ldots, 7$, and

$$K_m' = \frac{dK_m}{dz} \equiv \frac{\partial K_m}{\partial z} + \frac{\partial K_m}{\partial u_2} u_2' + \frac{\partial K_m}{\partial u_4} u_4' + \frac{\partial K_m}{\partial u_6} u_6'.$$

At this stage,

$$K_m \equiv K_m(z, u_2, u_4, u_6) = K_h \equiv K_h(z, u_2, u_4, u_6)$$

are functions of z and the solution components u_2, u_4, u_6, with u_z, v_z, θ_z replaced by u_2, u_4, u_6, respectively, and

$$I_1 = I_1(u_5(z_0))$$

and

$$H_0 = H_0(z_0, u_2(z_0), u_4(z_0), u_6(z_0))$$

given by (1.7), (1.8), with $\theta, u_z, v_z, \theta_z$, replaced by u_5, u_2, u_4, u_6, respectively.

An additional variable $u_7(z) = z$ is introduced in order to make the equations autonomous, as required by AUTO. Also, the overbar for θ_g is dropped for the sake of simplicity.

3. Bifurcation Analysis

3.1 Background

The set of equations both for the two-layer and multi-layer models listed above can be considered nonlinear dynamical systems with the nonlinearities entering through the turbulent diffusion terms. Such nonlinearities effectively prevent analytical solutions; however, the behavior of the solutions can be quantitatively examined as a function of external parameters using the methods of numerical bifurcation theory. Despite the lack of a closed-form solution, parameter dependence and other characteristics of the solution can be developed.

Techniques of nonlinear analysis have expanded in recent years to include numerical continuation methods. Sophisticated software packages are now available to carry out the analysis. For those not familiar with numerical continuation, SEYDEL (1988) and DOEDEL *et al.* (1991) provide excellent overviews of practical techniques of nonlinear analysis and continuation.

With numerical continuation the steady-state solution of the system can be traced out, thus providing an understanding of the dependence of the solution on the bifurcation parameter (DOEDEL and KERNEVÉZ, 1986). In our case, this parameter is imposed, similar to the geostrophic wind and surface roughness. Techniques of linear stability analysis can be used to determine the characteristics of the perturbed solution from the steady-state solution. By numerically calculating the eigenvalues, we can determine whether a portion of the curve can be characterized as stable or unstable to perturbations or whether the perturbed solution will exhibit oscillatory behavior.

MCN95 used numerical continuation to analyze the system and characterize the behavior of the system as a function of external parameters using the AUTO analysis system (DOEDEL and KERNEVÉZ, 1986). In the present analysis we choose the external parameters to be geostrophic wind, u_G, roughness length, z_0, layer heights, z_1 and z_2, and the Coriolis parameter, f_{co}. We first provide additional analysis of the two-layer system not provided in MCN95. Next, we provide the analysis of the multi-layer system.

The steady-state problem for (2.5)–(2.8) is a system of four algebraic equations. We take u_G, z_0, Z, and θ_m as bifurcation parameters and use AUTO to compute the corresponding bifurcation diagrams, since we are interested in the steady-state behavior of the wind field and the temperature field while these four quantities vary in appropriate domains. We have to provide AUTO with an approximate starting solution to the algebraic system for appropriately selected values of the parameters. We do this by applying a fourth-order Runge-Kutta scheme to produce the initial guess required by the Newton iteration scheme for solving nonlinear systems, and then use the Newton scheme to provide an initial guess for AUTO. Details are found in SHI (1997).

3.2 Additional Analysis of the Two-layer System

MCN95 reported on the first analysis of this system using AUTO. Figure 2, a bifurcation diagram from SHI (1997) (MCN95 showed the diagrams separately), illustrates that the system of equations supports hysteresis. This is indicated by the S shape in the bifurcation diagram. Because of hysteresis, multiple solutions exist in some parameter regimes. This is a classical case of the lack of predictability in the system with potential sensitivity to initial conditions. For light geostrophic winds the system goes to a light-wind, cold solution, and at large geostrophic winds the system goes to a warm, windy solution. This parameter dependence on wind speed is consistent with conventional thinking about the stable boundary layer. However, sharp variation occurs in the region between the warm and cold solutions; for larger roughness lengths the system supports three solutions in the foldback region. The upper and lower solutions are characterized by the eigenvalues (see below) as stable attracting solutions, while in the foldback region the solution is unstable. This

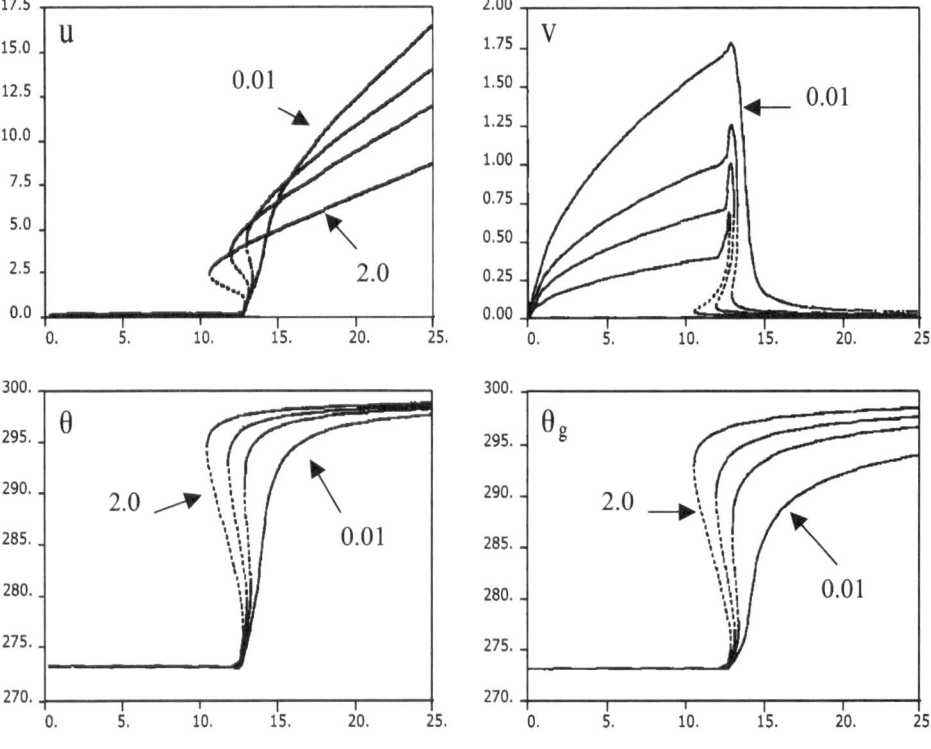

Figure 2

Bifurcation diagram for ODE systems after MCN95. Bifurcation parameter on the *x* axis is geostrophic wind. Four curves are given for roughness lengths – 0.01, 0.5, 1.0, and 2.0 m. The dashed portion of the curves indicates unstable solutions.

behavior is a classical case of the lack of predictability; for the same parameters three solutions are supported although only two are stable. Thus, the solution that evolves depends on the initial conditions.

The time-dependent behavior of the system in and near the regions of multiple solutions is interesting and illustrates both the sensitive dependence on initial conditions and the fact that the system can experience abrupt transitions through the night from warm to cold solutions. We carried out traditional time-dependent solutions of the ODE system using a fourth-order Runge-Kutta solver. Figures 3 and 4 present two examples of the time-dependent solutions also from MCN95.

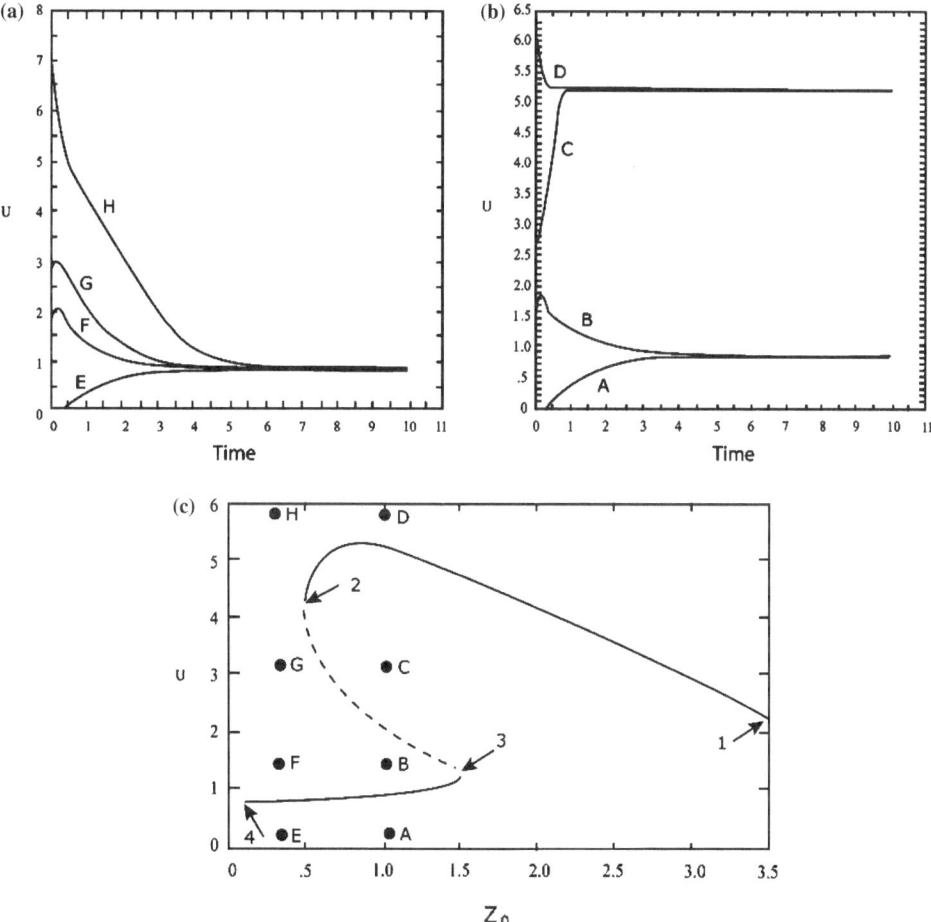

Figure 3
Time-dependent solutions for different initial conditions. Points for the initial conditions are shown on the bifurcation diagram (c) for (a) are indicated by E–H and for (b) by A–D. The roughness parameter z_0 is used as the bifurcation parameter.

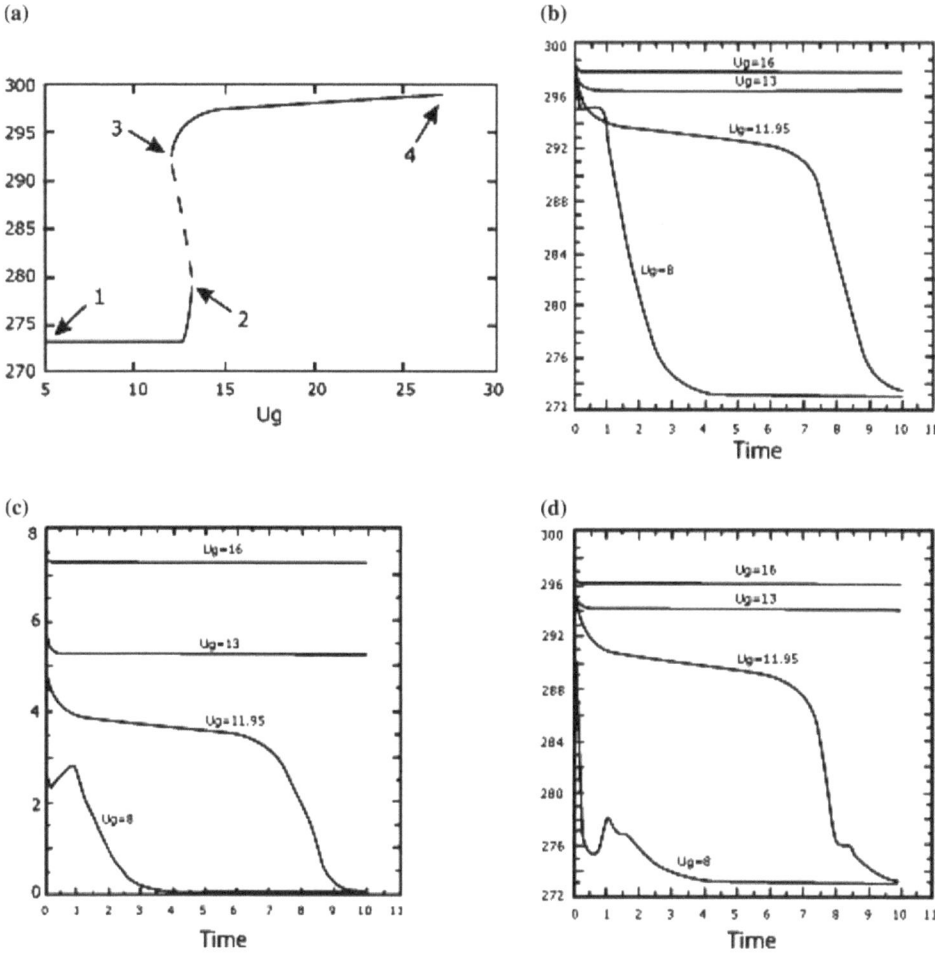

Figure 4
Time-dependent solutions for temperature different values of geostrophic winds. For reference the
bifurcation diagram for temperature is given in the upper left diagram.

Figure 3 shows that the model can be attracted to different solutions, depending on the initial conditions. Figure 4 shows that the solution can stay warm for most of the night, and then transition to a cold solution rather abruptly.

The relevance of these time-dependent solutions to real operational predictions of the nocturnal boundary layer was not discussed in MCN95. Some might regard the steady-state solutions shown in the bifurcation diagrams as a curiosity and not relevant to actual predictions. However, the time-dependent solutions of Figures 3 and 4 represent the time-dependent techniques used in weather forecast models. The bifurcation diagrams indicate where unpredictable behavior may exist and the time-dependent solutions demonstrate how this behavior is realized in a forecast setting.

The time-dependent results in Figures 3 and 4 establish that indeed the bifurcation diagrams have identified areas where predictability problems might exist.

Some analyses of the behavior of the stable boundary layer using nocturnal models similar to those presented here have indicated cyclic behavior (BLACKADAR, 1979; REVELLE,1993). Others, using a slightly different formulation, have found the existence of pure limit cycles (VAN DE WIEL *et al.*, 2002). It has been argued that such limit cycles might be responsible for the observed turbulent bursting or breakdown of the stable boundary layer. Besides the parameter dependence of the steady-state solution, the analysis by AUTO also provides a numerical calculation of the eigenvalues of the Jacobian matrix of the system. The eigenvalues are important because they characterize the behavior of the system of ODEs. If the real part of an eigenvalue is negative, then the system has a stable solution that attracts perturbations back to the steady-state solution. If the real part is positive, then the solution is unstable to perturbations. The existence of imaginary eigenvalues means oscillatory solutions can be supported (SEYDEL, 1988). Thus, we can examine whether the system might support cyclical behavior.

Figure 5 shows the four eigenvalues as computed by AUTO and their dependence on geostrophic wind speed. Three of the four eigenvalues have negative real parts over the entire range. However, one eigenvalue (5c) does have a positive real part in the narrow region $11.96 \text{ m s}^{-1} < u_G < 13.16 \text{ m s}^{-1}$. Imaginary values also exist over part of the domain. In most cases the imaginary part is paired with a negative real part (except for the third eigenvalue) causing any oscillations to be damped. Thus, the eigenvalue analysis reveals that pure limit cycles are not supported in this system for the parameter base studied. The numerical time-integration solutions in Figures 3 and 4 do show some transient but damped oscillations. While oscillations may exist, they appear to be highly damped.

Although this analysis does not support the existence of pure limit cycles, other investigators such as VAN DE WIEL *et al.* (2002) have shown that they may exist for other formulations of the problem. Van de Wiel *et al.* included a mulch layer at the ground with very scant heat capacity. We have carried out an analysis with heat capacity as a bifurcation variable to determine the effect of this parameter on the system.

Heat capacity as used in the surface energy budget in actual numerical modeling applications is relatively ill-defined, especially for composite surfaces comprising vegetation, bare soil, human structures, etc. MCNIDER *et al.* (2004) discuss the role of heat capacity in numerical models. Previous investigators who have applied models over mixed vegetated areas have used heat capacity values that vary over three orders of magnitude. Although some early mesoscale models employed a surface energy budget with zero heat capacity (MCNIDER and PIELKE, 1981), others used a prognostic equation for the surface temperature of the form employed in equation 2.4 (BLACKADAR, 1979). This includes a storage term, $C_b\theta\partial_g/\partial t$, so called because it

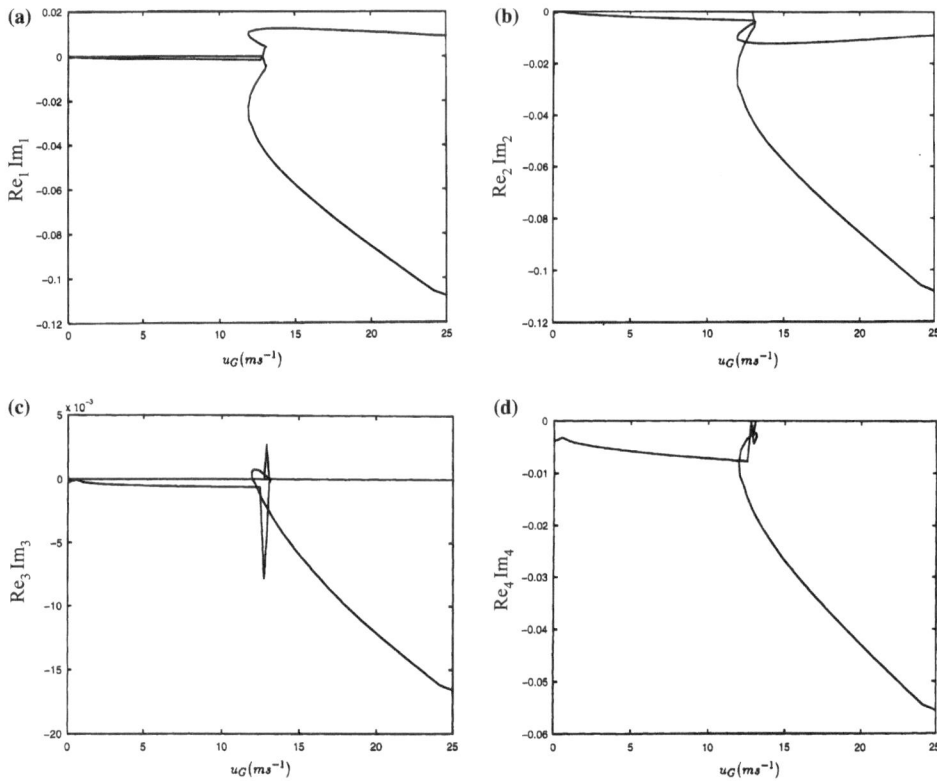

Figure 5
Diagrams for the real part (lower right curve) and imaginary parts (upper right curve) of the four eigenvalues of the Jacobian matrix of the two-layer system. (a)–(d) represent the 1st through 4th eigenvalues, respectively.

represents the imbalance in the forcing terms on the right-hand side. The definition and interpretation of C_b depends on the makeup of the surface.

Blackadar took the surface to be a uniform slab representative of bare soil. Thus C_b represented the heat capacity over some assumed depth d_s of the slab, i.e., $C_b = C_g d_s$. Blackadar also specified C_b to include the angular velocity of the earth, ω, and thermal conductivity, λ, so that the single-layer slab model would replicate the phase and amplitude of the surface temperature of a multi-layer, analytical soil model:

$$C_b = \left(\frac{\lambda C_g}{2\omega}\right)^{1/2}$$

This is the basic form of the force-restore model implemented here and in models such as MM5.

Heat capacity as used here can be viewed as equivalent to the bulk heat capacity, C_b, discussed above. The bifurcation diagrams of temperature versus bulk heat capacity in Figure 6 are interesting and instructive. We have covered a wide range of bulk heat-capacity values to span the parameter space discussed in the literature from the small values for vegetation to the larger values for wet soils and dense vegetation. VAN DE WIEL et al. (2002) considered values as small as 500 J m^{-2} K^{-1} and as large as 10000 J m^{-2} K^{-1}. MAHFOUF et al. (1995) used an equivalent vegetative heat capacity coefficient of 5×10^4 J m^{-2} K^{-1} for a vegetative layer. This value is more than an order of magnitude greater than the 1000 J m^{-2} K^{-1} of NOILHAN and PLANTON (1989), which was used for a vegetative layer. GIARD and BAZILE (2000) used a minimum value for a vegetative layer of 1.25×10^5 J m^{-2} K^{-1}. For comparison, dry sand has a value of approximately 2.5×10^4 J m^{-2} K^{-1} (VAN DE WIEL et al., 2002). This indicates the range and potential uncertainty in setting these parameters.

The bifurcation diagrams for the bulk heat capacity are provided for three different geostrophic winds (see Fig. 6). For the light wind case ($u_G = 8$ m s^{-1}) the air and slab temperature show a strong dependence on bulk heat capacity, especially at

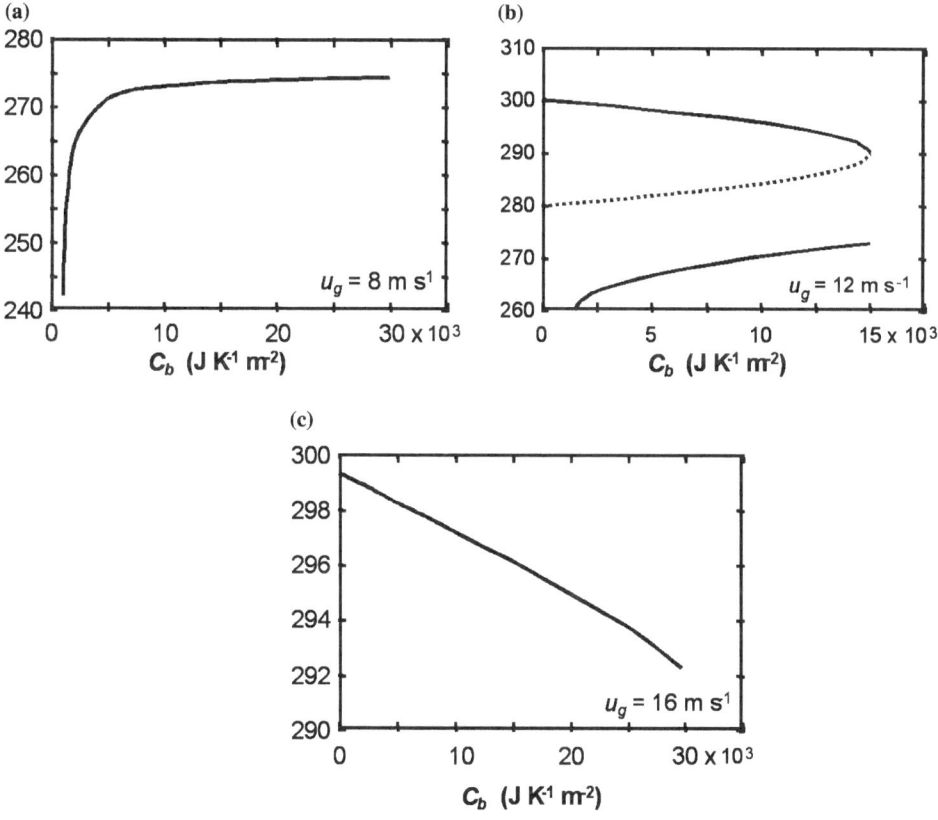

Figure 6

Bifurcation diagram with heat capacity as the bifurcation parameter for three different geostrophic wind speeds. Dashed lines indicate unstable solutions.

values between 3000 and 8000 J m^{-2} K^{-1}, where the air and surface temperature decrease as the bulk heat capacity decreases. In the highly stable regime turbulence is suppressed and the surface is dominated by longwave loss. Thus, the smaller the bulk heat capacity the larger the loss. For larger geostrophic wind speeds ($u_G = 16$ m s^{-1}) the relationship changes. Now surface and air temperatures decrease with increasing heat capacity. In this regime turbulent heat transfer is strong and warms the surface. The larger the heat capacity the less the surface is warmed; thus, temperatures decrease for larger heat capacities. The intermediate wind-speed case, which shows multiple solutions for a range of heat capacities, is the most interesting. The two stable attracting solutions and one unstable solution are highly sensitive to initial conditions. We might even expect possible transition between states if perturbations were large enough.

3.3 Analysis of the multi-layer model

We have now conducted a bifurcation analysis of the multi-layered system described above more representative of the full PDE system. The full analysis including details of the techniques for starting values is provided in the dissertation of SHI (1997).

Figure 7 displays the bifurcation diagrams for the PDE system which can be compared to Figure 2. Like the ODE system, the full PDE system also supports hysteresis and multiple solutions. The u, v and θ variables in the diagrams are now represented by the magnitude of their vectors. Bifurcation diagrams of the behavior of the PDE and the ODE systems cannot be directly compared since for the ODE system u, v and θ are scalars for the layer. However, the transition between the cold, calm solution and the warm, windy solution occurs at similar values of geostrophic winds. The variable that can be directly compared is the ground temperature since this is a scalar variable for both the two- and n-layer systems. As can be seen comparing Figures 2 and 7, the ground temperatures as functions of geostrophic wind are comparable. The transition from cold to warm solutions occurs at similar geostrophic wind. We conclude that the two-layer ODE system and its stability characteristics are a good surrogate for the full PDE system.

4. Summary and Conclusions

For weather prediction and climate modeling, understanding the behavior and predictability of the stable boundary layer is critical. This paper has extended the work of MCN95 by using techniques of nonlinear analysis to examine the behavior of the nocturnal boundary layer as represented in formulations of the type currently used in weather-forecast and climate models. These analyses were performed using a two-layer spatial truncation of the original full PDE system.

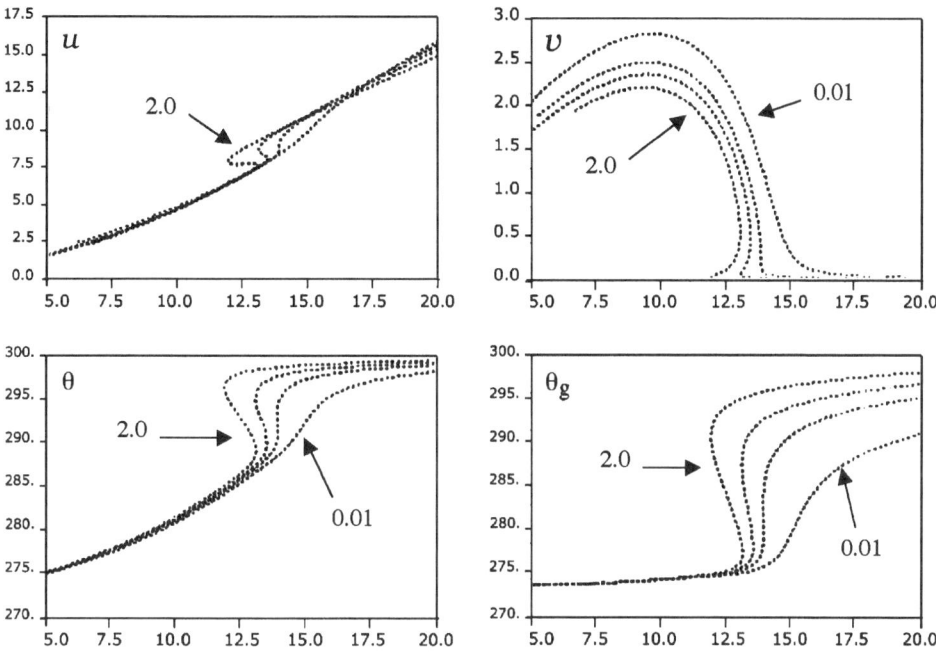

Figure 7
Same as Figure 2 except for the PDE system. Values for u, v and θ are the magnitude norm for the z vector.
Roughness lengths range from 0.01 to 2.0 m.

The analysis has shown that models can make a rapid transition from a cold solution to a warm solution as a function of imposed parameters such as geostrophic wind, surface roughness, and heat capacity. In some parameter spaces the model has multiple solutions. In these spaces the model exhibits a lack of predictability since the solution depends not on the imposed parameters but on slight changes in the initial conditions. The heat capacity or thermal inertia parameter, which is highly uncertain in weather and climate forecasting, can also change the behavior and characteristics of the solutions.

It has been speculated, with some confirmation from modeling results, that the nocturnal boundary-layer formulations of the type used here can support chaotic bursting or limit cycle behavior (BLACKADAR, 1979; REVELLE, 1993; VAN DE WIEL et al., 2002). In examining the eigenvalues of this system we do not find purely complex eigenvalues that would support undamped oscillations. We do find a narrow region where the eigenvalues are complex and the real part is positive, indicating an unstable solution. In carrying out time-dependent integrations of the systems, we selected the parameter values that allowed us to examine the space where the foldback or multiple solutions exist in the bifurcation diagram. We found evidence of transient oscillations in the solution but not pure undamped oscillations or chaotic behavior. We have not carried out an eigenvalue analysis for cases of very small heat capacities. These might support pure oscillations such as found by VAN DE WIEL *et al.* (2002).

The original analysis by MCN95 and the additional analyses of the eigenvalues with heat capacity as a bifurcation parameter were done using a truncated two-layer system. Such a truncation allowed the PDE boundary-layer equations to be represented by four ODEs. We were able to analyze the behavior of the system but were concerned that the behavior might be an artifact of the truncation. To investigate this possibility we carried out an analysis of an *n*-layer system representing the PDE system. The results show that the basic characteristics of the original two-layer system are preserved. That is, regions of multiple solutions and lack of predictability are also found in the more complete system.

Acknowledgements

This work has been partially supported by the following grants — National Science Foundation, Grant ATM-9120321; U.S. EPA Star Grant, R 826770-01-0; NASA GEWEX Grant, NCC8-200; Southern Oxidant Study, U.S. EPA Cooperative Agreement R-82897701-0; and NASA Polar Program Grant—NCC8-200. The results in this study do not necessarily reflect policy or science positions by the funding agency. We also wish to acknowledge and thank an anonymous reviewer whose detailed comments significantly improved the structure and quality of the manuscript.

REFERENCES

BLACKADAR, A. K., *High resolution models of the planetary boundary layer* In *Advances in Environmental and Scientific Engineering*, Vol. I. (Gordon and Breach, New York, 1979).

DERBYSHIRE, S. H., (1999), *Boundary-layer Decoupling over Cold Surfaces as Physical Boundary Layer Instability,* Bound. Layer Meteorol. *90*, 297–325

DOEDEL, E. J. and KERNEVÉZ, J. P., *AUTO: Software for Continuation and Bifurcation Problems in Ordinary Differential Equations*, User Manual "Auto 86" (California Institute of Technology, Appl. Math. Program Report, 1986).

DOEDEL, E. J., KELLER, H. B., and KERNEVÉZ, J. P. (1991), *Numerical Analysis and Control of Bifurcation Problems*, Int. J. Bif. and Chaos: (I) Bifurcation in Finite Dimensions *1*(3), 493–520; (II) Bifurcation in Infinite Dimensions *1*(4), 745–772.

ENGLAND, D. E. and MCNIDER, R. T. (1995), *Stability Functions Based upon Shear Functions,* Bound. Layer Meteorol. *74*, 113–130.

GIARD, D. and BAZILE, E. (2000), *Implementation of a New Assimilation Scheme for Soil and Surface Variables in a Global NWP Model,* Mon. Wea. Rev. *120*, 997–1015.

MAHRT, L. (1999), *Stratified Atmospheric Boundary Layers.* Bound. Layer Meteorol. *90*, 375–396.

MCNIDER, R. T. and PIELKE, R. A. (1981), *Diurnal Boundary Layer Development over Sloping Terrain,* J. Atmos. Sci. *38*, 2198–2212.

MCNIDER, R. T., ENGLAND, D. E., FRIEDMAN, M. J., and SHI, X. (1995), *Predictability of the Stable Atmospheric Boundary Layer,* J. Atmos. Sci. *52,* 1602–1614.

MCNIDER, R. T., BIAZAR, ARASTOO, and LAPENTA, W. M. (2004), *Role of Surface Heat Capacity in Grid Scale Models and Retrieval of Bulk Heat Capacity from Satellite Skin Brightness Temperatures,* to be submitted to J. Appl. Meteorol.

MAHFOUF, J-F, MANZI, A., NOILHAN, J., GIORDANI, H., and DEQUE, M. (1995), *The Land Surface ISBA Scheme within Meteo-France Climate Model ARPEGE,* J. Climate, *8,* 2039–2057.

NOILHAN, J. and PLANTON, S. (1989), *A Simple Parameterization for Land Surfaces in Meteorological Models,* Mon. Wea. Rev., 536–549.

PIELKE, R. A., *Mesoscale Meteorological Modeling (Academic Press, 1984).*

REVELLE, D. O. (1993), *Chaos and Bursting in the Planetary Boundary Layer.* J. Appl. Meteorol., 32, 1169–1180.

SEYDEL, R., *From Equilibrium to Chaos: Practical Bifurcation and Stability Analysis* (Elsevier Science Publishing Co., Inc., 1988).

SHI, X, *Numerical Investigation of the Stable Nocturnal Boundary Layer,* Ph.D. Dissertation. (Dept. of Mathematical Sciences, University of Alabama in Huntsville, 1997.

STALEY, D. O. and JURICA, G. M. (1972), *Effective Atmospheric Emissivity under Clear Skies,* J. Appl. Meteorol., *11,* 349–356.

STULL, ROLAND B., *An Introduction to Boundary Layer Meteorology* (Kluwer Academic Publishers, Dordrecht, 1988).

VAN DE WIEL, B., RONDA, R. J., MOENE, A. F., DE BRUIN, H. A. R., and HOLTSLAG, A. A. M. (2002), *Intermittent Turbulence and Oscillations in the Stable Boundary Layer over Land. Part I: A Bulk Model,* J. Atmos. Sci., *59,* 942–958.

VITERBO, PEDRO, BELJAARS, A., MAUHOFF, J. F., and TEIXEIRA, J. (1999), *The Representation of Soil Moisture Freezing and its Impact on the Stable Boundary Layer,* Quart. J. Roy. Meterol. Soc. *125,* 2401–2426.

(Received November 14, 2003; accepted May 14, 2004)
Published Online First

 To access this journal online:
http://www.birkhauser.ch

Pure appl. geophys. 162 (2005) 1831–1857
0033–4553/05/101831–27
DOI 10.1007/s00024-005-2695-6

© Birkhäuser Verlag, Basel, 2005

┃ **Pure and Applied Geophysics**

An Asymptotic Analysis of a Simple Model for the Structure and Dynamics of the Ramdas Layer

A. S. VASUDEVA MURTHY[1], RODDAM NARASIMHA[2], and SAJI VARGHESE[1]

Abstract—This paper presents a complete asymptotic analysis of a simple model for the evolution of the nocturnal temperature distribution on bare soil in calm clear conditions. The model is based on a simplified flux emissivity scheme that provides a nondiffusive local approximation for estimating longwave radiative cooling near ground. An examination of the various parameters involved shows that the ratio of the characteristic radiative to the diffusive timescale in the problem is of order 10^{-3}, and can therefore be treated as a small parameter (μ). Certain other plausible approximations and linearization lead to a new equation whose asymptotic solution as $\mu \to 0$ can be written in closed form. Four regimes, consisting of a transient at nominal sunset, a radiative-diffusive boundary ('Ramdas') layer on ground, a boundary layer transient and a radiative outer solution, are identified. The asymptotic solution reproduces all the qualitative features of more exact numerical simulations, including the occurrence of a lifted temperature minimum and its evolution during night, ranging from continuing growth to relatively sudden collapse of the Ramdas layer.

Key words: Ramdas layer, lifted temperature minimum, nocturnal temperature distribution.

1. Introduction

The climate of air near the earth's surface is a subject of considerable interest in agricultural meteorology (especially in connection with the occurrence of frost which can have a strong adverse effect on various crops), remote sensing (for determinations of true surface temperatures), and in general as part of the microclimate in which most animals and insects live. A recent treatment of the subject is available in GEIGER *et al.* (1995). In the present study we are particularly interested in profiles of air temperature near ground. Under clear skies and windless conditions the air temperature has generally been taken to be a monotonic function of the height z from the surface, based on data of the kind shown in Figure 1. During daytime the temperature decreases with increasing z, ground being the level of highest

---

[1] TIFR Centre, Indian Institute of Science, Bangalore, 560 012, India
[2] Jawaharlal Nehru Centre for Advanced Scientific Research, Bangalore 560 064, India. E-mail: roddam@caos.iisc.ernet.in

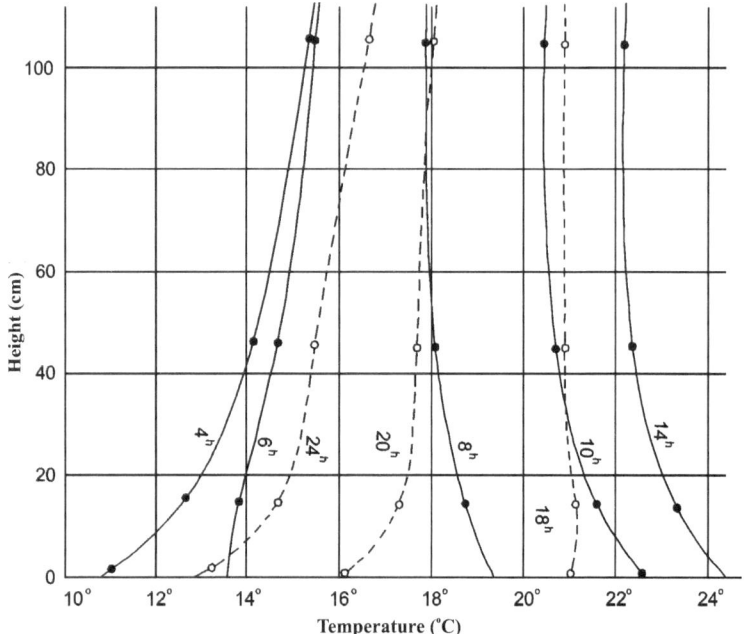

Figure 1
Daily evolution of temperature profiles below 1 m, after GEIGER (1965).

temperature. This is because of the low absorptive power of air for shortwave solar radiation: the bulk of incoming solar energy is absorbed by the earth's surface. On the contrary during nights the temperature is generally considered (not always with justification, as we shall see below) to increase with z upto what is called the inversion height z_i. This is attributed to the rapid cooling of the surface through the infrared radiation to 'space', i.e., to atmospheric layers cooler than ground. This rapid cooling of the surface also cools the air layers above. It has generally been considered that this cooling effect diminishes with height, thus resulting in a monotonically increasing temperature profile up to z_i. Indeed, GEIGER (1965, p. 81) proposed that a good fit to temperature profiles near ground (for $z < 10$ m) is Az^B where A and B are constants; during daytime B is negative while during nights B is positive. What this implies is that near the ground the thermal interaction between atmosphere and the earth's surface may be basically seen as a diffusive phenomenon. In fact KUO (1968) successfully simulated such temperature profiles by a model consisting of a linear diffusion equation supplemented by a Newtonian cooling term (we shall return to this model below). There is no doubt that the temperature profile depends upon local conditions and various other factors; however, at least under the clear-sky and no-wind conditions considered here, the general view for a long time was that the behavior of near-surface temperature profiles was well understood.

This, however, is not correct and ignores the surprising observations of RAMDAS and ATMANATHAN (1932), who reported that the nighttime temperature profile under calm clear conditions was not monotonically increasing from the ground up to z_i but contained a kink (Fig. 2) at heights of order 10 cm. This means that there was a lower air layer where temperature *decreased* from the surface, as during daytime. As it emerges from the detailed account of the history of the phenomenon by NARASIMHA (1994), these results were first accepted with reservations, as pointed out by GEIGER (1965, p. 98). The main objection was that such a cold layer of air above warm ground should be subject to Rayleigh-Benard instability and should quickly overturn. However the observations of Ramdas indicated that the cool layer near ground often persisted until dawn. This layer is what we refer to as the Ramdas layer. In spite of further confirmatory work by Ramdas, these results were attributed, for more than twenty years, to either instrumentation errors in the measurements or to advection from colder regions in the neighborhood. It was perhaps the eventual confirmation by LAKE (1956) and Raschke (1957) that removed all doubts about the existence of the Ramdas layer. The phenomenon is variously known as the lifted or elevated temperature minimum, and was highlighted by LETTAU (1979) as the Ramdas paradox in micrometeorology. Such singular or anomalous temperature

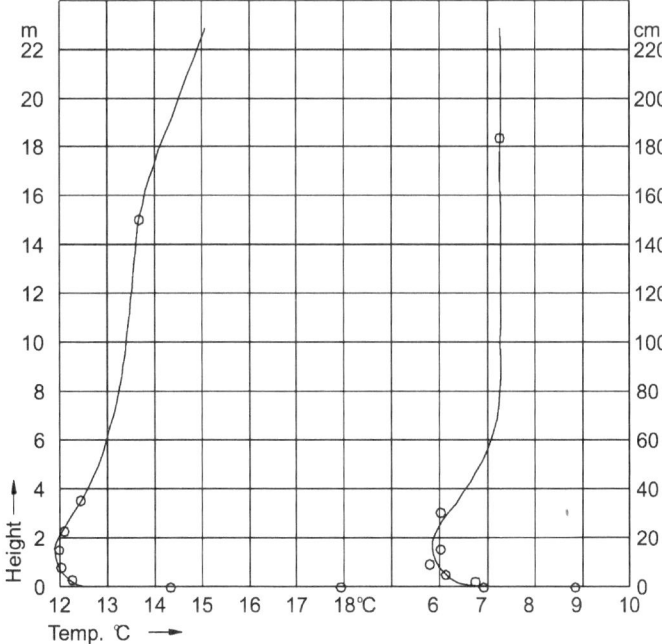

Figure 2
Temperature profiles (RAMDAS and ATMANATHAN, 1932) observed at Poona (left) and Agra (right) on Nov. 29, 1931 at 0600 h and 0536 h local time, respectively.

distributions near ground continue to be of considerable current interest; see e.g., the recent study of YAMADA and TAKAHASHI (2004). A schematic of such a singular temperature distribution, showing both a lifted minimum and the inversion, is displayed in Figure 3.

VASUDEVA MURTHY *et al.* (1993, hereafter refered as VSN) have proposed a model that successfully simulates the Ramdas layer. They also advanced a heuristic argument showing how the Rayleigh-Benard convection can be circumvented by radiative stabilization. The proposed model was a nonlinear diffusion equation with nonlocal terms. This was solved numerically for a wide range of ground cooling parameters. To illustrate the evolution of the lifted temperature minimum through night, and its response to certain kinds of turbulence episodes or gusts, detailed numerical solutions were provided by RAGOTHAMAN *et al.* (2001, 2002, hereafter R01, R02). An energy balance analysis by NARASIMHA and VASUDEVA MURTHY (1995, hereafter NV) indicated the terms in the model that were crucial for the

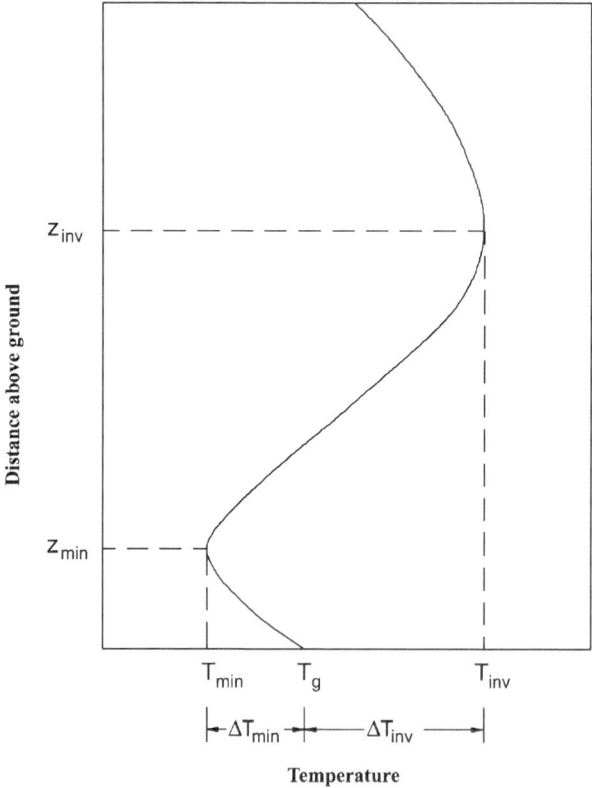

Figure 3

Schematic showing the lifted temperature minimum and inversion, and the notation used in this paper.

formation of the Ramdas layer. A survey of much of this work is available in
NARASIMHA (1997).

Given the counter-intuitive nature of the phenomenon studied, it is important to
understand the dominant mechanisms that are responsible for the origin and
evolution of the layer through night. For this purpose we derive here a simple
equation from the VSN model by linearization and localization of the 'conspiring'
terms. We then obtain an approximate analytical solution of this simple equation
using singular perturbation techniques. This solution reproduces qualitatively the
structure and dynamics of the Ramdas layer, and helps to illuminate the nature of the
lifted temperature minimum phenomenon. Our purpose here, it must be emphasized,
is to obtain insight into the physics and mathematics of the problem. More accurate
treatments of the problem are best sought through computer simulations of the kind
described in R01, R02.

2. Mathematical Formulation

We first briefly describe the VSN model. We consider the temperature
distribution over bare soil in flat terrain, on calm clear nights with no advective
changes. The temperature T can then be assumed to be homogeneous in the
horizontal and so is only a function of the vertical height z and time t. Under these
conditions T is governed by the one-dimensional energy equation

$$\rho_a c_p \frac{\partial T}{\partial z} = -\frac{\partial Q}{\partial z}, \quad z > 0, \quad t > 0. \tag{2.1}$$

Here ρ_a is the density of air, c_p is the specific heat of air at constant pressure and Q is
the total energy flux, conveniently split into three components representing,
respectively the contributions of molecular conduction (Q_m), turbulent convection
(Q_t) and radiation (Q_r),

$$Q = Q_m + Q_t + Q_r. \tag{2.2}$$

The first two are given by

$$Q_m = -k_m \frac{\partial T}{\partial z}, \quad Q_t = -k_t \frac{\partial (T + \Gamma z)}{\partial z}, \tag{2.3}$$

where k_m is the thermal conductivity of air, $T + \Gamma z$ is the so-called potential
temperature, Γ is the adiabatic lapse rate (of order 6 K/km) at which temperature
drops with altitude in the free atmosphere and k_t is a suitable eddy conductivity.

As the lifted temperature phenomenon we discuss here is most strongly seen when
there is little or no turbulent transport, we shall be setting $Q_t = 0$. This incidentally
limits the altitudes upto which the present analysis will apply, because even in calm
conditions the contribution of eddy transport *relative* to radiation may not be

negligible at large z. The detailed numerical simulations reported in R02 (see Fig. 8) show the effect of turbulent transport. At a friction velocity of 0.1 ms^{-1}, for example, a lifted minimum does occur but the layer slowly fades away beginning about 2 hours after sunset.

The radiative flux divergence is written as

$$\frac{1}{\rho_a c_a}\frac{\partial Q_r}{\partial z} = \frac{q}{c_p}\frac{\partial F}{\partial u}, \tag{2.4}$$

where

$$q(z) = \rho_w(z)/\rho_a(z) \tag{2.5}$$

is the water vapor mixing ratio (ρ_w is the density of water vapor in the atmosphere) and

$$u = \int_0^z \rho_w(z')\,dz' \tag{2.6}$$

is the water vapor mass path length. The flux F in (2.4) is split into four components (see NV for more details),

$$F = F_{eg}^{\uparrow} + F_{rg}^{\uparrow} + F_a^{\uparrow} - F_a^{\downarrow}, \tag{2.7}$$

where, using the flux-emissivity scheme for parameterizing radiation following VSN, we can write

$$F_{eg}^{\uparrow} = \varepsilon_g(1 - E(u))\sigma T_g^4, \tag{2.8}$$

$$F_{rg}^{\uparrow} = (1 - \varepsilon_g)(1 - E(u))F_a^{\downarrow}(0), \tag{2.9}$$

$$F_a^{\uparrow} = \int_0^u \sigma T^4(u',t)\dot{E}(u - u')\,du', \tag{2.10}$$

$$F_a^{\downarrow} = \int_0^{u_{\infty}} \sigma T^4(u',t)\dot{E}(u - u')\,du'. \tag{2.11}$$

These terms represent respectively emission and reflection from ground and emission by air, arrows denoting upwelling and downwelling fluxes. The radiative transport scheme we adopt consists in the use of the broadband flux emissivity function $E(u)$, whose values for water vapor are provided by ZDUNKOWSKI and JOHNSON (1965); $\dot{E}(u) \equiv dE/du$ is the derivative of $E(u)$. The function $E(u)$ is discussed below in section 3. Finally σ is the Stefan-Boltzmann constant, ε_g is the emissivity of ground, and T_g is the ground temperature, whose variation with time will be considered known or prescribed in the present analysis (see (2.13) below; a more complete treatment, with air-soil coupling, is provided by VSN).

A more elaborate band-model scheme for fast, accurate computation of the fluxes (2.7) has recently been proposed and implemented by VARGHESE *et al.* (2003), but for our present purposes the flux emissivity scheme (2.8)–(2.11) is adequate.

Equation (2.1) is supplemented with the initial/boundary conditions

$$T(z,0) = T_{g_0} - \Gamma z, \tag{2.12}$$

$$T(0,t) \equiv T_g(t) = T_{g_0} - \beta\sqrt{t}, \tag{2.13}$$

$$\frac{\partial T}{\partial z}(\infty, t) = -\Gamma, \tag{2.14}$$

where the function $T_g(t)$ in (2.13) is from an expression due to BRÜNT (1941) and T_{g_0} is the ground temperture at time $t = 0$ (which we can think of as the time of nominal sunset: see R01). Boundary condition (2.14) is weaker than prescribing

$$T(z,t) = T_{g_0} - \Gamma z,$$

which would be valid at heights greater than those considered here. The system (2.1), (2.8)–(2.11) is a nonlinear parabolic integro-differential equation. Its solution, satisfying the conditions (2.12) – (2.14) for various illustrative values of ε_g and β, was computed by VSN numerically using the method of lines. It was found (see also NV) that if

$$k_t \approx 0, \beta < 15\mathrm{K}/\sqrt{\mathrm{h}} \quad \text{and} \quad \varepsilon_g < 1, \tag{2.15}$$

the temperature profiles exhibit a lifted minimum. In fact the numerical solution was able to predict closely the temperature distributions observed by Ramdas.

The physical explanation for the conditions given in (2.15) have been discussed in detail by VSN and NV. It was particularly pointed out there that the strong dependence of the lifted minimum on ε_g is due to the rapid variation of $E(u)$ near the origin. The radiative flux divergence in (2.4) near ground includes the product

$$(1 - \varepsilon_g)\dot{E}(0), \tag{2.16}$$

which is chiefly responsible for the heavy radiative cooling near ground. As VSN point out $\dot{E}(0)$ is large (due to the large absorption coefficients in the vibration-rotation bands of the H_2O molecule), and so the product in (2.16) can be significant even if ε_g is close to unity. However, this fact is not immediately evident from a casual look at the fluxes in (2.7). In fact equation (2.1) is masked by too many formulae for the radiative fluxes, making it hard to visualize the structure of the solution.

The formation of the Ramdas layer is basically due to the unusual balance between this radiative cooling and thermal diffusion. The temperature distribution in the air column is therefore like that in a conducting rod that experiences a loss of heat (i.e., cooling) on its lateral surface — a situation in which a temperature minimum can occur at some section along the rod, as pointed out by RASCHKE (1957).

The main aim of this work is therefore to present a simpler version of (2.1) which will enable us to infer the structure of the solution by more transparent analysis of the unusual balance between radiation and conduction mentioned above. This also has the advantage of providing greater physical insight into the problem. We achieve the simplification by approximating the integrals in (2.10)–(2.11) and linearizing the fourth powers of T; we then show that this simplified equation simulates the qualitative behavior of the Ramdas layer to a reasonable approximation.

Before we conclude this section it is noteworthy to mention why Kuo's model (which was able to simulate clear weather temperature profiles successfully) cannot simulate the lifted minimum for the boundary conditions (2.12)–(2.14). Kuo represented the radiative flux divergence (2.4) by

$$k_r \frac{\partial^2 T}{\partial z^2} + a^2 (T - \overline{T}),$$

where the diffusion term (with radiative diffusivity k_r) is supposed to represent the effect of the strongly absorbing regions in the spectrum, while the Newtonian cooling term represents the weakly absorbing regions. It is well known (ÖZISIK, 1973, p. 319) that the diffusion approximation breaks down near the boundary (in the present case $z = 0$) and one has to use a boundary condition which can accommodate the associated discontinuity (slip). For a derivation of such a boundary condition see LARSEN *et al.* (1983). It was shown rigorously in VASUDEVA MURTHY (1986) that such a model for (2.4) cannot predict the lifted minimum. Also a radiation model that does not include the surface emissivity effect cannot simulate the Ramdas layer either, because it does not allow for the mechanism indicated in (2.16). In fact most authors (save for a few like GARRATT and BROST, 1981, who use a detailed radiation model) assume $\varepsilon_g = 1$. The point that we are emphasizing here is that the energy balance near the ground is rather delicate and radiation plays a subtle role; this is the principal reason why the phenomenon long remained without a convincing explanation (see VSN for a discussion of this point).

3. *Towards a Local, Linear Model*

Our first approximation is, as already mentioned, to ignore eddy diffusion (cf.,2.15), so we put

$$Q_t = 0. \tag{3.1}$$

We can now write (2.1) as

$$\frac{\partial T}{\partial t} = -\frac{1}{\rho_a c_p} \frac{\partial Q_m}{\partial z} - \frac{q}{c_p} [F_{eg}'^{\uparrow} + F_{rg}'^{\uparrow} + F_a'^{\uparrow} - F_a'^{\downarrow}], \tag{3.2}$$

where the prime denotes derivatives with respect to u of the various fluxes in (2.8)–(2.11). Note that it is the *derivatives* (or divergences) of the fluxes, rather than the fluxes themselves, that matter. These terms were computed for various cases in NV so as to assess their contribution to the total cooling rate. For the so-called baseline case ($\varepsilon_g = 0.8, \beta = 2K/\sqrt{h}$ and $q = 0.005$) which shows a pronounced lifted temperature minimum, NV showed that the cooling rates due to the downwelling and upwelling fluxes near the surface ($u \to 0$) can be approximated by

$$F_a'^{\downarrow} \approx 0, \tag{3.3}$$

$$F_a'^{\uparrow} \approx \sigma T^4(u)\dot{E}(u), \tag{3.4}$$

Using (3.1) and (3.3)–(3.4) in (2.2) and (2.7) we obtain

$$\frac{\partial T}{\partial t} = K_m \frac{\partial^2 T}{\partial z^2} + \frac{q}{c_p}\dot{E}(u)\left[\varepsilon_g \sigma T_g^4(t) + (1 - \varepsilon_g)F_a^{\downarrow}(0) - \sigma T^4\right], \tag{3.5}$$

an equation which does not contain integrals and so is local (but not diffusive) *if* $F^{\downarrow}(0)$ can be considered known or prescribed (as we shall in fact do below). Note that the presence of q here is due to the fact that $\partial u/\partial z = \rho_w$ (cf. 2.4–2.5); and $K_m = K_m/\rho_a c_p$.

We now proceed to nondimensionalize this equation. Make the change of variables

$$\bar{z} = z/\ell_e; \ell_e = 1/\rho_{wo}\dot{E}(0), \tag{3.6}$$

$$\bar{t} = t/t_d; t_d = \ell_e^2/K_m, \tag{3.7}$$

$$\bar{T} = \frac{T - T_{g0}}{T_{g0}} + \bar{\Gamma}\bar{z}, \tag{3.8}$$

where $\bar{\Gamma}$ is defined in (3.17) below. The order of magnitude of these variables can be estimated by choosing the typical values (cf VSN) $T_{g0} = 300$ K, $\rho_{w_0} = 0.012$ kg/m³ (water vapor density at the surface), $\dot{E}(0) = 63$ m²/kg and $K_m = 2.5 \times 10^{-5}$m²/s. We thus obtain

$$\ell_e \approx 1.3\,\text{m}, \quad t_d \approx 7 \times 10^4\,\text{s}.$$

The length scale here is characteristic of what VSN have called the 'emissivity sublayer'; t_d is the order of time required for heat to diffuse over the height ℓ_e by pure molecular conduction.

Using (2.13) and (3.8) we approximate the nonlinear terms by

$$\sigma T_g^4 \approx \sigma T_{g0}^4[1 - 4\bar{\beta}\sqrt{\bar{t}}], \tag{3.9}$$

$$\sigma T^4 \approx \sigma T_{g0}^4[1 + 4(\bar{T} - \bar{\Gamma}\bar{z})], \tag{3.10}$$

where

$$\bar{\beta} = \frac{\beta\sqrt{t_d}}{T_{g0}}. \tag{3.11}$$

Note that the approximations in (3.9, 3.10) assume that $\bar{\beta}\sqrt{\bar{t}}$ is small, which limits the range of validity of the present theory to relatively low ground cooling.

Finally, using the expressions

$$\rho_a(z) = \rho_{a_0}\exp(-z/H_a), \quad \rho_w(z) = \rho_{w_0}\exp(-z/H_w), \tag{3.12}$$

where H_a, H_w are respectively the atmospheric scale heights for air and water vapor, we obtain

$$q(z) = q_0 \exp(-z/H_q); \quad q_0 = \rho_{w_0}/\rho_{a_0}, \quad \frac{1}{H_q} = \frac{1}{H_w} - \frac{1}{H_a}.$$

Typical values for the scale heights in the tropics (ANANTHASAYANAM and NARASIMHA, 1989) are $H_w \approx 2.7$ km, $H_a \approx 8$ km, giving $H_q \approx 4.1$ km.

Using (3.9)–(3.10) in (3.5), we obtain the governing linear equation

$$\mu\left[\frac{\partial^2 \bar{T}}{\partial \bar{t}} - \frac{\partial^2 \bar{T}}{\partial^2 \bar{z}^2}\right] = \exp(-z/H_q)\bar{E}(u)[\lambda + \varepsilon_g\bar{\beta}\sqrt{\bar{t}} + \bar{T} - \bar{\Gamma}\bar{z}], \tag{3.13}$$

with the transformed initial/boundary conditions (2.12)–(2.14),

$$\bar{T}(\bar{z},0) = 0, \tag{3.14}$$

$$\bar{T}(0,\bar{t}) = -\bar{\beta}\sqrt{\bar{t}}, \tag{3.15}$$

$$\frac{\partial \bar{T}}{\partial \bar{z}}(\infty,\bar{t}) = 0. \tag{3.16}$$

Here

$$\begin{cases} \mu \equiv t_r/4q_0 t_d, t_r \equiv c_p/\sigma T_{g0}^3 \dot{E}(0), \bar{E}(u) \equiv \dot{E}(u)/\dot{E}(0), \\ \lambda \equiv (1-\varepsilon_g)(1-\varepsilon_s)/4, \varepsilon_s \equiv F_a^\downarrow(0)/\sigma T_{g0}^4 \approx E(u_\infty), \bar{\Gamma} \equiv \Gamma \ell_e/T_{g0}. \end{cases} \tag{3.17}$$

The parameter μ is basically a ratio of the radiative to the diffusive time scale in the problem. Substitution for t_d from (2.7) gives

$$\mu = \frac{(K_m \Delta T_{g0}/\ell_e)}{4\sigma T_{g0}^3 \Delta T_{g0}},$$

showing μ can be alternatively interpreted as the ratio of the [change in] heat flux to that in the radiative energy flux due to a change of order ΔT_{g0} in air temperature. The fact that μ is very small (see below) plays a crucial role in the present theory.

From (2.11) we have

$$F_a^\downarrow(0) = \int_0^{u_\infty} \sigma T^4(u',t)\dot{E}(u')\,du'.$$

Using (3.10) in this equation we obtain

$$\varepsilon_s \approx E(u_\infty) + 4 \int_0^\infty (\overline{T} - \overline{\Gamma z})\dot{E}(u(\bar{z})) \, du(\bar{z}). \tag{3.18}$$

Fortunately the integral term here is relatively small, and in any case not crucial for the formation of the Ramdas layer, as will be evident from the results we shall report below. The integrand is significant only near the ground. An *a posteriori* check shows that omitting the integral does not introduce significant error (and if necessary an iteration procedure can be adopted). For our present purposes, therefore, we will, as a first approximation take

$$\varepsilon_s \simeq E(u_\infty). \tag{3.19}$$

We shall now estimate the various parameters appearing in (3.17). Using $\Gamma \sim 0.01$ K/m, $F^\downarrow(0) \sim 200$ W/m^2, $\beta \sim 2K\sqrt{h}$ and $\varepsilon_g = 0.8$, we obtain the following typical values:

$$\mu \approx 4 \times 10^{-3}, t_r \approx 10\text{s}, \lambda \approx 0.03, \varepsilon_s \approx 0.44, \overline{\Gamma} \approx 4 \times 10^{-5}, \overline{\beta} \approx 0.03. \tag{3.20}$$

4. A Simplified Flux-emissivity Scheme

We now propose an approximation for the factor $\exp(-z/H_q)\overline{E}(u)$ that appears on the right of (3.13). To do this we shall consider a widely used expression for $E(u)$ proposed by ZDUNKOWSKI and JOHNSON (1965, cf. (3.7) in VSN),

$$E(u) = a \ln(1 + bu), \tag{4.1}$$

where

$$a = 0.04902, \; b = 1263.5 \text{ m}^2/\text{kg} \quad \text{for } u < 0.01 \text{ kg/m}^2,$$

$$a = 0.05624, \; b = 875 \text{ m}^2/\text{kg} \text{ otherwise.}$$

As already mentioned, it is the flux derivatives rather than the fluxes themselves that are important; similarly it is the derivative $\dot{E}(u)$ that is crucial rather than the function itself. However the Zdunkowski-Johnson expression for $\dot{E}(u)$ has a discontinuity at $u = 0.01$ kg/m^2. For the construction of an approximate analytical solution it is preferable to avoid this discontinuity. So we adopt a single value for a and b and choose the first set of values because of their relevance near the ground ($u \approx 0$). Figure 4 shows that the error committed in the derivative $\dot{E}(u)$ due to this choice of a and b is small.

We therefore get, from the definition of $\overline{E}$ given in (3.17),

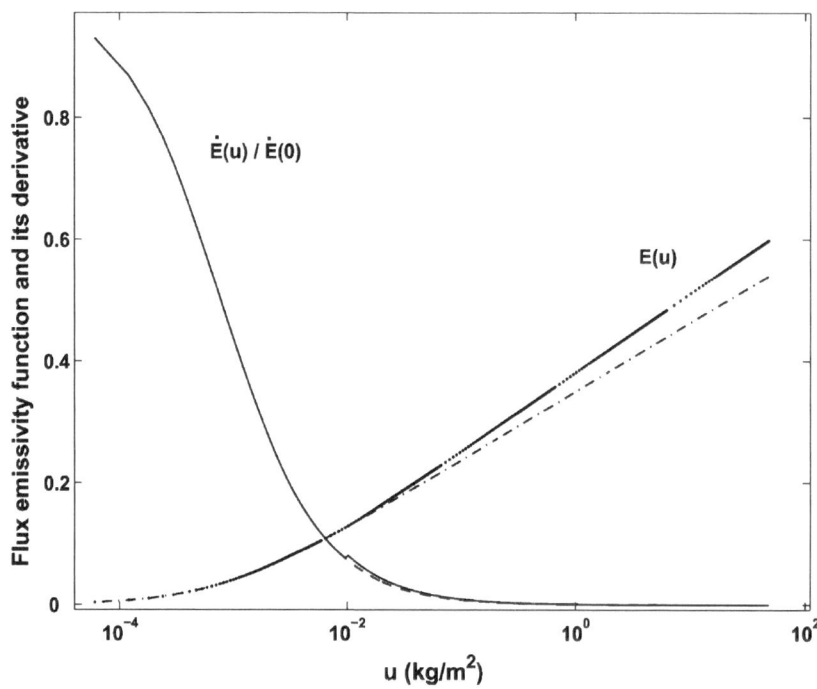

Figure 4
Flux emissivity function (4.1) and its derivative. ZDUNKOWSKI and JOHNSON (1965): E, solid line; $\dot{E}$, dotted
lines. Present approximation: E, dashed line; $\dot{E}$, dash-dot line.

$$\overline{E}(u) = \frac{1}{1 + bu}. \tag{4.2}$$

Using (3.12) in (2.6) we obtain

$$u(z) = u_\infty(1 - \exp(-z/H_w)), \quad u_\infty \equiv \rho_{w_0} H_w. \tag{4.3}$$

Consequently

$$\overline{E}(z) = [1 + bu_\infty\{1 - \exp(-z/H_w)\}]^{-1}. \tag{4.4}$$

For $z = O(\ell_e) \ll H_w < H_q$ we have

$$\overline{E}(z) \approx \frac{1}{1 + bu_\infty z/H_w}. \tag{4.5}$$

For $z \gg H_w$, (4.1) gives

$$\overline{E}(z) \approx \frac{1}{1 + bu_\infty} \approx O(10^{-5}). \tag{4.6}$$

Qualitatively $\overline{E}(z)$ starts from unity and drops rapidly over a length scale $\ell_e = O$ (1 m) and decays to a small value (cf. 4.5) over a length scale $H_w = O$ (10^3 m).

Thus we can approximate the product $\exp\left(-z/H_q\right)\overline{E}(u(z))$ by

$$e_m(z) = \frac{\exp(-\delta z)}{1 + \alpha z}, \tag{4.7}$$

where $\delta^{-1} = H_q \approx 4.1$ km is the largest length scale in the problem, while $\alpha = (b\rho_{w_0})^{-1} \approx 0.1$ m is the smallest length scale. Expression (4.7) combines the effect of these two length scales. Figure 5 shows that replacing the product $\exp(-z/H_q)\overline{E}(u(z))$ in (3.13) by the function $e_m(z)$ of (4.7) is a good approximation.

5. The Proposed Model

The model we propose to study below is now rewritten with the following (cosmetic) changes in notation: replace $\overline{T}$ by θ, remove the bars on the independent variables $\overline{z}$ and $\overline{t}$ and replace the product $\exp(-z/H_q)\overline{E}(u(z))$ with $e_m(z)$ given in (4.7). The governing equation for the present model, (3.13), now takes the form

$$\mu\left[\frac{\partial\theta}{\partial t} - \frac{\partial^2\theta}{\partial z^2}\right] = -e_m(z)[g(t) + \theta - \Gamma z] \quad \text{for } z > 0, t > 0, \tag{5.1}$$

with the initial boundary conditions (3.14)–(3.16) transformed to

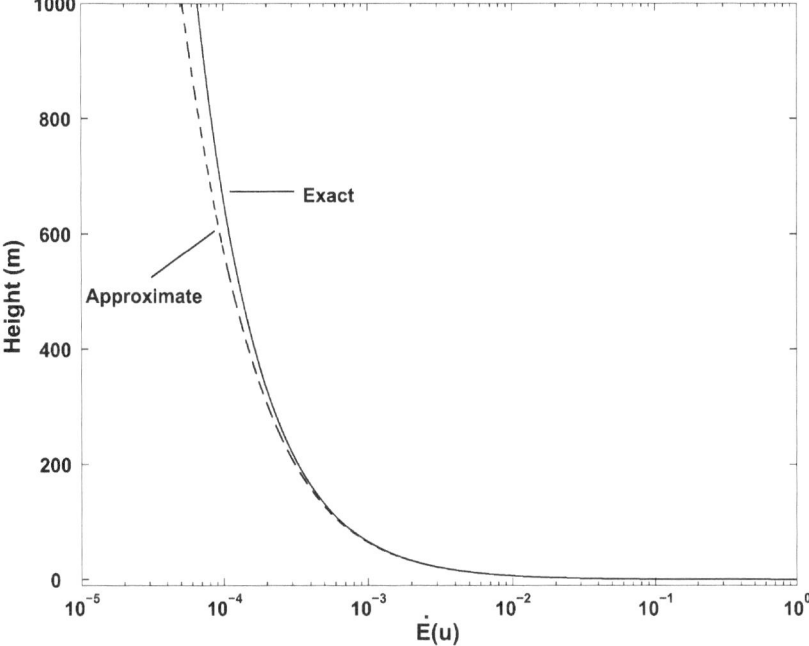

Figure 5
Function $e_m(z)$, solid line, compared with values of $\exp(-z/H_q)\overline{E}(u(z))$, dashed line.

$$\theta(z,0) = 0, \tag{5.2}$$

$$\theta(0,t) = -\beta\sqrt{t}, \tag{5.3}$$

$$\frac{\partial\theta}{\partial z}(\infty,t) = 0. \tag{5.4}$$

Here

$$g(t) = \lambda + \varepsilon_g\beta\sqrt{t}.$$

The above system has seven parameters, namely

$$\mu, \delta, \alpha, \Gamma, \beta, \lambda, \varepsilon_g. \tag{5.5}$$

From estimates previously given we find that $\mu \approx 10^{-3}$, $\delta \approx 10^{-3}$, $\Gamma \approx 10^{-5}$. We therefore find it useful to consider the limit $\mu, \delta, \Gamma \ll 1$, $\mu \sim \delta \gg \Gamma$. The other parameters have the approximate values $\alpha \approx 0.1$, $\beta \approx 0.32$, $\lambda \approx 0.2$, $\varepsilon_g \approx 0.8$. A small value for μ implies that (5.1) is a singular pertubation problem; a small δ means that $e_m(z)$ decays (algebraically) over a length scale α^{-1}. A small Γ means that the temperature gradient at large heights is small compared to that near ground. The parameter ε_g plays a subtle and important role in the formation of the lifted minimum (note that, from (3.17), $\lambda = 0$ when $\varepsilon_g = 1$).

6. A Simple Limit

We now describe heuristically how the Ramdas layer is formed. For small values of time the temperature evolution will be governed by radiation because the diffusive time scale is considerably longer than the radiative time-scale (as indicated by the small value of μ, see (3.20)). Dropping the diffusion term in (5.1) we derive the purely radiative problem (denoted by subscript ra)

$$\mu\frac{\partial\theta_{ra}}{\partial t} = e_m(z)[-\lambda + \Gamma z - \theta_{ra}], \quad \text{for } z > 0, t > 0, \tag{6.1}$$

$$\theta_{ra}(z,0) = 0.$$

The solution to (6.1) is

$$\theta_{ra}(z,t) = (-\lambda + \Gamma z)\left[1 - \exp\left\{-\frac{t}{\mu}e_m(z)\right\}\right]. \tag{6.2}$$

Note that this solution has the limits

$$\theta_{ra}(0+,t) = -\lambda[1 - e^{-t/\mu}], \quad \theta_{ra}(\infty,t) = 0; \tag{6.3}$$

therefore the actual temperature at $z = 0+$ (just above ground) will be (cf. 3.8)

$$T_{g_0}[(1 - \lambda) + \lambda e^{-t/\mu}], \tag{6.4}$$

while for large z it will be essentially $T_{g_0} - \Gamma z$.

The crucial role of parameter ϵ_g is now clearly seen. For $\epsilon_g = 1$, i.e., $\lambda = 0$, we have $\theta_{ra}(0+, t) = 0$. For $\lambda \neq 0$, on the other hand, the value given in (6.3) is different from $T_g(t)$ which is the value of T at $z = 0$ (cf. 2.13). This means that the radiative solution (5.2) has a discontinuity at $z = 0$; there is thus what may be called a radiative slip, given by

$$T(0+, t) - T_g(t) = -\lambda T_{g_0}(1 - e^{-t/\mu}) + \beta\sqrt{t}.$$

The smearing of this discontinuity by the diffusion term causes the formation of the Ramdas layer. The precise way that this happens is considered next.

7. The Asymptotic Solution

We now construct an approximate solution to the singularly perturbed system (5.1)–(5.4) in the limit $\mu \to 0$ following the procedure of VASILÉVA et al. (1995, p. 95). First introduce the stretched variables

$$\varsigma = \mu^{-1/2}z, \tau = \mu^{-1}t, \tag{7.1}$$

and look for a solution in the form

$$\theta = \theta_r(z, t) + \theta_i(z, \tau) + \theta_b(\varsigma, t) - \theta_c(\varsigma, \tau). \tag{7.2}$$

Here θ_r is the regular solution valid for z and t distant from the origin, i.e., in the bulk of the (z, t) domain; θ_i is the initial layer solution or the "transient" valid in the domain $t \in (0, \mu)$ (already briefly discussed in the previous section); θ_b is the boundary layer solution (slow diffusive smoothing mentioned in the previous section) valid near the ground, $z \in (0, \sqrt{\mu})$; and θ_c is the initial/boundary layer correction in the corner near the origin $(0, 0)$, required to make the ansatz (7.2) satisfy the initial/ boundary values (5.2)–(5.4). The domain of validity of the solutions in each of the four distinguished limits is schematically shown in Figure 6.

Each term in (7.2) is now expanded in powers of $\mu^{1/2}$:

$$\begin{aligned}
\theta_r(z, t) &= \theta_0(r, t) + \mu^{\frac{1}{2}}\theta_1(z, t) + \cdots \\
\theta_i(z, \tau) &= \hat{\theta}_0(r, \tau) + \mu^{\frac{1}{2}}\hat{\theta}_1(z, t) + \cdots \\
\theta_b(\varsigma, t) &= \tilde{\theta}_0(\varsigma, t) + \mu^{\frac{1}{2}}\tilde{\theta}_1(\varsigma, t) + \cdots \\
\theta_c(\varsigma, \tau) &= \breve{\theta}_0(\varsigma, \tau) + \mu^{\frac{1}{2}}\breve{\theta}_1(\varsigma, \tau) + \cdots
\end{aligned} \tag{7.3}$$

We substitute (7.2) into (5.1) and group equations according to their independent variables. For instance θ_r is a function of z and t, and therefore all functions with z

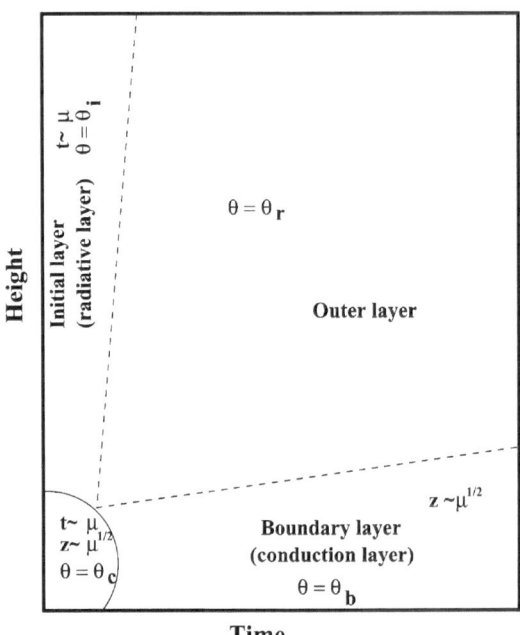

Figure 6
Distinguished subdomains in the height-time (zt) plane.

and t as independent variables are grouped together; while θ_i is a function of z and τ and thus will be grouped with functions of z and τ and so on. We thus obtain

$$\mu\left[\frac{\partial \theta_r}{\partial t} - \frac{\partial^2 \theta_r}{\partial z^2}\right] = -e_m(z)[g(t) + \theta_r - \Gamma z], \tag{7.4}$$

$$\frac{\partial \theta_i}{\partial \tau} - \mu\frac{\partial^2 \theta_i}{\partial z^2} = -e_m(z)\theta_i, \tag{7.5}$$

$$\mu\frac{\partial \theta_b}{\partial t} - \frac{\partial^2 \theta_b}{\partial \varsigma^2} = -e_m(\varsigma\sqrt{\mu})\theta_b = -\theta_b\left[e_m(0) + O(\sqrt{\mu})\right], \tag{7.6}$$

$$-\left[\frac{\partial \theta_c}{\partial \tau} - \frac{\partial^2 \theta_c}{\partial \varsigma^2}\right] = e_m(\varsigma\sqrt{\mu})\theta_c = e_m(0)\theta_c + O(\sqrt{\mu}). \tag{7.7}$$

Note that when we sum (7.4) to (7.7) we obtain (5.1). In order for θ to satisfy (5.2)–(5.4), the various terms in (7.2) have to satisfy the initial/boundary conditions

$$\text{(a)} \quad \theta_r(z,0) + \theta_i(z,0) = 0, \qquad \text{(b)} \quad \theta_b(\varsigma,0) - \theta_c(\varsigma,0) = 0, \tag{7.8}$$

$$\text{(a)} \quad \theta_r(0,t) + \theta_b(0,t) = -\beta\sqrt{t}, \quad \text{(b)} \quad \theta_i(0,\tau) - \theta_c(0,\tau) = 0, \tag{7.9}$$

$$\frac{\partial \theta_b}{\partial \varsigma}(\infty,t) = \frac{\partial \theta_c}{\partial \varsigma}(\infty,t) = 0. \tag{7.10}$$

Now substituting the series (7.3) in (7.4)–(7.10) and considering only the leading order terms in μ, we obtain simplified equations for $\theta_0, \hat{\theta}_0, \tilde{\theta}_0$ and $\check{\theta}_0$. For θ_0 we need to solve

$$e_m(z)[g(t) + \theta_0 - \Gamma z] = 0. \tag{7.11}$$

If $e_m \neq 0, \theta_0$ is uniquely given by

$$\theta_0 = -g(t) + \Gamma z. \tag{7.12}$$

But $e_m(z)$ tends to zero as $z \to \infty$, so at very large heights θ_0 may have to be determined differently. The dynamics at very large heights is however beyond the scope of the present analysis.

For $\hat{\theta}_0$ we obtain an ordinary differential equation from (7.5) with the spatial variable z being just a parameter,

$$\frac{\partial \hat{\theta}_0}{\partial \tau} = -e_m(z)\hat{\theta}_0, \tag{7.13}$$

whose solution is

$$\hat{\theta}_0(z, \tau) = \hat{\theta}_0(z, 0)e^{-\tau e_m(z)}.$$

Using (7.8a) and (7.12), we find

$$\hat{\theta}_0(z, 0) = -\theta_0(z, 0) = (\lambda - \Gamma z).$$

Thus the first order approximation $\hat{\theta}_0$ is

$$\hat{\theta}_0(z, \tau) = (\lambda - \Gamma z)e^{-\tau e_m(z)}. \tag{7.14}$$

Next consider the term $\tilde{\theta}_0$ in (7.6). This requires us to solve

$$\frac{\partial^2 \tilde{\theta}_0}{\partial \varsigma^2} = \tilde{\theta}_0, \tag{7.15}$$

for which boundary conditions, obtained from (7.9a), (7.10) and (7.12), are

$$\tilde{\theta}_0(0, t) = -\beta\sqrt{t} - \theta_0(0, t) = -\beta\sqrt{t} + g(t), \tag{7.16}$$

$$\frac{\partial \tilde{\theta}_0}{\partial \varsigma}(\infty, t) = 0. \tag{7.17}$$

The solution to (7.15)–(7.17) is

$$\tilde{\theta}_0(\varsigma, t) = (g(t) - \beta\sqrt{t})e^{-\varsigma} = (\lambda + (\varepsilon_g - 1)\beta\sqrt{t})e^{-\varsigma}. \tag{7.18}$$

From (7.7) the equation for the correction term $\check{\theta}_0$ is

$$\frac{\partial \breve{\theta}_0}{\partial \tau} - \frac{\partial^2 \breve{\theta}_0}{\partial \varsigma^2} = -\breve{\theta}_0. \tag{7.19}$$

Initial and boundary conditions for (7.19) are obtained from (7.8b), (7.9b) and (7.10):

$$\begin{aligned}
\breve{\theta}(\varsigma, 0) &= \breve{\theta}_0(\varsigma, 0) = \lambda e^{-\varsigma}, \\
\breve{\theta}_0(0, \tau) &= \hat{\theta}_0(0, \tau) = \lambda e^{-\tau}, \\
\frac{\partial \breve{\theta}_0}{\partial \varsigma}(\infty, \tau) &= 0.
\end{aligned} \tag{7.20}$$

The solution (7.19)–(7.20) is

$$\breve{\theta}_0(\varsigma, \tau) = \lambda e^{-\tau} \operatorname{erfc}\left(\frac{\varsigma}{2\sqrt{\tau}}\right) + \frac{\lambda}{2}\left[e^{-\varsigma}\operatorname{erfc}\left\{\frac{2\tau - \varsigma}{2\sqrt{\tau}}\right\} - e^{\varsigma}\operatorname{erfc}\left\{\frac{2\tau - \varsigma}{2\sqrt{\tau}}\right\}\right] \tag{7.21}$$

where

$$\operatorname{erfc} x = 1 - \operatorname{erf} x, \quad \operatorname{erf} x \equiv \frac{2}{\sqrt{\pi}} \int_0^x e^{-s^2} ds.$$

Note that $\operatorname{erf}(-x) = -\operatorname{erf}(x)$, and hence $\operatorname{erfc}(-\infty) = 2$.

Thus our first order asymptotic solution to (5.1)–(5.4) is, from (7.12), (7.14), (7.18) and (7.21),

$$\begin{aligned}
\Theta_0 &= \theta_0 + \hat{\theta}_0 + \tilde{\theta}_0 - \breve{\theta}_0 = -\lambda - \varepsilon_g \beta \sqrt{t} + \Gamma z + [\lambda - \Gamma z]e^{-r e_m(z)} \\
&\quad + [\lambda + (\varepsilon_g - 1)\beta\sqrt{t}]e^{-\varsigma} - \lambda e^{-\tau}\operatorname{erfc}\left(\frac{\varsigma}{2\sqrt{\tau}}\right) \\
&\quad - \frac{\lambda}{2}\left[e^{-\varsigma}\operatorname{erfc}\left\{\frac{2\tau - \varsigma}{2\sqrt{\tau}}\right\} - e^{\varsigma}\operatorname{efrc}\left\{\frac{2\tau + \varsigma}{2\sqrt{\tau}}\right\}\right].
\end{aligned} \tag{7.22}$$

The most important property of Θ_0 is that it is asymptotic to θ_0, in the sense that

$$\theta - \Theta_0 = O(\mu^{1/2})$$

everywhere. That the method gives a uniformly valid asymptotic solution has been shown by VASILÉVA *et al.* (1995) for a bounded domain. Their proof can be easily extended to the present case of unbounded domain because $e_m(z)$ decays exponentially to zero for large z. Thus Θ_0 provides a first-order approximate solution to (5.1)–(5.4) in the limit of small μ. We can therefore study the qualitative properties of θ through Θ_0; this is done in the next section.

8. Results and Discussion

In order to understand the structure of the approximate solution (7.22) we first rewrite the solution Θ_0 in the original dimensional variables through the relation (cf. 3.8)

$$T = T_{g_0}[1 + \overline{T}] - \Gamma z \approx T_{g_0}[1 + \Theta_0] - \Gamma z. \tag{8.1}$$

Firstly it is easy to see that T satisfies all the initial/boundary conditions (2.12)–(2.14).

The first term θ_0 in (7.22) is the outer solution valid for $z \gg l_e\sqrt{\mu}$ and $t \gg t_d\mu$, taking the value $-g(t) + \Gamma z$ in this region. The initial or radiative layer solution $\hat{\theta}_0$ is valid for short times ($t \approx t_d\mu$) but for nearly all heights, while the boundary or conduction layer solution $\tilde{\theta}_0$ is valid for low heights ($z \approx l_e\sqrt{\mu}$) but for nearly all times. The corner solution $\check{\theta}_0$ is valid for very short times and heights in the sense mentioned above. This is essentially a correction added to the sum of outer, initial layer and boundary layer solutions so that they satisfy the initial/boundary conditions and can be neglected away from the small corner shown in Figure 6. Outside this corner we can therefore consider an approximation that ignores this term, and write

$$\begin{aligned}
T_\theta &\equiv T_{g_0}[1 + \Theta_0 + \tilde{\theta}_0] - \Gamma z \\
&= \left[(1 - \lambda)T_{g_0} - \varepsilon_g\beta\sqrt{t} + (\lambda T_{g_0} - \Gamma z)\exp(-4q_0 e_m(z)t/t_r)\right] \\
&\quad + (1 - \varepsilon_g)\left[\lambda_0 T_{g_0} - \beta\sqrt{t}\right]\exp(-2z(q_0/t_r K_m)^{1/2}).
\end{aligned} \tag{8.2}$$

from (7.12), (7.14) and (7.18). Here $\lambda_0 = (1 - \varepsilon_s)/4$.

We consider two cases: (i) $\varepsilon_g = 0.8$, $\beta = 2$ K/$\sqrt{h}$, $q_0 = 0.01$ and (ii) $\varepsilon_g = 1$, β and q_0 being the same.

The solution (8.2), for $F_a^\downarrow(0) = 200$ W/m² and 300 W/m² respectively, is shown in Figures 7–8 along with the full numerical solution described in VSN. Note that for the latter $F_a^\downarrow(0)$ is obtained as a part of the solution and not (as in the present approach) prescribed. It is clear that the full numerical solution of VSN in case (i) is well approximated by the present analytical solution taking $F_a^\downarrow(0) = 300$ W/m², whereas in case (ii) there is only a slight change in the present analytical solution at $\varepsilon_g = 1$ between the two downwelling flux values considered.

We now analyze the expression (8.2) to see under what conditions a lifted minimum will occur or be destroyed.

From (4.7) it is clear that $e_m(z)$ is a monotonically decreasing function of z. Therefore the first exponential term in (8.2) (which is the radiative contribution (7.14)) is a monotonically increasing function of z, and hence cannot contribute to the negative gradient at $z = 0$ which is a necessary condition for the profile to exhibit a minimum. The minimum actually arises from the second exponential term, which corresponds to the diffusion solution (7.18). Therefore if a minimum in T_θ is to occur at some value of z, then a necessary and sufficient condition is that the factor multiplying the diffusion solution is positive so that the gradient at ground is negative. This requires that

$$(1 - \varepsilon_g)(\lambda_0 T_{g_0} - \beta\sqrt{t}) > 0.$$

which can be achieved only if

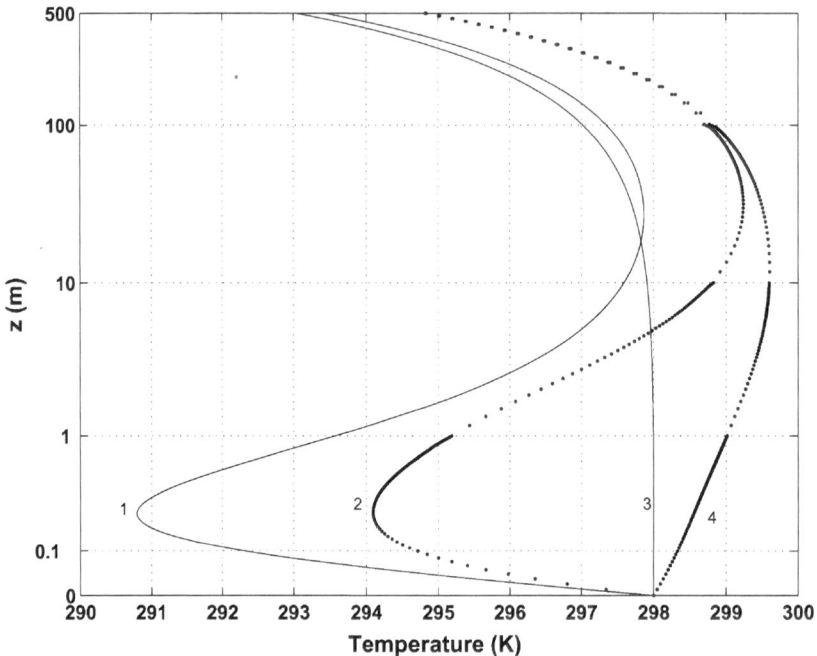

Figure 7
Solutions for temperature profile. (1) Numerical solution of VSN model, $F_a^\downarrow = 200$ W/m^2, $\varepsilon_g = 0.8$; (2) present analytical solution (8.1) with $F_a^\downarrow(0) = 200$ W/m^2, $\varepsilon_g = 0.8$; (3) same as 1 but with $\varepsilon_g = 1$. (4) same as 2 but with $\varepsilon_g = 1$.

$$\varepsilon_g < 1 \qquad (8.3)$$

and

$$\beta\sqrt{t} < \lambda_0 T_{g0} \qquad (8.4)$$

Of course it is also achieved if the signs are reversed in (8.3)–(8.4) but $\varepsilon_g > 1$ is not possible.

The condition (8.3) was pointed out in VSN, however the result (8.4) is new. It can be recast in the form of an inequality for a nondimensional number, namely

$$\overline{\overline{\beta}} \equiv \frac{\beta\sqrt{t}}{\lambda_0 T_{g0}} = \frac{\overline{\beta}}{\lambda_0}(t/t_d)^{1/2} < 1, \qquad (8.5)$$

where we have used (3.11). Note that the above condition is always fulfilled at $t = 0$, so that a lifted temperature minimum appears immediately after $t = 0$ (provided of course $\varepsilon_g < 1$). However it can be destroyed quickly if $\beta/\lambda_0 T_{g0}$ is large; Figures 9 and 10 show examples of this behaviour.

Although the present model and solution are based on certain approximations (including in particular one about β not being large), we may ask if, under more general

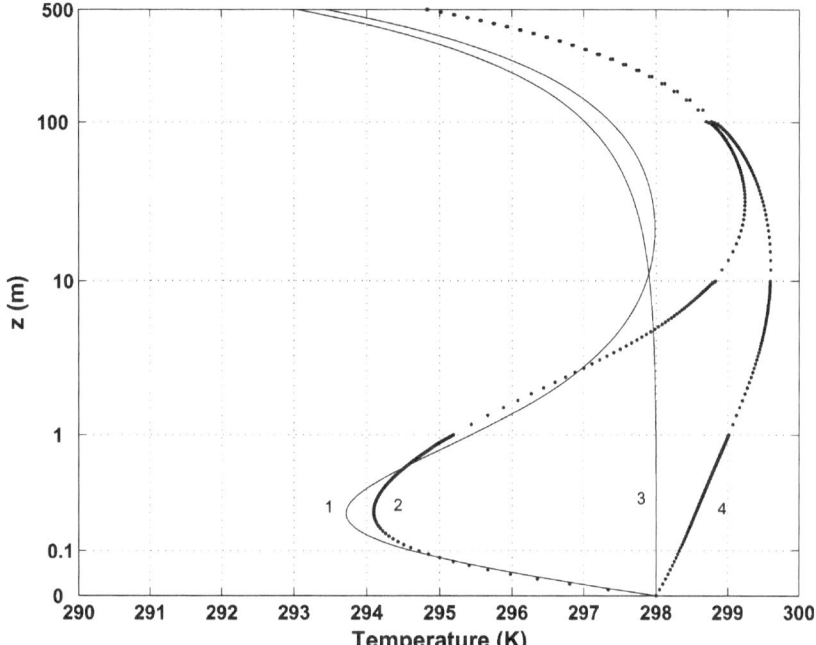

Figure 8
Same as Figure 7 but with $F_d^\downarrow(0) = 300$ W/m^2.

conditions, the value of the nondimensional parameter $\bar{\bar{\beta}}$ provides the criterion for the occurrence of the lifted minimum. A precise answer to the question is difficult, as in general we may expect a critical value of $\bar{\bar{\beta}}$ to depend on other nondimensional parameters, e.g., $\bar{\beta}$, ε_g. We may nevertheless examine the more exact computer simulations (e.g., VSN, R01) to throw light on the issue, although they were not carried out with the specific objective of answering this question. Thus consider the data in Figure 7 of R01. Here the intensity of the minimum vanishes around $t \simeq 10$ h for $\beta = 12\text{K}/\sqrt{\text{h}}$. Assuming $F^\downarrow(0) \simeq 300$ W/m^2, we find $\lambda_0 = 1$ and hence the critical value of $\bar{\bar{\beta}}$ to be $\bar{\bar{\beta}}_{cr} \simeq 1.3$. However for the values given in NV for the nonexistence of a minimum we obtain $\bar{\bar{\beta}}_{cr} \simeq 0.4$ for $\beta = 15$ K$/\sqrt{\text{h}}$, $t = 1$ h, $\varepsilon_g = 0.8$. Therefore all we can conclude at present is that the critical value of $\bar{\bar{\beta}}$ is of order unity for $\varepsilon_g = 0.8$.

From (8.2) it is clear that if $K_m \to \infty$ (with t_r bounded away from zero), or $t_r \to \infty$ (with K_m bounded away from zero) then the diffusion term will collapse to a constant destroying the Ramdas layer.

Also, from (8.2), for heights $l_e \ll z < O(H_q)$ the first exponential term is near unity and the second exponential term is near zero leading to the expression

$$T_\theta(z, t) \approx T_{g_0} - \varepsilon_g \beta \sqrt{t} - \Gamma z.$$

This shows that, under calm clear conditions, the temperature is still affected by ground cooling parameters like ε_g, β at heights well above the emissivity sublayer.

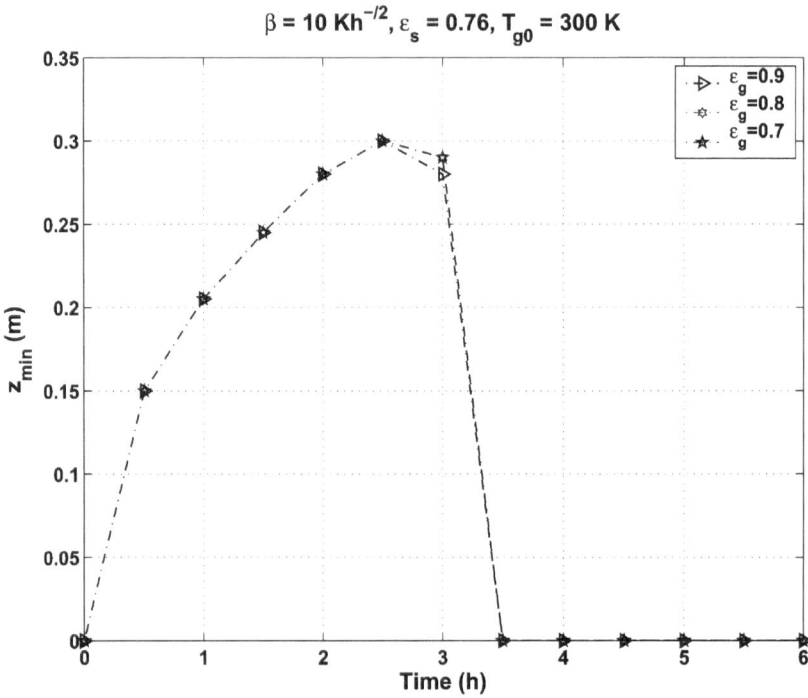

Figure 9

Evolution of $z_{\min}$, showing growth followed by collapse, as predicted by the present analytical solution (8.1) for various surface emissivities.

In particular this conclusion holds upto inversion height, as first pointed out by R02 based on the full numerical solution of (2.1). We now pursue this further based on our analytical solution (8.1). Let the inversion height be denoted by z_{inv} and the intensity of the inversion defined as

$$\Delta T_{z_{\text{inv}}}(t) = T_{z_{\text{inv}}}(t) - T_g(t), \tag{8.6}$$

where $T_{z_{\text{inv}}}$ is the temperature at z_{inv} (Fig. 3). Two typical results are given in Figures 11–12, which show the evolution of the above-mentioned inversion parameters for various values of ground emissivity. Except for $\varepsilon_g = 1$ the inversion height grows steadily; from the analytical expression it is clear that it grows like $\sqrt{t}$, following the classical proposal of TAYLOR (1915). However, $\Delta T_{z_{\text{inv}}}$ decreases with increase in ε_g. These results are qualitatively very similar to those of R02, except that, in the latter, the inversion height showed a steady growth even for $\varepsilon_g = 1$. Also $\Delta T_{z_{\text{inv}}}$ was always positive in R02 which means that the ground temperature is always less than the temperature at inversion height, unlike in the present model. Incidentally the Taylor $\sqrt{t}$ growth does not always occur; R02 shows that in the presence of even slight wind, growth of inversion height may be completely suppressed.

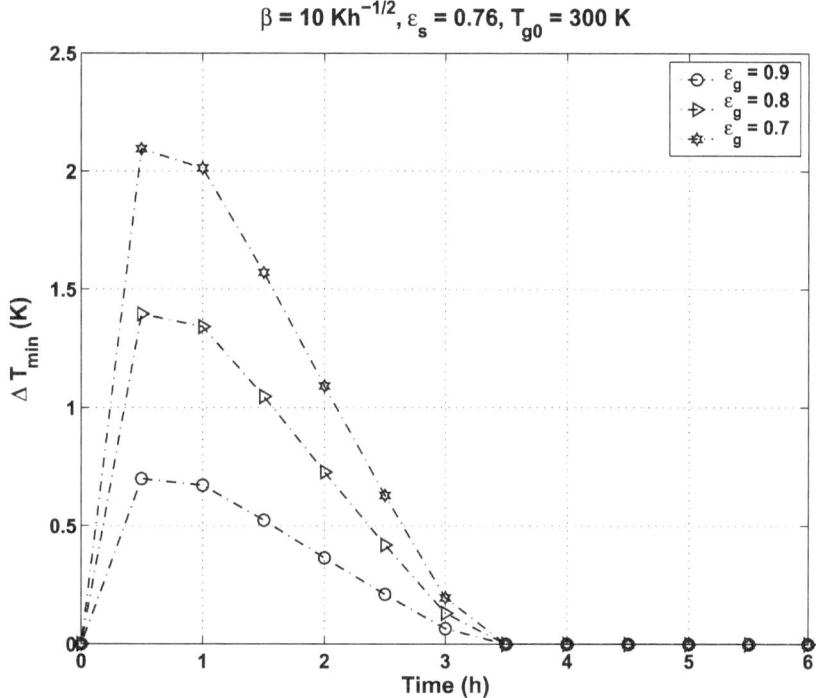

Figure 10
Same as Figure 9 but for $\Delta T_{\min}$.

9. Conclusion

Let us first recapitulate the arguments that have led to the formulation of the equation studied in this paper.

We start with an energy equation that expresses a balance between thermal conduction and radiation (turbulence being ignored as we consider only calm conditions). We then note that the radiative time constant is much less than a typical diffusion time, the ratio μ being of order 10^{-3}. (The response of the layer of air we are considering to any disturbance, or even to the initial conditions at (nominal) sunset, therefore consists of a quick radiative adjustment followed by a slow diffusive relaxation.) In terms of formulating an appropriate asymptotic theory, we therefore have a ready small parameter in μ.

This limit is however still not sufficient to unravel the structure of the solution of the original nonlinear integro-differential equation that governs energy balance. Fortunately two further major simplifications are possible.

The first derives from the analysis of NV, showing that near the surface the radiative fluxes lend themselves to a local approximation if the flux-emissivity model is adopted. It is important to note that this local approximation is *not* the obvious

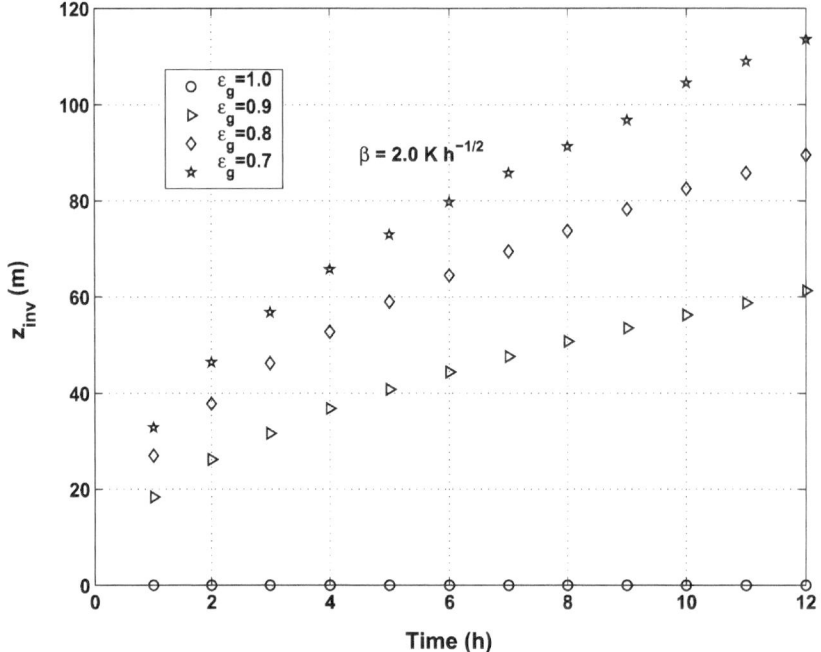

Figure 11

Evolution of inversion height z_{inv} for various surface emissivities.

diffusive one, which is in general not valid for radiation near ground as photon mean free paths vary widely (wildly!) with the frequency of radiation. (This variation mirrors that in the absorption spectrum of the H_2O molecule.)

The second simplification involves linearizing the equations. Radiation involves the fourth power of the temperature; if temperature differences are small we can replace the fourth power by a linear approximation. This seems reasonable near the ground, as long as temperature differences are of the order of a few degrees, but limits the analysis to relatively low ground cooling.

Two other approximations, both excellent and harmless, complete the formulation. The first exploits the fact that what VSN call the emissivity sublayer is only about a meter thick, and so is considerably smaller than a mixed air-water vapor scale height of the atmosphere (about 4 km). The second recognizes that the lapse rate in the free atmosphere (of order 6 K/km) is far smaller than the temperature gradients we observe near the surface (of order 10^5 K/km).

Armed with these approximations we can derive a new diffusion equation which is linear but has radiative sources which are both time- and space-dependent, and boundary conditions which are time-dependent. Nevertheless in the limit $\mu \to 0$ this initial/boundary value problem needs to be solved in four distinct space-time (zt) domains, respectively a transient at sunset, a boundary layer near ground, a radiative

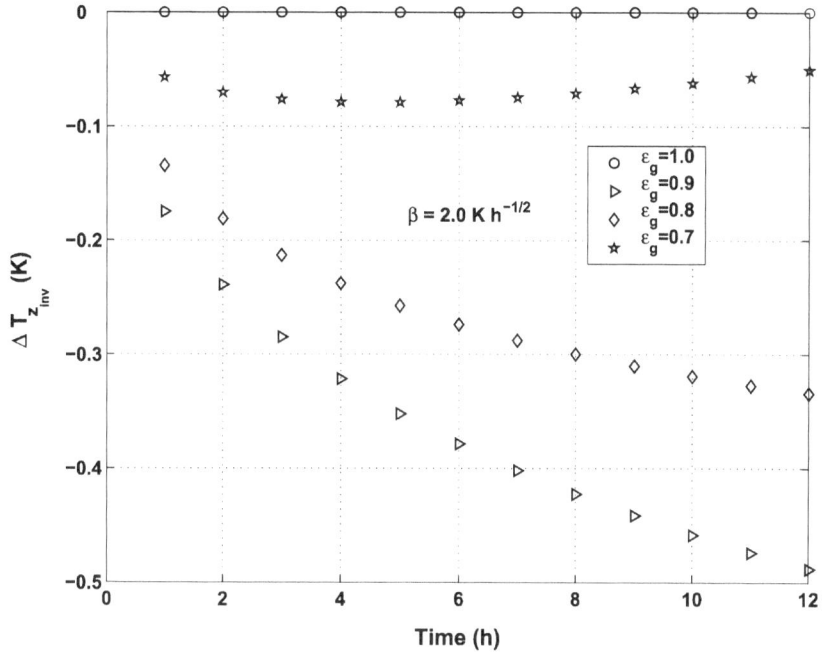

Figure 12
Same as Figure 11 but for ΔT_{inv}.

outer solution, and a correction for the transient boundary layer at nominal sunset. With the solution of the limiting equations in the different domains a uniformly valid solution can also be constructed, which reveals the structure and dynamics of the Ramdas layer.

The physical picture that emerges is briefly as follows. Firstly the radiative cooling near ground can be very large, especially when ground is not radiatively black, because of the extremely rapid variation of the flux emissivity of the air at short optical paths : it is the *gradient* of the emissivity which determines the cooling. The high gradients are due to the presence of the strongly absorbing vibration-rotation bands of the H_2O molecule, as pointed out by VSN. This radiative cooling provides the analogue of transverse side losses in the problem of heat conduction along a rod, and it is clear that such losses, when heavy, can lead to temperatures *within* the rod lower than either at the top or the bottom. This mechanism leads to the lifted minimum in the air.

The thickness of the cold layer is in general time-dependent because ground temperature varies with time. It is largely determined by (molecular) thermal diffusion. For diffusivities K_m of order 10^{-5} m²/s, the layer thickness after some three hours should be of order $\sqrt{K_m t} \approx 0.3$ m, which is about right if the ground is cooling slowly. However, if the ground cools very rapidly it can dominate the effects of radiative cooling above, and consequently the lifted minimum can entirely disappear.

Another way of viewing this is as follows. The inner limit at the surface of the outer radiative solution is a temperature that will in general differ from the actual ground temperature. If the latter is higher, a Ramdas layer develops; if it is lower the layer collapses.

Acknowledgement

ASVM and SV thank the Council of Scientific and Industrial Research, New Delhi for their financial support through the NMITLI project. RN thanks the Defense Research and Development Organization for support of his work, and the Centre for Atmospheric and Oceanic Sciences, Indian Institute of Science, for their kind hospitality.

REFERENCES

ANANTHASAYANAM, M.R. and NARASIMHA, R. (1989), *Standard for the Indian Tropical Atmosphere (SITA)*, Final Consolidated Technical Report to AR and DB Part II, Dept. of Aeronautics, Indian Inst. of Science, Bangalore, India.

BRÜNT, D., *Physical and Dynamical Meteorology*, 2nd ed. (Cambridge University Press, New York, 1941).

GARRATT, J.R. and BROST, R.A. (1981), *Radiative Cooling Effects within and above the Unstructural Boundary Layer*, J. Atmos. Sci. *38*, 2730–2746.

GEIGER, R., *The Climate Near the Ground*, (Harvard University Press, 1965).

GEIGER, R., ARON, R.H., and TODHUNTER, P., *The Climate Near the Ground*, 5th ed. (Vieweg, Braunschweig, 1995).

KUO, H.L. (1968), *The Thermal Interaction between the Atmosphere and the Earth and Propagation of Diurnal Temperature Waves*, J. Atmos. Sci. *25*, 682–717.

LAKE, J.V. (1956), *The Temperature Profile above Bare Soil on Clear Nights*, Q. Jl. Roy. Met. Soc. *82*, 187–197.

LETTAU, H.H. (1979), *Wind and Temperature Profile Prediction for Diabatic Surface Layers Including Strong Inversion Cases*. Boundary Layer Meteorology *17*, 443–464.

LARSEN, E., POMRANNING, G., and BADHAM, V. (1983), *Asymptotic Analysis of Radiative Transfer Problems*, J. Quant. Spect. and Rad. Transfer *29*, 285–310.

NARASIMHA, R. (1994), *The Dynamics of the Ramdas Layer*, Curr. Sci. *66*, 16–23.

NARASIMHA, R. and VASUDEVA MURTHY, A.S. (1995), *The Energy Balance in the Ramdas Layer*, Boundary Layer Meteorology *76*, 307–321.

NARASIMHA, R., *Down-to-earth temperatures: The mechanics of the environment*. In *Theoretical and Applied Mechanics, ICTAM96)* (eds. Tatsumi, T., Watanabe, E., and Kambe, T.) (Elsevier, Amsterdam, 1997) pp. 567–582.

ÖZISIK, M.N., *Radiative Transfer and Interactions with Conduction and Convection* (Wiley, New York, 1973).

RASCHKE, K. (1957), *Über das nächtliche Temperature Minimum über nackten Boden im Poona*, Met. Rundschau *10*, 1–11.

RAGOTHAMAN, S., NARASIMHA, R., and VASUDEVA MURTHY, A.S. (2001), *The Dynamical Behaviour of the Lifted Temperature Minimum*, Il Nuovo Cimento. *24C*, 353–375.

RAGOTHAMAN, S., NARASIMHA, R., and VASUDEVA MURTHY, A.S. (2002), *Evolution of Nocturnal Temperature Inversions: A Numerical Study*. Il Nuovo Cimento *25C*, 147–163.

RAMDAS, L.A. and ATMANATHAN, S. (1932), *The Vertical Distribution of Air Temperature Near the Ground at Night*, Beit. Geophys. *37*, 116–117.

TAYLOR, G.I. (1915), *Eddy Motion in the Atmosphere*, Phil. Trans. Roy. Soc. *A215*, 1–16.

VARGHESE, S., VASUDEVA MURTHY, A.S., and NARASIMHA, R. (2003), *A Fast, Accurate Method of Computing Near-surface Longwave Fluxes and Cooling Rates in the Atmosphere*, J. Atmos. Sci. *60*, 2869–2886.

VASILÉVA, A.B., BUTUZOV, V.F., and KALACHÉV, L.V. (1995), *The Boundary Function Method for Singular Perturbation Problems*, SIAM Studies in Applied Mathematics, 14, SIAM, Philadelphia, *14* pp. 221.

VASUDEVA MURTHY, A.S. (1986), *Heat Transfer Processes in the Lowest Layers of the Atmosphere II*, Report 86 AS2, Center for Atmospheric Sciences, Indian Institute of Science, Bangalore.

VASUDEVA MURTHY, A.S., SRINIVASAN, J., and NARASIMHA, R. (1993), *A Theory of the Lifted Temperature Minimum on Calm Clear Nights*, Phil. Trans. Roy. Soc. (London), *A344*, 183–206.

YAMADA, M. and TAKAHASHI, H. (2004), *Frost Damage to Hemerocallis Esculenta in Mire Relationship between Flower Bud Height and Air Temperature Profile during Calm, Clear Nights*, Can. J. Bot. *82*, 409–419.

ZDUNKOWSKI, W. and JOHNSON, F.G. (1965), *Infrared Flux Divergence Calculations with Newly Constructed Radiation Tables*, J. App. Met. *4*, 371–377.

(Received June 30, accepted August 11, 2004)
Published Online First June 17, 2005

To access this journal online:
http://www.birkhauser.ch

D. Air Quality

Pure appl. geophys. 162 (2005) 1861–1892
0033–4553/05/101861–32
DOI 10.1007/s00024-005-2696-5

Pure and Applied Geophysics

An Analytical Study for the Dispersion of Pollutants in a Finite Layer under Low Wind Conditions

MAITHILI SHARAN[1], and MANISH MODANI[1]

Abstract—A steady-state three-dimensional analytical model for the dispersion of a pollutant from a continuously emitting point source in a finite inversion layer has been proposed. The advection along the mean wind and the diffusion in all three directions have been accounted. A closed-form analytical solution of the proposed problem has been obtained using the method of integral transform and the same happens out to be non-Gaussian. Various limiting cases including the crosswind integrated concentration and the standard Gaussian-plume formula have been deduced from the solution obtained.

The present model has been evaluated using Hanford diffusion experiment (DORAN *et al.*, 1984) in stable conditions and IIT diffusion experiment (SINGH *et al.*, 1991) in unstable conditions. The dispersion parameters have been computed using the various sigma schemes in accordance with the available observations. The proposed model has performed reasonably well with the data from both the diffusion experiments considered here.

Key words: Mathematical model, Low wind dispersion, Inversion/mixing layer, Dispersion parameters.

1. Introduction

The importance of dispersion modeling in low wind conditions ($U \leq 2$ ms^{-1}) lies in the fact that these conditions occur frequently (VAN DER HOVEN, 1976) and are crucial for air pollution episodes (SHARAN *et al.*, 1995a, 2003). The situation becomes considerably more risky (potentially hazardous) particularly during winter periods associated with stable conditions, for example, the infamous Bhopal gas leak (SHARAN *et al.*, 1995a). In case of low wind conditions, (i) pollutants are not able to travel far from the source, (ii) concentration distribution is not cone-shaped, non-Gaussian and has the multiple peaks (SAGENDORF and DICKSON, 1974; SHARAN *et al.*, 2003). The wind direction fluctuations are also observed to be large under these

---

[1]Centre for Atmospheric Science, Indian Institute of Technology, Hauz Khas, New Delhi, 110016, India.

conditions (HANNA et al., 2003). These factors have generated considerable interest in recent years to undertake modeling and observational studies to gain an insight in the low wind dispersion. The various aspects of dispersion in low wind conditions have recently been reviewed extensively (SHARAN and GOPALAKRISHNAN, 2003; SHARAN et al., 2003).

The classical models such as Gaussian puff/plume with suitable assumptions are being used extensively (ZANNETTI, 1990; ARYA, 1999) to explain the diffusion data (SEINFELD, 1986; HANNA et al., 2003). However, their applicability may be questionable in low winds because of (i) the neglect of downwind diffusion in comparison to advection and (ii) the non-availability of appropriate dispersion parameters. ARYA (1995) has shown that the assumption of neglecting the downwind diffusion works well in the presence of moderate wind ($U > 2$ ms^{-1} at 10 m) and moderate turbulent intensity ($\sigma_u/U \leq 0.3$, in which σ_u is the standard deviation of u-component of wind velocity) whereas in weak wind and high turbulence intensity ($\sigma_u/U > 0.3$), the longitudinal diffusion becomes important. SHARAN et al. (1996a) have shown that the errors may reach 25% in predicting the near-source concentrations without longitudinal diffusion during the weak wind situations. Recently, PHILIP (1997) has also shown the importance of diffusion in the downwind direction to estimate the upwind diffusive flux.

SIRAKOV and DJOLOV (1979) proposed a mathematical model to account for the dispersion of pollutants from a continuous source in the absence of wind. However, the parameterization for its practical application was not done. ARYA (1995) discussed a mathematical model for the near-source diffusion in weak winds which was only valid for the convective boundary layer over a flat and homogeneous terrain. SHARAN et al., (1996a) presented a three-dimensional mathematical model by explaining longitudinal diffusion and validated it with the data obtained at IIT (Indian Institute of Technology) Delhi in low wind conditions. The variable K model has also been proposed (SHARAN et al., 1996b; SHARAN and YADAV, 1998) to deal with low wind dispersion. However, this model does not explain the upstream diffusion and the parameterization used is limited to non-zero wind speed. Further, CIRILLO and POLI (1992) and SHARAN et al., (1996c) proposed models to deal with the dispersion in non-homogeneous and nonstationary conditions to account for the variability in the wind direction.

In low wind conditions, the appropriate estimation of dispersion parameters becomes a challenging task because (i) the turbulence structure of the atmospheric boundary layer, particularly in low wind conditions is not well defined; (ii) plume spread in the horizontal direction is increased because of meandering and (iii) non-availability of adequate observations (SHARAN et al., 2003). SHARAN et al., (1995b) and YADAV and SHARAN (1996) have analyzed, both qualitatively and quantitatively, various sigma schemes for the estimation of dispersion parameters in low wind conditions. All these schemes are essentially based on seven versions of a K-theory based model. ARYA (1995) has suggested a parameterization for the eddy diffusivities based on Taylor's statistical theory (1921). Recently, OETTL et al., (2001) suggested new parameterizations for turbulent intensities based on observations in low wind stable conditions. More recently, SHARAN et al., (2002) used the turbulence parameterization based on friction velocity to explain the data observed in low-wind unstable conditions at IIT Delhi. However, the general parameterization scheme for low wind dispersion is still lacking.

Most of the pollution emissions take place in the atmospheric boundary layer (ABL) and this plays a vital role in the dispersion of pollutants because the pollutants emitted near the surface are largely diluted and confined within it. It is recognized that ABL is often capped by an inversion which reflects back the material reaching the inversion base (BEYRICH 1997; ARYA, 1999). Surface, based temperature inversion in stable boundary layer (SBL) tends to suppress the dispersion of pollutants in the atmosphere. Recently, Narasimha and his coworkers (VASUDEVA MURTHY et al., 1993; RANGOTHAMAN et al., 2003) have examined the evolution of surface-based inversion in the SBL. Models have also been developed (ROBSON, 1983; LIN and HILDEMANN, 1996) for the dispersion in the finite layer. However, these may not be appropriate to deal with the low wind dispersion. The earlier models (ARYA, 1995; SHARAN et al., 1996a; SHARAN and YADAV, 1998) for the dispersion in low wind allow unrestricted diffusion of plume in the vertical direction. This does not normally happen in the real atmosphere where a finite inversion layer/mixing layer of vanishing turbulence at the top of ABL restricts vertical diffusion.

Thus, in the present study, a model for the dispersion of pollutants released from an elevated source in a finite inversion/mixing layer under low wind conditions has been formulated. The mathematical description of the model is discussed in the next section. The estimation of dispersion parameters is described in section 3. Validation of the model with the observations from field experiments has been described in section 4, followed by conclusions in section 5.

2. Mathematical Description

The steady-state concentration (c) of a pollutant is governed by the advection-diffusion equation (SHARAN et al., 1996a):

$$U\frac{\partial c}{\partial x} = \frac{\partial}{\partial x}\left(K_x\frac{\partial c}{\partial x}\right) + \frac{\partial}{\partial y}\left(K_y\frac{\partial c}{\partial y}\right) + \frac{\partial}{\partial z}\left(K_z\frac{\partial c}{\partial z}\right) + S, \qquad (1)$$

where U is the mean wind speed along x axis, K_x, K_y and K_z are the eddy diffusion coefficients along x, y and z directions, respectively and S is the source term. Equation (1) forms the basis for most of the air-quality models based on K-theory. A point source of strength q is assumed to be located at the point $(0, 0, H)$ and accordingly $S = q\delta(x)\delta(y)\delta(z - H)$ in which $\delta\,(.)$ is the Dirac's delta function and H is the effective stack height. Assuming U and Ks are constant, equation (1) reduces to:

$$U\frac{\partial c}{\partial x} = K_x\frac{\partial^2 c}{\partial x^2} + K_y\frac{\partial^2 c}{\partial y^2} + K_z\frac{\partial^2 c}{\partial z^2} + q\delta(x)\delta(y)\delta(z - H). \tag{2}$$

Equation (2) is subject to the following boundary conditions:
(i) Concentration of the pollutant tends to zero remotely from the source i.e.,

$$c \to 0 \text{ as } |x|, |y| \to \infty. \tag{3}$$

(ii) The pollutant is not absorbed by the ground surface and therefore there is no diffusive flux at the surface, i.e.,

$$-K_z\frac{\partial c}{\partial z} = 0 \text{ at } z = 0. \tag{4}$$

(iii) The pollutant dispersion remains confined to a layer capped by an inversion lid at the top which serves as an impermeable upper boundary for the pollutant (DOBBINS, 1979; BEYRICH, 1997; SHARAN and GUPTA, 2002). Thus, the boundary condition at the top (h) of the inversion/mixed layer can be prescribed as:

$$-K_z\frac{\partial c}{\partial z} = 0 \quad \text{at } z = h. \tag{5}$$

An analytical solution of equation (2) with the boundary conditions (3–5) is obtained using methods of eigenfunction expansion and Fourier transforms (Appendix A):

$$c(x, y, z) = \frac{q}{2\pi h\sqrt{K_xK_y}}\exp\left(\frac{Ux}{2K_x}\right)\left[K_0\left\{\frac{U}{2\sqrt{K_x}}\left(\frac{x^2}{K_x} + \frac{y^2}{K_y}\right)^{1/2}\right\}\right.$$

$$+ \sum_{n=1}^{\infty}K_0\left\{\left(\frac{U^2}{4K_x} + \frac{n^2\pi^2}{h^2}K_z\right)^{1/2}\left(\frac{x^2}{K_x} + \frac{y^2}{K_y}\right)^{1/2}\right\}$$

$$\left. \times\left\{\cos\left(\frac{n\pi}{h}(z - H)\right) + \cos\left(\frac{n\pi}{h}(z + H)\right)\right\}\right] \tag{6}$$

where K_0 is the modified Bessel function of the second kind of order zero.

The solution (6) essentially represents the dispersion of a pollutant in the presence of the inversion/mixed layer and accounting for the diffusion in all three coordinate directions. It may be interpreted as the contribution from a source term located at $z = H$ and its successive images by reflection from two parallel boundaries at $z = 0$ and $z = h$ respectively (HANNA et al., 1982; SEINFELD, 1986). Notice that the concentration distribution represented by (6) is not Gaussian.

2.1. Special Cases

Case I: When source is located at the ground, the expression for the concentration is obtained from the relation (6) by taking $H \to 0$ and is

$$c(x,y,z) = \frac{q}{2\pi h\sqrt{K_xK_y}}\exp\left(\frac{Ux}{2K_x}\right)\left[K_0\left\{\frac{U}{2\sqrt{K_x}}\left(\frac{x^2}{K_x}+\frac{y^2}{K_y}\right)^{1/2}\right\}\right.$$

$$\left. + 2\sum_{n=1}^{\infty}K_0\left\{\left(\frac{U^2}{4K_x}+\frac{n^2\pi^2}{h^2}K_z\right)^{1/2}\left(\frac{x^2}{K_x}+\frac{y^2}{K_y}\right)^{1/2}\right\}\cos\left(\frac{n\pi}{h}z\right)\right]. \quad (7)$$

Case II: Slender plume approximation:

The slender or thin plume approximation represents the plume emitted from a point source with its crosswind spread small compared to the downwind distance it has traveled. The expression for concentration in this case is obtained by taking the limit $K_x \to 0$ in equation (6) and using the fact that the argument of the modified Bessel function becomes considerably larger than its order as K_x tends to zero. Accordingly we use the property for the modified Bessel function $K_0(X) \approx \frac{\pi}{\sqrt{2\pi X}}\exp(-X)$ for every $X >> 0$ (GRADSHTEYN and RYZHIK, 1980). The resulting expression is

$$c(x,y,z) = \frac{q}{2h\sqrt{\pi x U K_y}}\exp\left(-\frac{Uy^2}{4xK_y}\right)\left[1+\sum_{n=1}^{\infty}\exp\left(-\frac{n^2\pi^2}{h^2}\frac{x}{U}K_z\right)\right.$$

$$\left. \times\left\{\cos\left(\frac{n\pi}{h}(z-H)\right)+\cos\left(\frac{n\pi}{h}(z-H)\right)\right\}\right]. \quad (8)$$

This is analogous to the well-known Gaussian plume solution in the presence of an inversion layer at the top (SEINFELD, 1986).

Case III: When there is near absence of inversion (i.e., $h \to \infty$):

This case essentially represents the dispersion of a pollutant in the absence of the inversion lid (HANNA *et al.*, 1982). The limit $h \to \infty$ in equation (6) has been obtained using the property of Riemann integral to convert the summation series into an integral (Appendix B). The resulting expression can be written as

$$c(x,y,z) = \frac{q}{4\pi\sqrt{K_xK_yk_z}}\exp\left(\frac{U_x}{2K_x}\right)\times\left[\frac{\exp\left\{-\frac{U}{2\sqrt{K_x}}\left(\frac{x^2}{K_x}+\frac{y^2}{K_y}+\frac{(z-H)^2}{K_z}\right)^{1/2}\right\}}{\left(\frac{x^2}{K_x}+\frac{y^2}{K_y}+\frac{(z-H)^2}{K_z}\right)^{1/2}}\right.$$

$$\left. +\frac{\exp\left\{-\frac{U}{2\sqrt{K_x}}\left(\frac{x^2}{K_x}+\frac{y^2}{K_y}+\frac{(z+H)^2}{K_z}\right)^{1/2}\right\}}{\left(\frac{x^2}{K_x}+\frac{y^2}{K_y}+\frac{(z+H)^2}{K_z}\right)^{1/2}}\right]. \quad (9)$$

The solution (9) is the same as obtained by SHARAN *et al.*, (1996a) for the dispersion of a pollutant in the absence of surface-based inversion.

To facilitate the solution for practical applications, the eddy diffusion coefficients K_i' s are parameterized in terms of standard deviations of concentration using the following relations (BATCHELOR, 1949):

$$K_i = \frac{1}{2}\frac{d\sigma_i^2}{dt}, \quad i = x, y \text{ or } z, \tag{10}$$

where t is the travel time and σ_i is the standard deviation or dispersion parameter in the ith direction. In case of constant K and expressing travel time in terms of mean-wind speed and downwind distance, the equation (10) reduces to:

$$K_i = \frac{U}{2x}\sigma_i^2, \quad i = x, y \text{ or } z. \tag{11}$$

The solution (6) in terms of dispersion parameters is:

$$c(x, y, z) = \frac{q}{\pi h U \sigma_x \sigma_y} \exp\left(\frac{x^2}{\sigma_x^2}\right)\left[K_0\left\{\frac{x}{\sigma_x}\left(\frac{x^2}{\sigma_x^2} + \frac{y^2}{\sigma_y^2}\right)^{1/2}\right\}\right.$$

$$+ \sum_{n=1}^{\infty} K_0\left\{\left(\frac{x^2}{\sigma_x^2} + \frac{n^2\pi^2}{h^2}\sigma_z^2\right)^{1/2}\left(\frac{x^2}{\sigma_x^2} + \frac{y^2}{\sigma_y^2}\right)^{1/2}\right\}$$

$$\left. \times \left\{\cos\left(\frac{n\pi}{h}(z - H)\right) + \cos\left(\frac{n\pi}{h}(z + H)\right)\right\}\right]. \tag{12}$$

The expression for the Gaussian plume model is obtained by rewriting the equation (8) in terms of σ's using equation (11) i.e.,

$$c(x, y, z) = \frac{q}{\sqrt{2\pi}U\sigma_y h}\exp\left(-\frac{y^2}{2\sigma_y^2}\right)$$

$$\times \left[1 + \sum_{n=0}^{\infty}\exp\left\{-\left(\frac{n\pi}{h}\right)^2\frac{\sigma_z^2}{2}\right\}\left\{\cos\left(\frac{n\pi}{h}(z + H)\right) + \cos\left(\frac{n\pi}{h}(z - H)\right)\right\}\right]. \tag{13}$$

3. Estimation of σ's

The standard deviations (σ_x, σ_y, σ_z) are seldom measured by instruments. Thus it is essential to estimate σ's in terms of measured quantities. These dispersion parameters can be estimated (SHARAN et al., 2003) from (i) empirical formulations (SMITH, 1968; TURNER, 1970; BRIGGS, 1973; GREEN et al., 1980), (ii) the parameterizations in terms of turbulence measurements (ERBRINK, 1991; WEBER, 1998; DEGRAZIA et al., 2000), (iii) similarity relationships (NIEUWSTADT, 1979; GRYING et al., 1987) and (iv) surface layer parameters (VENKATRAM, 1980, 1996; TIRABASSI and RIZZA, 1997). In the present study, while considering the following schemes, the dispersion parameters in x and y directions are taken equal (i.e., $\sigma_x = \sigma_y$) unless otherwise specified.

3.1 Power Law Formula:

In the absence of measurements, the empirical formulations proposed by BRIGGS (1973) and GREEN et al., (1980) are commonly used in estimating the dispersion parameters. They are of the form

$$\sigma_y = \frac{a_1 x^{p_1}}{(1 + b_1 x^{q_1})^{r_1}},$$ (14)

$$\sigma_z = \frac{a_2 x^{p_2}}{(1 + b_2 x^{q_2})^{r_2}},$$ (15)

where a_1, a_2, b_1, b_2, p_1, p_2, q_1, q_2, r_1, r_2 are constants depending on atmospheric stability. These formulations require basic minimum information on atmospheric stability.

3.2. Methods Based on Standard Deviation of Velocity Fluctuations

3.2.1. TAYLOR's (1921) theory for plume dispersion in homogeneous turbulence relates σ_y and σ_z to the standard deviations of the corresponding velocity fluctuations in the lateral (σ_v) and the vertical (σ_w) directions (GRYNING et al., 1987).

$$\sigma_y = \sigma_v t f_y \left(\frac{t}{T_y}\right),$$ (16)

$$\sigma_z = \sigma_w t f_z \left(\frac{t}{T_z}\right),$$ (17)

where t is the travel time, T_y and T_z are the Lagrangian time scales in the lateral and the vertical directions, respectively. For practical use, the following functional forms for f_y and f_z are taken:

$$f_i(X) = \left(1 + \sqrt{\frac{X}{2}}\right)^{-1} \quad i = y, z.$$ (18)

In (16) and (17), T_z = 300 s for L (Monin - Obukhov length) < 0 and T_z = 30 s for L > 0 are taken from GRYNING et al., (1987). IRWIN (1983) has suggested T_y = 600 s for ground level as well as elevated sources.

In the absence of turbulence measurements, σ_w appearing in relation (17) is obtained for H/h < 1 from

$$\left(\frac{\sigma_V}{u_*}\right)^2 = \begin{cases} 1.5\left[\frac{H}{(-kL)}\right]^{-2/3} \exp\left(-\frac{2H}{h}\right) + \left(1.7 - \frac{H}{h}\right) & L < 0 \\ 1.7\left(1 - \frac{H}{h}\right)^{3/2} & L > 0 \end{cases}$$ (19)

in which u_* is the friction velocity and k is von Karman constant. Similarly, σ_v can be estimated for unstable and stable conditions as

$$\left(\frac{\sigma_v}{u_*}\right)^2 = \begin{cases} 0.35\left(-\frac{h}{kL}\right)^{2/3} + \left(2 - \frac{H}{h}\right) & L < 0, \\ 2\left(1 - \frac{H}{h}\right)^{3/2} & L > 0. \end{cases} \tag{20}$$

Recently, OETTL et al., (2001) suggested the following formulations for σ_u, σ_v and σ_w for low wind stable conditions in terms of friction velocity

$$\sigma_u = 2u_*\left(1 - \frac{z}{h}\right), \tag{21}$$

$$\sigma_v = 1.3u_*\left(1 - \frac{z}{h}\right), \tag{22}$$

$$\sigma_w = 1.07u_*. \tag{23}$$

3.3. Methods Based on Standard Deviation of Wind Direction Fluctuations

3.3.1. Let σ_θ and σ_ϕ be the root-mean-square value of horizontal wind and vertical wind directions, respectively. YOKOHAMA (1979) proposed the following formulations for σ_y and σ_z

$$\sigma_y = 0.017\sigma_\theta x F(x), \tag{24}$$

$$\sigma_z = 0.0054\sigma_\phi \omega x, \tag{25}$$

where ω varies with Pasquill's stability category, which is equal to 2.67 for A, 2.04 for B, 1.0 for C and 0.83 for D, E, F stability categories. $F(x)$ is a function of downwind distance from the source (Table 1). Here σ_θ and σ_ϕ are taken in degree.

CIRILLIO and POLI (1992) proposed the following relations to estimate σ_x and σ_y from σ_θ as

$$\sigma_x = x\left[\cosh\left(\sigma_\theta^2\right) - 1\right]^{-1/2}, \tag{26}$$

$$\sigma_y = x\left[\sinh\left(\sigma_\theta^2\right)\right]^{-1/2}. \tag{27}$$

Table 1

Values of F as a function of downwind distance x
(HANNA et al., 1977)

X	F(x)
0.0	1.0
0.1	0.8
0.2	0.7
0.4	0.65
1	0.6
2	0.4
4	0.4
10	0.33
> 10	$0.33\,(10/x)^{1/2}$

These relationships are valid even for larger σ_θ. The parameter σ_z in terms of σ_ϕ can be estimated from the relation (25). In the absence of σ_ϕ, CIRILLIO and POLI (1992) and SHARAN et al., (1995b) estimated σ_z from BRIGGS' formulations. In this scheme, the horizontal stability is accounted through σ_θ for estimating σ_x and σ_y and the vertical stability is taken on the basis of either temperature gradient for estimating σ_z from BRIGGS formula or σ_ϕ. Resulting, this scheme is also termed as "split sigma-theta" scheme (SHARAN et al., 1995b).

4. Results and Discussion

4.1. Role of Downwind Diffusion

In order to show the importance of downwind diffusion, we define the ratio R of the concentration obtained from the Gaussian model (without downwind diffusion, eq. 8) to that from the present model (eq. 6). The value of R larger/smaller than 1 indicates overestimation/underestimation of the concentration from the Gaussian model. The ratio can be written as

$$R = \sqrt{\frac{\pi K_x}{Ux}}$$

$$\times \frac{\exp\left(-\frac{Ux}{2K_x} - \frac{Uy^2}{4xK_y}\right)\left[1 + 2\sum_{n=1}^{\infty}\exp\left(-\frac{n^2\pi^2}{h^2}\frac{x}{U}K_z\right)\cos\left(\frac{n\pi}{h}z\right)\cos\left(\frac{n\pi}{h}H\right)\right]}{\left[K_0\left\{\frac{U}{2\sqrt{K_x}}\left(\frac{x^2}{K_x} + \frac{y^2}{K_y}\right)^{1/2}\right\} + 2\sum_{n=1}^{\infty}K_0\left\{\left(\frac{x^2}{K_x} + \frac{y^2}{K_y}\right)^{1/2}\left(\frac{U^2}{4K_x} + \frac{n^2\pi^2}{h^2}K_z\right)^{1/2}\right\}\cos\left(\frac{n\pi}{h}z\right)\cos\left(\frac{n\pi}{h}H\right)\right]}.$$

$$(28)$$

In general, the plume is interpreted physically relative to the plume centerline. Thus, we transform the coordinates in such a way that the plume centerline coincides with the x axis taken in the direction of the mean wind passing through the location of the source. For this purpose we substitute $z^* = z - H, h^* = h - H$ and accordingly the ratio R is rewritten in the form

$$R = \sqrt{\frac{\pi}{\beta}}\frac{\exp\left\{-\frac{\beta}{2}\left(1 + \frac{1}{2}\frac{Y^2}{X^2}\right)\right\}\left[1 + 2\sum_{n=1}^{\infty}\exp\left(-n^2\pi^2\frac{X^2\lambda_1^2}{\beta Z^2}\right)\cos(n\pi(\lambda_1 + \lambda_2))\right]}{\left[K_0\left\{\frac{\beta}{2}\left(1 + \frac{1}{2}\frac{Y^2}{X^2}\right)^{1/2}\right\} + 2\sum_{n=1}^{\infty}K_0\left\{\frac{\beta}{2}\left(1 + \frac{1}{2}\frac{Y^2}{X^2}\right)^{1/2}\left(1 + 4n^2\pi^2\frac{X^2\lambda_1^2}{\beta^2 Z^2}\right)^{1/2}\right\}\cos(n\pi(\lambda_1 + \lambda_2))\right]},$$

$$(29)$$

where

$$X = \frac{x}{\sqrt{K_x}}, \quad Y = \frac{y}{\sqrt{K_y}}, \quad Z = \frac{z^*}{\sqrt{K_z}}$$

and

$$\beta = \frac{Ux}{K_x};$$

$$p = \frac{K_x}{x^2}\left(\frac{y^2}{K_y} + \frac{z^{*2}}{K_z}\right) = \frac{Y^2}{X^2} + \frac{Z^2}{X^2};$$

$$\lambda_1 = \frac{z^*}{h^* + H}$$

and

$$\lambda_2 = \frac{H}{h^* + H}.$$

are dimensionless parameters.

Recognize that β resembles the well-known Peclet number and it represents the ratio of advective (convective) transport to diffusive transport. The dimensionless parameter p shows the region of interest relative to the plume centerline. Both β and p assume positive values. The magnitude of β indicates the atmospheric conditions in terms of strength of winds. Small values of β represent the low wind condition, in which the downwind diffusion is expected to contribute and the region of interest remains close to the source, whereas larger values of β refer to the moderate and strong winds when the downwind diffusion is neglected in comparison to the advection and the region of interest extends to a greater distance from the source. p close to zero represents the region in the proximity to the plume centerline. For practical purposes, the values of p are restricted in the range $(0,1)$. λ_1 is the dimensionless distance measured from the plume central line and λ_2 is the dimensionless source height referred to the thickness of the layer.

Figure 1 represents the variation of R with β for various values of p for the typical values of $\lambda_1 = 0.001$, $\lambda_2 = 0$ and $\frac{Z^2}{X^2} = \frac{p}{4}$. The figure reveals that for a given value of β, the value of R decreases as p decreases and it approaches 1 as p tends to zero. This implies that both models provide similar concentration along the plume centerline and this is true for all finite values of β.

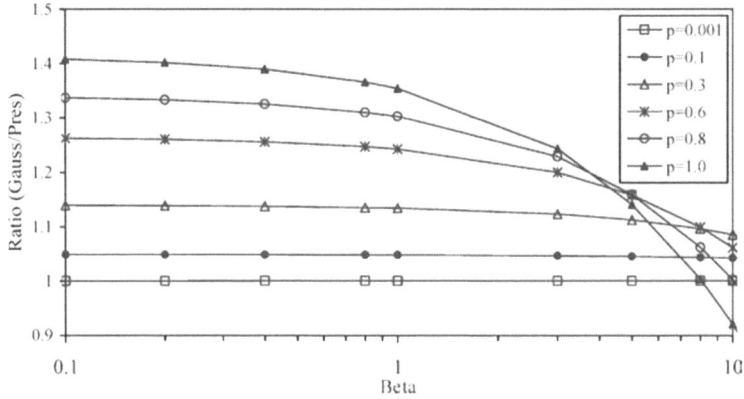

Figure 1
Variation of the ratio R with the parameter β for various of the parameter p.

Figure 1 also reveals that for β less than one, the ratio R increases with the increase in p. That is, under weak wind conditions (when the downwind diffusion is expected to be influential) the Gaussian formula leads to overestimation, for example, to the extent of about 25% when $p = 0.6$. The overpredicting trend is obtained even for slightly higher values of β also. However, the condition with higher values of β and p is physically uncommon and the variation of R is found to be reverse order. This is consistent with the theoretical considerations (SHARAN et al., 1996a) for the inclusion of the downwind diffusion term under a low wind condition in the absence of inversion lid.

4.2. Model Validation

The present model is validated by comparing computed crosswind integrated concentrations (as the concept of crosswind integrated concentration has been widely used in air pollution studies (DU, 2002)) and overall concentrations with the observed concentrations.

4.2.1. Crosswind Integrated Concentrations

The expression for crosswind integrated concentration is obtained by integrating equation (12) with respect to y from $-\infty$ to $+\infty$ and is given by

$$
c_y(x,z) = \frac{q}{hU} \left[1 + \exp\left(\frac{x^2}{\sigma_x^2}\right) \right.
$$

$$
\left. \sum_{n=1}^{\infty} \frac{\exp\left\{ -\frac{x^2}{\sigma_x^2}\left(1 + \frac{n^2\pi^2}{h^2}\frac{\sigma_x^2\sigma_z^2}{x^2}\right)^{1/2} \right\}}{\left(1 + \frac{n^2\pi^2}{h^2}\frac{\sigma_x^2\sigma_z^2}{x^2}\right)^{1/2}} \left\{ \cos\left(\frac{n\pi}{h}(z-H)\right) + \cos\left(\frac{n\pi}{h}(z+H)\right) \right\} \right].
$$

$$(30)$$

For validating crosswind integrated concentrations, data obtained at Hanford diffusion grid in stable conditions (DORAN et al., 1984) are used.

The diffusion experiment was conducted in May-June, 1983 at Hanford in a semi-arid region of southeastern Washington ($46°34'$N, $119°36'$W) on generally flat terrain. A detailed description of the experiment was provided by DORAN and HORST (1985). A total of six test runs was conducted (Table 2). In each Run release time was 30 min except Run #5 in which it was 22 min. Samplers were placed at an angular distance of $8°$, $4°$, $4°$, $2°$ and $3°$ on the concentric circles (center at release point) of radii 100, 200, 800, 1600 and 3200 m, respectively. The release height of SF_6 was 2 m and the average release rate was around 0.3 g/s. Meteorological inputs have been provided by the measurements taken at levels 1, 2, 4, 8, 16 and 32 m on a 122 m tower located at approximately 100 m to the north of the release point (DORAN et al., 1984). The values of parameters such as h, U, σ_θ, σ_ϕ, u_*, L and atmospheric stability

Table 2

The meteorological data observed at 2 m height during the Hanford diffusion grid experiment

Run No.	Date (DD/MM/YY)	P-G Stability	U (ms^{-1})	σ_θ (deg.)	σ_ϕ (deg.)	u_* (ms^{-1})	h (m)	L (m)
1.	18-05-83	E	3.63	13.0	4.1	0.40	325	165
2.	26-05-83	E	1.42	12.8	3.5	0.26	135	44
3.	05-06-83	E	2.02	13.1	3.7	0.27	182	77
4.	12-06-83	E	1.50	13.7	3.8	0.20	104	34
5.	24-06-83	E	1.41	15.8	4.6	0.26	157	59
6.	27-06-83	E	1.54	14.9	4.7	0.30	185	71

are given in Table 2. Values of U, σ_θ and σ_ϕ are at 2 m level. The atmospheric stability is determined on the basis of temperature gradient whereas h, u_* and L have been taken from GRYNING *et al.* (1987). The terrain was considered as an urban terrain (DORAN and HORST, 1985).

The crosswind integrated concentrations from Hanford diffusion data set under stable conditions (DORAN and HORST, 1985) have been simulated using the formula (30), with the following five schemes for the estimation of dispersion parameters using the present model.

Scheme I: BRIGGS formulations (equations 14–15).

Scheme II: Taylor's formulations (equations 16–17) for σ_y and σ_z in which σ_v and σ_w are computed using Gryning's formulations (equations 19–20).

Scheme III: In analogy to the Taylor's formulations for σ_y and σ_z in terms of σ_v and σ_w (equations 16–17), the formulation for σ_x has been written in terms of σ_u. The required σ_u and σ_w have been obtained from Oettl's formulations (equations 21 and 23) and Lagrangian time scale in x direction is assumed to be the same as in y direction.

Scheme IV: Formulations (equations 24–25) in terms of σ_θ and σ_ϕ.

Scheme V: Equation (26) for σ_x and equation (15) for σ_z.

Formula (30) does not have σ_y in view of being a crosswind concentration. In all these schemes σ_x is approximated by σ_y unless otherwise specified. Concentrations computed from the model using all five schemes are given in Table 3 and Figure 2 (a). In general, concentrations from all the schemes except Scheme IV are reasonably close to those observed. Scheme IV shows an overpredicting trend whereas the remaining schemes predict most of the concentrations within a factor of 2 (Fig. 2(a)).

Figure 2(b) reveals that the number of cases predicted within a factor of 2 are 93%, 90%, 80%, 33% and 93% for Schemes I, II, III, IV and V, respectively. This implies that the number of predictions within a factor of 2 are maximum from Scheme I and V and are minimum from Scheme IV. This may be due to the fact that the empirical constants appearing in the formulation (24) and (25) used in Scheme IV are site-specific and may vary in the Hanford experimental site.

Table 3

Observed and predicted crosswind-integrated concentrations C_y/Q ($10^{-3}sm^{-2}$) at Hanford using various parameterization schemes

Run No.	Distance (m)	OBS	Scheme I	Scheme II	Scheme III	Scheme IV	Scheme V	GRYNING et al. (1987)
1	100	19.5	26.3	24.7	29.3	65.6	26.4	42.5
1	200	11.7	13.7	14.8	17.8	48.4	13.8	24.8
1	800	3.7	3.6	5.6	6.8	14.7	3.6	7.7
1	1600	2.1	1.9	3.6	4.3	7.4	1.9	4.4
1	3200	1.3	1.1	2.3	2.8	3.7	1.1	2.5
2	100	51.9	67.3	48.1	57.1	185.1	67.4	72.3
2	200	36.7	35.1	29.7	35.5	135.9	35.2	44.3
2	800	12.9	9.3	12	14.5	43.8	9.3	15.5
2	1600	9.1	5.5	7.9	9.5	22.2	5.5	9.2
2	3200	7.2	5.2	5.7	6.5	11.2	5.2	5.6
3	100	27.1	47.3	41.9	49.6	125.8	47.4	66
3	200	18.1	24.7	25.7	30.8	92.6	24.7	39.5
3	800	5.9	6.5	10.2	12.2	29.2	6.5	13.1
3	1600	3.3	3.5	6.6	8	14.8	3.5	7.6
3	3200	1.8	2.7	4.4	5.3	7.4	2.7	4.5
4	100	91.8	63.7	60.8	71.4	168.4	63.8	96.9
4	200	48.6	33.3	37.9	45.1	122.7	33.3	60.1
4	800	20.1	8.8	15.4	18.4	38.3	8.8	21.5
4	1600	13.1	6.4	10.1	12.1	19.4	6.4	12.9
4	3200	9.2	6.4	7.2	8.2	9.8	6.4	7.9
5	100	83.9	67.8	48.2	57.2	165.2	67.9	70.2
5	200	42.4	35.4	29.7	35.6	115.3	35.4	42.4
5	800	10.5	9.3	12.1	14.5	33.8	9.3	14.4
5	1600	8.6	5.2	7.9	9.5	17.1	5.2	8.5
5	3200	6.6	4.5	5.5	6.4	8.6	4.5	5.1
6	100	88.4	62.1	40.8	48.6	149.9	62.1	59.8
6	200	61.1	32.4	25.1	30.1	104	32.4	35.9
6	800	13.4	8.6	10.1	12.1	30.3	8.6	12
6	1600	6.2	4.6	6.6	8	15.3	4.6	7
6	3200	3.1	3.5	4.5	5.3	7.7	3.5	4.2

The statistical measures (YADAV and SHARAN, 1996) such as the mean absolute error (MAE), fractional bias (FB), fractional variance (FS), root-mean-square error (RMSE) and normalized mean square error (NMSE) between the observed and the predicted concentrations from all schemes are computed (Table 4). The smaller the value of the measures, the better the performance. The values in Table 4 reveal that Scheme IV based on σ_θ-σ_ϕ is not performing well. The statistical measures of Scheme I (Briggs formulations) and Scheme V (Split-sigma theta) are mutually closed. Scheme V has low values of MAE, FS, RMSE and NMSE. Thus, the performance of the split-sigma theta scheme (Scheme V) is found to be better than the other schemes.

To evaluate the performance of the model on the basis of an unpaired analysis, the quantile-quantile (Q-Q) plot has been drawn for the predicted and observed concentration distributions (OLESEN, 1995; VENKATRAM, 1999). The usefulness and

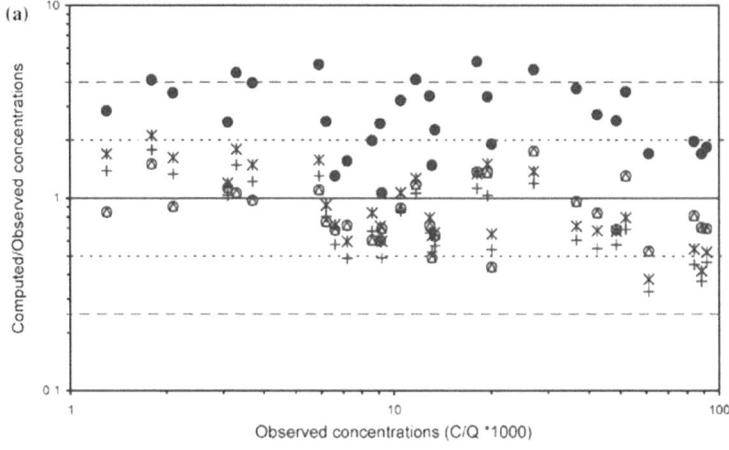

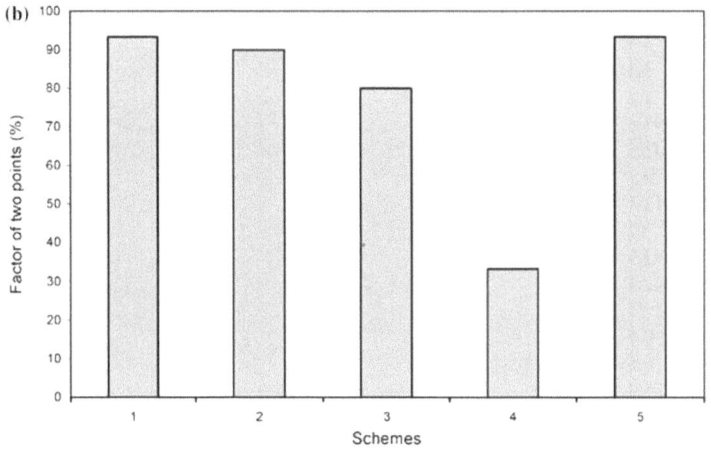

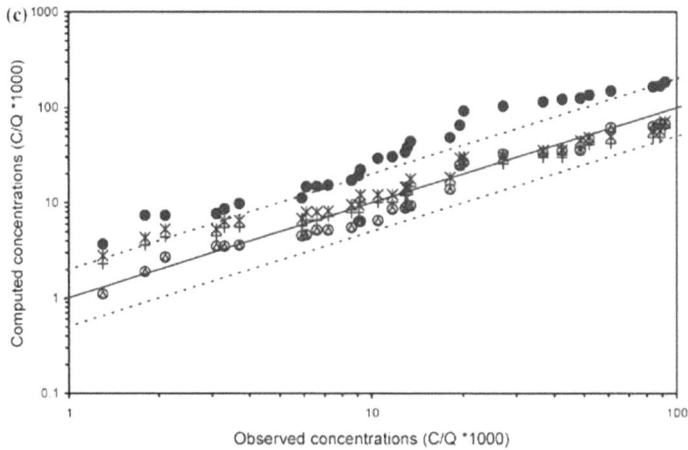

◄

Figure 2
(a) Ratio of predicted and observed concentrations with observed concentration. The lines with dashes—are the lines with factor two and the line with dashes — — — are the lines with factor four. Solid line is the one-to-one line. Δ Scheme I; + Scheme II;* Scheme III; • Scheme IV and o Scheme V. (b) Percentage of cases predicted within a factor 2 using the present model. (c) Q-Q diagram between the observed and predicted concentrations. Middle line is the one-to-one line and the outer lines are the lines with factor two to observations. Δ Scheme I; + Scheme II;* Scheme III; • Scheme IV and o Scheme V.

importance of Q-Q plot has been discussed by VENKATRAM (1999). In this approach, first the predictions and the observations are ranked from the highest to the lowest and then both ranked predictions and ranked observations are plotted (SHARAN and GUPTA, 2002). Figure 2(c) is the Q-Q plot for all the Schemes, in which the middle line is the one-to-one line and the outer lines with a factor of two to observations. Figure 2(c) reaffirms the conclusion drawn earlier that Scheme IV has the overpredicting trend and the performance of Scheme V is relatively better than the others for predicting crosswind integrated concentrations.

We have also computed the crosswind integrated concentrations (Table 3) from the model used by GRYNING et al., (1987) which was originally developed by VAN ULDEN (1978). Results computed from our model using Scheme V i.e., split-sigma theta scheme are closer to the observations than those predicted form the model used by GRYNING et al., (1987). The RMSE from the present model is $0.0113 \ \mathrm{sm}^{-2}$ whereas it is about $0.0145 \ \mathrm{sm}^{-2}$ from the model used by GRYNING et al. (1987). The RMSE obtained from the present model is found to be comparable with the standard Gaussian model because σ_z is computed by the same formulations in both the models and it appears to play a dominant role in computing the crosswind integrated concentrations.

4.2.2. Concentration Distribution

In order to validate the model (equation 12) for overall concentration distribution, two sets of data, one representing stable and the other convective conditions, have been considered. The data from Hanford diffusion grid experiments have been utilized for stable conditions whereas IIT-SF$_6$ tracer data have been considered for convective conditions.

Table 4

Quantitative analysis with Hanford data

Statistical Measures	Scheme I	Scheme II	Scheme III	Scheme IV	Scheme V
MAE	0.00733	0.00837	0.00745	0.0363	0.00733
FB	0.1713	0.240	0.0674	−0.861	0.1701
FS	0.41514	0.929	0.647	−1.3026	0.4123
RMSE	0.0113	0.0148	0.0125	0.0510	0.01130
NMSE	0.2647	0.483	0.291	1.7912	0.2638

4.2.2.1. Stable Conditions

The description of the Hanford diffusion experiment has been given in section 4.1.1 and the observed meteorological parameters are reported in Table 2. The release rate has been computed by dividing the amount of SF_6 released by the average release time. The σ's have been computed using all five Schemes, namely Scheme I (BRIGGS' formulations i.e., equations 14 and 15), Scheme II (Taylor and GRYNING formulations i.e., equations 16–17 and 19–20), Scheme III (Taylor and OETTL formulations i.e., equations 16–17 and 21–23), Scheme IV (equations 24–25) and Scheme V (Split sigma Scheme i.e., equations 26, 27 and 15).

The model results with each scheme have been computed: (i) to analyze peak concentrations (observed and computed) as the assessment of performance for high ground-level concentrations in compliance with air-quality regulations is necessary (WEIL et al., 1992) and (ii) to examine overall distribution.

Peak Concentrations:

Table 5 provides the ratio of computed concentrations to the corresponding observed peak concentrations with each of the schemes. For each arc the values in the Table are obtained by taking the arithmetic mean of the ratios over all the test runs. In addition, the averaged ratios from the Gaussian model with Scheme I are given in the Table. Ratios from the present model are close to those obtained from the Gaussian model with Scheme I. This is in conformity with the theoretical considerations in which both models are expected to produce corresponding results along the plume central line.

The scatter diagram (Fig. 3 (a)) between the ratios of predicted to observed peak concentrations and observed peak concentrations reveals that Scheme IV overpredicts the peak in most of the cases whereas Scheme V shows the underpredicting trend. The Q-Q plot (Fig. 3(b)) also confirms a similar trend. The majority of the cases are predicted within a factor of 2 by Schemes I, II and III. The cases predicted within a factor of two are 83%, 77%, 90%, 53% and 23% from Schemes I, II, III, IV and V, respectively. The relatively poor performance of Scheme IV and Scheme V may be due

Table 5

Ratio of computed to observed peak concentrations for all the schemes and averaged over all the test runs. O—observed; a—Scheme I, b—Scheme II, c—Scheme III, d-Scheme IV, e—Scheme V and g— Gaussian model with Briggs formulations

Arc (m)	a/o	b/o	c/o	d/o	E/o	g/o
100	1.28067	0.764368	0.97892	2.126541	0.687002	1.282935
200	0.811215	0.616513	0.78379	2.263932	0.489844	0.812457
800	0.761006	0.774165	1.013851	2.064543	0.313995	0.761379
1600	0.763521	0.889037	1.169346	2.124095	0.27854	0.769484
3200	1.109256	1.076485	1.381218	1.958724	0.353554	1.109947

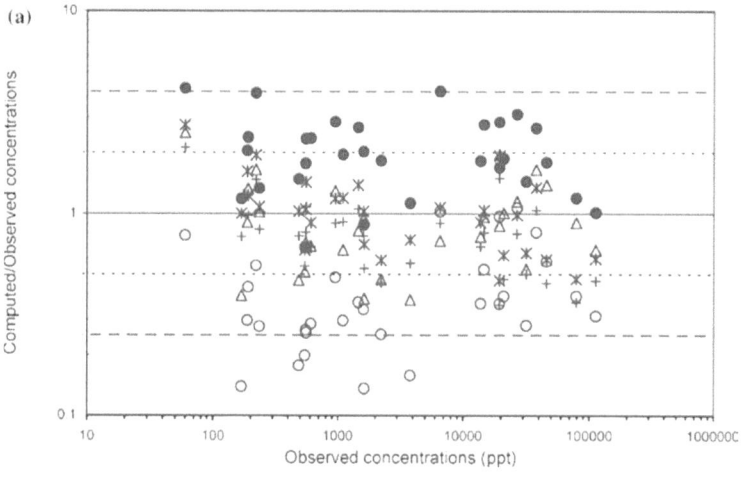

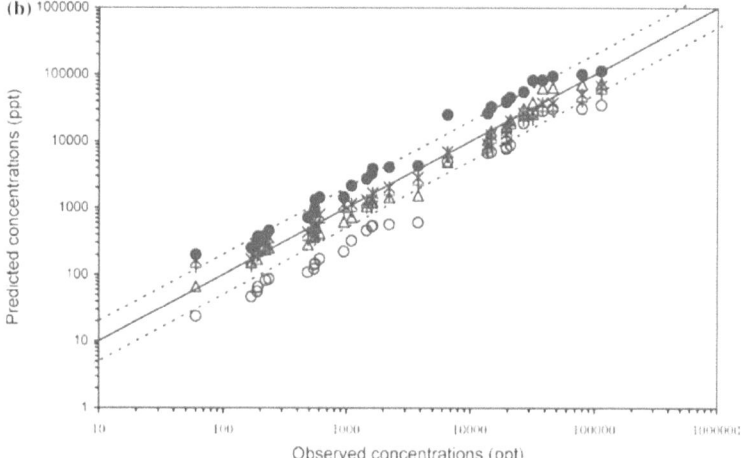

Figure 3
(a) Scatter diagram between the ratio of peaks predicted to observed and observed peak concentrations Δ Scheme I; + Scheme II;* Scheme III; Scheme IV and o Scheme V. (b) Q-Q plot for observed peaks to corresponding predicted peak concentrations. Middle line is the one-to-one line and the outer lines within a factor of two to observations. Δ Scheme I; + Scheme II;* Scheme III; • Scheme IV and o Scheme V.

to the non availability of detailed observations for σ_θ - σ_ϕ. This is in contrast to the performance of Scheme V for the crosswind integrated concentrations as seen in section 4.2.1, primarily because of the dominant role played by σ_z in comparison to σ_x.

For quantitative analysis we have computed the various statistical measures such as MAE, FB, FS, RMSE, NMSE, MG (geometric mean bias) and VG (geometric mean variance). When the range of predicted and observed values is sufficiently large, one prefers to use statistical measures MG and VG, as the large values of the computed and observed concentrations strongly influence the FB and

NMSE (HANNA *et al.*, 1991). Since the range of computed and observed concentrations here is large, we have used the performance measures MG and VG which are defined as

$$MG = \exp\left(\overline{\ln\ C_o} - \overline{\ln\ C_p}\right)$$

$$VG = \exp\left(\overline{\left(\ln\ C_o - \ln\ C_p\right)^2}\right)$$

where overbar indicates average values and C_o and C_p are the observed and the corresponding predicted concentrations, respectively. The ideal value for MG and VG is one. The magnitude of MG less than one indicates overestimation whereas a magnitude greater than one represents underestimation of the concentrations. The measures in Table 6 confirm the earlier inference that the performance of Schemes I, II and III is better as compared to Schemes IV and V. In general, the performance of the first three Schemes for the peak concentrations is comparable to each other (Table 6, Fig. 3(b)). Consequently, these three Schemes have been used for the analysis of overall distribution.

Overall Distribution:

As an illustration, the computed concentrations from the model for all three Schemes, I (Briggs formulations), Scheme II (Taylor and Gryning formulations) and Scheme III (Taylor and Oettl formulations) for test Run #4 are shown in Figure 4. In general, the direction of the predicted peak matches that of the observed peak (Fig. 4). In this test Run, the model with each of the schemes underpredicts the peak concentration on all the arcs except Scheme II on 1600 m arc. The scatter diagram (Fig. 5 (a)) for overall concentration distribution shows an almost similar trend for all three Schemes while Fig. 5 (b) provides a more objective picture regarding the performance of the schemes.

It is clear from Figure 5(b) as well as from statistical measures in Table 6 that for overall distribution the model with Scheme I based on Briggs formulations shows a better performance than Schemes II and III (least FS, least VG and MG closed to unity). This is also confirmed from the Q-Q plot (Fig. 5 (c)) as the predicted concentrations from Scheme I are nearer one-to-one line to the observation. These

Table 6

Quantitative analysis with Hanford data for peak and over all concentrations

Statistical Measures	Peak Concentrations					Overall Concentrations		
	Scheme I	Scheme II	Scheme III	Scheme IV	Scheme V	Scheme I	Scheme II	Scheme III
FS	0.1857	1.1549	0.7726	−0.6245	1.3496	0.0899	0.9364	0.5505
MG	1.19076	1.3291	1.0290	0.5212	2.7592	0.8853	0.5468	0.4805
VG	1.297015	1.2948	1.1965	1.8614	3.8056	4.3849	17.7436	18.5096

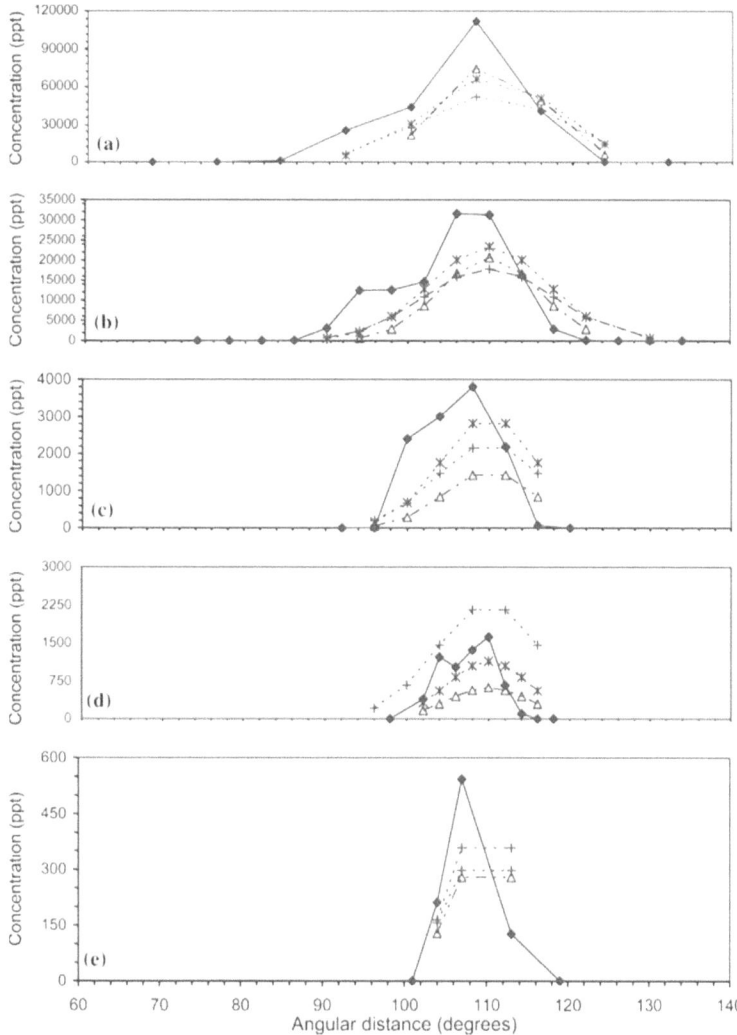

Figure 4

Observed and Predicted concentrations for Run #4 on (a) 100 m, (b)200 m, (c) 800 m, (d) 1600 m, and (e) 3200 m, arcs respectively. Observed (—●—), Scheme I (–Δ–); Scheme II (–+–) and Scheme III (–*–).

findings establish the robustness of Briggs formulations for peak as well as overall distribution even in low wind conditions. Also, we have seen that the overall distribution is predicted well by a scheme based on the Oettl formulation specifically for low wind stable conditions.

4.2.2.2. Unstable Conditions

A diffusion experiment was conducted at IIT Delhi (28°52′N, 77°18′E) in low wind conditions during February 1991. The release rate of SF_6 tracer varied between

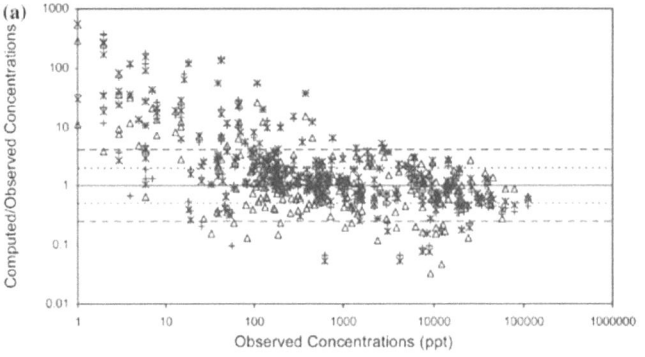

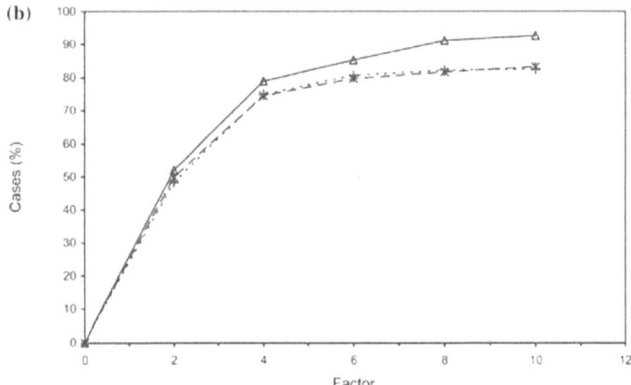

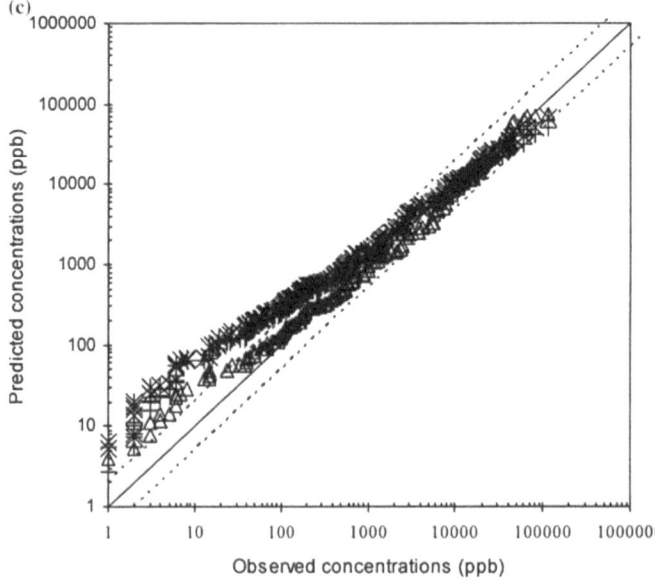

◄

Figure 5
(a) Ratio of predicted and observed concentrations with observed concentration. The lines with dashes—are the lines with factor two and the lines with dashes — — — are the lines with factor four. Solid line is the one-to-one line to observations. Δ Scheme I; + Scheme II;* Scheme III. (b) Percentage of cases predicted within a given factor using present model. Scheme I (—Δ—); Scheme II (–+–) and Scheme III(–*–). (c) Q-Q diagram between the observed and predicted concentratins. Middle line is the one-to-one line and the outer lines are the lines within a factor two to observations. Δ Scheme I; + Scheme II; * Scheme III.

30 to 50 ml min^{-1}. The sampling period for each run was 30 min. Wind and temperature measurements were obtained at four levels (2, 4, 15 and 30 m) from a 30-meter micrometeorological tower. Details of the experiment including the atmospheric stability are given in SINGH et al., (1991) and SHARAN et al. (1996a). The values of wind speed and atmospheric stability are given in Table 7. In all, 14 test runs were conducted. Out of these, 4 (runs #4, 5, 9 and 10) correspond to stable conditions, 2 (Runs #3 and 14) are in neutral conditions and the rest are in unstable conditions. Here we have chosen runs corresponding to the unstable conditions only because the adequate data were not available in stable runs. In all the cases the wind speed was less than 2 ms^{-1}. In the absence of onsite sounding data, the mixing heights were obtained from solar measurements done by the National Physical Laboratory at a nearby location in Delhi (SHARAN et al., 1996b). The samplers were located on arcs of 50 m and 100 m radii. However, in some of the test runs a few samplers were also placed on arcs of 150 m and 200 m radii. The angular distance between samplers, in general, was 45°. Concentrations have been computed here in two ways: (i) applying formula (12) for every 3 min during the sampling period and then averaging over the entire test period which is termed as 3-min averaging and (ii) using formula (12) for half hourly average of the wind data which is known as 30-min averaging.

As an illustration, results computed from these two approaches using BRIGGS formulations for test Run #11 on the arcs of radii 50 m and 100 m are shown in Figure 6. The predicted concentrations using both the approaches are reasonably close to the observations on the 50 m arc (Fig. 7a) as well as on the 100 m arc (Fig. 7b). On the 100 m arc, observed concentrations by the samplers located at 45° and 225° are not shown in the Figure as they were unreliable (SINGH et al., 1991). This is because the sampler on 225° was close to the wall and the concentration at the sampler on 45° is higher on the 100 m arc than the 50 m arc (SINGH et al., 1991). The 3-min averaging is expected to cover a greater number of receptors and predict better results due to the fact that the variations in the wind direction during the test run are taken into account by dividing the sampling period into smaller intervals. Here the role of short-term averaging (3 min in this study) could not be captured properly due to large scatter in the data and low variability in the wind direction. However, it has been found in literature that segmenting of the plume is useful in dealing with weak

Table 7

Meteorological conditions during the IIT experiment (SHARAN et al., 1996a)

Run No.	Time	Wind Speed (ms^{-1})	Mixing Height (m)	Atmospheric Stability
1	1200–1230	1.56	1570	A-B
2	1530–1600	0.74	1240	B
6	1000–1030	1.34	1070	B
7	1245–1315	1.54	1240	B
8	1645–1715	0.89	943	B
11	1000–1030	1.07	1070	A-B
12	1215–1245	1.55	1325	B
13	1530–1600	1.08	1070	B

and variable wind conditions (SAGENDORF and DICKSON, 1974; SHARAN et al., 1996c; SHARAN and YADAV, 1998). In the rest of the analysis, results for 30-min averaging are presented.

In view of the available observations for this dataset, model computations are tenable only with Schemes I and II. The peak concentration at the plume centerline is obtained by taking $y = 0$ and $z = 0$ in equation (12). The computed peak concentrations from the present model using Scheme I are mostly found to be closer to the observations than those from Scheme II (Table 8). The computed peak concentrations with the Scheme I are within a factor of two to the observations in most of the runs.

The scatter diagram (Fig. 7(a)) between the ratios of predicted-to-observed concentrations and observed concentrations reveals that the majority of the points are predicted within a factor of two line with Scheme I. Scheme I predicts 64 % cases within a factor of two whereas Scheme II predicts 48 % cases. The values of RMSE and NMSE with Scheme I are 127.92 ppt and 0.38236 whereas with Scheme II the values are 193.27 ppt and 1.729, respectively. Thus Scheme I shows a better performance in comparison to Scheme II. The Q-Q plot (Fig. 7(b)) shows that in general both schemes show an overpredicting trend for the low concentrations whereas an underpredicting trend is indicated for the higher concentrations. The concentrations predicted from Scheme I, i.e., Briggs formulations, are relatively closer to the one-to-one line than those obtained from Scheme II. The computed value of RMSE from the present model with IIT diffusion data is 126.7 ppt, whereas it is 142 ppt with the Gaussian model.

It can be concluded that widely used and tested BRIGGS formulation (Scheme I) has given reasonably good results for peak as well as overall concentration distribution. At the same time another formulation by Oettl used in Scheme III has produced the best results for peak concentration in low wind conditions. It could very well be used in other studies with confidence in similar conditions as the formulations are specifically meant to be used in a low wind stable condition.

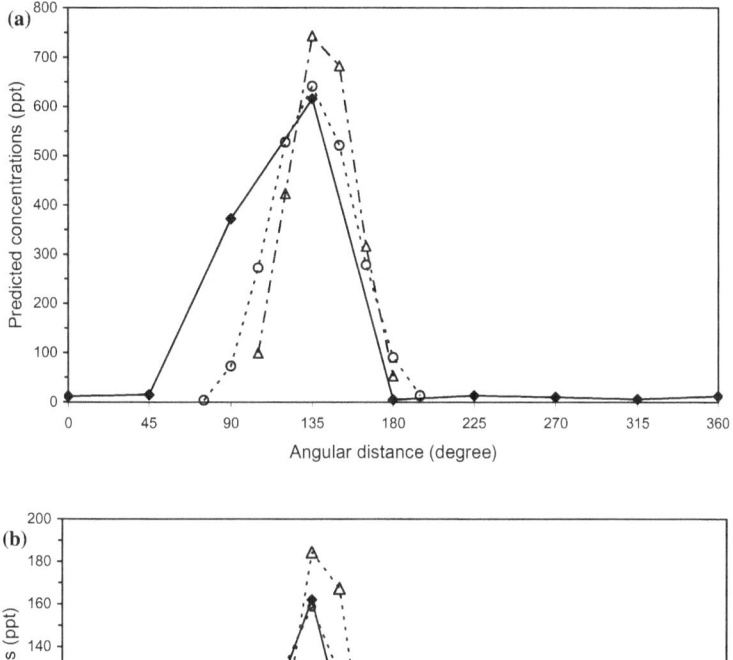

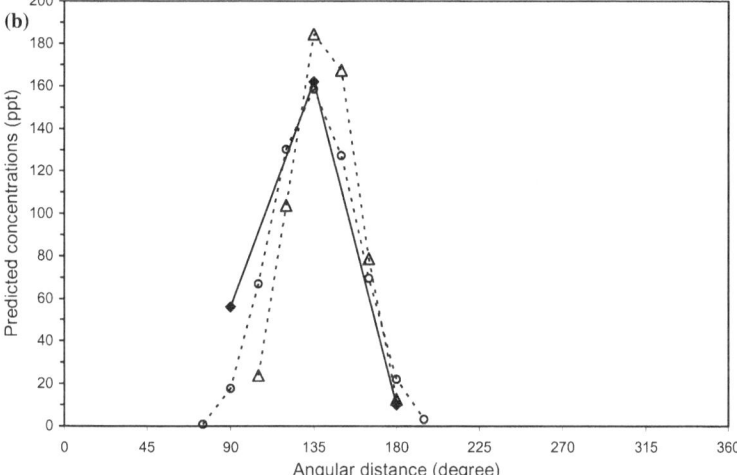

Figure 6
Observed and predicted concentrations with 3-min and 30-min averaging for Run # 11 on (a) 50 m, (b)
100 m, arcs. Observed (—◆—), 30 min (–Δ–) and 3 min (–o–).

In the present model, eddy diffusivities are assumed constant to find the solution
of the resulting advection-diffusion equation and subsequently on, they are expressed
in the solution obtained as a function of downwind distance through dispersion
parameters. The assumption of considering eddy diffusivities constant in the
formulation and relaxing it later, although mathematically inconsistent, is an
approach that has been widely accepted from the application point of view
(LLEWELYN, 1983; SHARAN *et al.*, 1996a). The traditional conversion from eddy
diffusivities to dispersion parameter (Eq. (10)) is valid for extended travel time, i.e.,

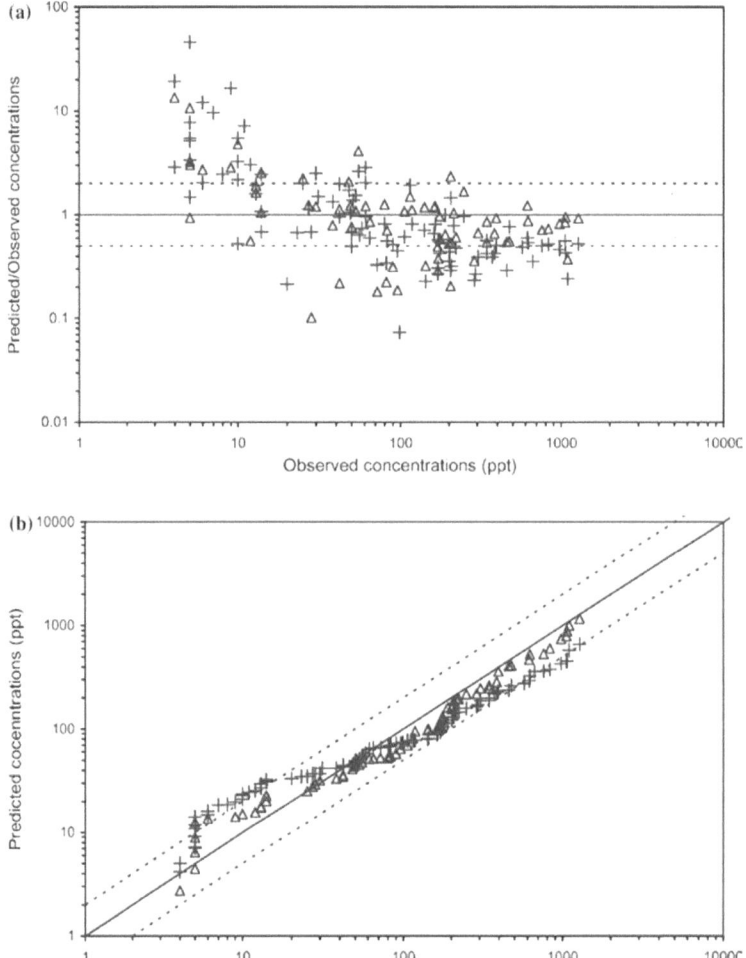

Figure 7
(a) Ratio of predicted and observed concentrations with observed concentration. The lines with dashes —
are the lines with factor two. Solid line is the one-to-one line. Δ Scheme I; + Scheme II. (b) Q-Q diagram
between the observed and predicted concentratins. Middle line is the one-to-one line and the outer lines are
the lines within a factor two to observations. Δ Scheme I; + Scheme II.

greater than the Lagrangian time scale typically of the order of 100–200 s (ZANNETI, 1990) and this may not be appropriate in dealing with near-source dispersion. The model presented here is valid for nonzero wind speed only and it is likely that the steady-state solution for the advection-diffusion equation which describes the transport and diffusion of pollutants released from an elevated point source in the surface-based inversion in calm wind ($\sim U = 0$) may not exist. This aspect is being investigated in detail using the mathematical analysis. We wish to point out that the

Table 8

Peak values of tracer concentration (ppt) observed and predicted by the two schemes at 50 m and 100 m downwind distance of the source

Run No.	50 m arc			100 m arc		
	Observed	Predicted		Observed	Predicted	
		Scheme I	Scheme II		Scheme I	Scheme II
1	832	600	419	345	185	143
2	1068	1006	429	460	250	126
6	1101	407	258	176	102	73
7	248	410	231	288	103	65
8	1282	1169	654	345	293	189
11	616	745	355	162	188	103
12	759	535	367	222	133	103
13	1060	883	564	215	221	162

steady-state solution of the corresponding problem for an elevated release in the absence of inversion exists (SHARAN *et al.*, 1996a) for zero wind speed.

In the section 4 an attempt has been made to validate the present model with the available data in the literature for crosswind integrated concentrations as well as for overall concentration distribution. It has been found that the present model simulated the observations well. The observational studies are lacking for elevated release under low wind conditions. The meteorological observations averaged over the released period in Hanford are used for the simulation. However, the on-site turbulence measurements provide the improved parameterization of dispersion to be used in the dispersion model, resulting in the realistic computation of the concentration of the tracer. Further, turbulence data may facilitate the explanation of some of the observed phenomena such as plume meandering, (SHARAN *et al.*, 1996c) puddling and recirculation, etc. (SHARAN *et al.*, 1996a). The limitations with the IIT diffusion experiment such as a lesser number of samplers, angular spacing between the samplers being too large, poor response of the wind measuring instruments during light winds, etc. have already been pointed out in detail by SHARAN *et al.* (1996a).

5. Conclusions

A steady-state three-dimensional mathematical model for the dispersion of a pollutant from a continuously emitting elevated point source in a finite layer has been described. Besides advection along the mean wind, the model takes into account the diffusion in all three directions, i.e., downwind, crosswind and vertical directions, respectively. The closed form analytical solution of the proposed problem has been obtained using the method of integral transforms. The solution proves to be non-Gaussian. Various cases including crosswind integrated concentration and standard Gaussian-plume formula have been deduced from the obtained solution. The ratio R between the Gaussian and present model has been expressed as a function of the dimensionless parameters p, β, λ_1 and λ_2, and computationally it is shown that the Gaussian approach results in overestimation (up to 25%) during low wind conditions.

To facilitate the solution for practical applications, the conventional conversion from eddy diffusivities to Gaussian dispersion parameters has been used. The present model has been validated in both stable (Hanford diffusion experiment) and unstable (IIT diffusion experiments) conditions. The dispersion parameters have been computed using the various sigma schemes in accordance with the available observations.

In general, the present model performs well in both the diffusion experiments considered here. However, the datasets used are primarily for the surface-based releases. Thus, the model needs to be further validated with the availability of diffusion data in elevated release under low wind conditions.

Acknowledgements

Observational and modeling studies on low wind dispersion were, in fact, initiated by Professor M. P. Singh at the Indian Institute of Technology, Delhi and the authors are grateful for his continuous support and encouragement. The authors wish to thank Dr. Anil Kumar Yadav for his valuable comments on the manuscript. They also acknowledge the support extended by Professor J. C. Doran in the analysis of Hanford data. The Reviewers are acknowledged for their valuable comments.

Appendix A

In view of the boundary conditions (4) and (5), the solution of equation (2) can be written as:

$$c(x, y, z) = A_0(x, y) + \sum_{n=1}^{\infty} A_n(x, y) \cos \frac{n\pi}{h} z, \qquad (A1)$$

where $\{A_0, A_1, A_2 \ldots \ldots\}$ is the set of unknown coefficients. Here $\frac{n\pi}{h}, n = 0, 1, 2 \ldots$ are the eigenvalues and $\cos \frac{n\pi}{h} z$ are the corresponding eigenfunctions.

Estimation of Unknown Coefficients:

Differentiating equation (A1) with respect to x, y and z and substituting them in equation (2) we obtain:

$$\{K_xA_{0xx} - UA_{0x} + K_yA_{0yy}\} + \sum_{n=1}^{\infty}\left\{K_xA_{nxx} - UA_{nx} + K_yA_{nyy} - \frac{n^2\pi^2}{h^2}K_z\right\}$$

$$\cos\frac{n\pi}{h}z + q\delta(x)\delta(y)\delta(z-H) = 0, \tag{A2}$$

where the subscripts "x" and "y" to A denote the partial derivatives with respect to x and y.

Using orthogonal properties of eigenfunctions (i.e., multiplying equation (A2) by $\cos\frac{j\pi}{h}z, j = 0, 1, \ldots$ and integrating it from 0 to h with respect to z) we derive a set of partial differential equations:

$$\{K_xA_{0xx} - UA_{0x} + K_yA_{0yy}\} + \frac{q}{h}\delta(x)\delta(y) = 0, \tag{A3}$$

$$\left\{K_xA_{nxx} - UA_{nx} + K_yA_{nyy} - \frac{n^2\pi^2}{h^2}K_z\right\} + \frac{2q}{h}\delta(x)\delta(y)\cos\left(\frac{n\pi}{h}H\right) = 0, n = 1, 2, \ldots. \tag{A4}$$

The boundary conditions (3) become

$$A_n(x, y) \to 0; \quad |x|, |y| \to \infty \quad n = 0, 1, 2, \ldots\ldots \tag{A5}$$

Notice that equation (A3) may be deduced from (A4) by taking $n = 0$ and q in place of $2q$. Thus we will present briefly the solution of (A4) with the boundary conditions (A5).

Using the double Fourier transform with respect to x and y

$$\tilde{A}_n(\lambda_1, \lambda_2) = \int_{-\infty}^{+\infty}\int_{-\infty}^{+\infty} A_n(x, y)e^{2\pi i(\lambda_1 x + \lambda_2 y)}\, dxdy \tag{A6}$$

in equation (A4) to obtain an algebraic equation

$$\tilde{A}_n(\lambda_1, \lambda_2) = \frac{q}{2\pi^2hK_x}\frac{\cos\left(\frac{n\pi}{h}H\right)}{\left\{\left(\lambda_1 - \frac{iU}{4\pi K_x}\right)^2 + \lambda_2^2\frac{K_y}{K_x} + \frac{U^2}{16\pi^2K_x^2} + \frac{n^2}{4h^2}\frac{K_z}{K_x}\right\}} \tag{A7}$$

On inverting (A7) with respect to λ_1, using $\oint 632, 207$ (CAMPBELL and FOSTER, 1957, pp, 74, 39), we obtain

$$A_n^*(x, \lambda_2) = \frac{q\cos\left(\frac{n\pi}{h}H\right)\exp\left(\frac{Ux}{2Kx}\right)\exp\left[-2\pi x\sqrt{\frac{K_y}{K_x}}\left\{\lambda_2^2 + \frac{U^2}{16\pi^2K_xK_y} + \frac{n^2}{4h^2}\frac{K_z}{K_y}\right\}^{1/2}\right]}{2\pi hK_x\sqrt{\frac{K_y}{K_x}}\left\{\lambda_2^2 + \frac{U^2}{16\pi^2K_xK_y} + \frac{n^2}{4h^2}\frac{K_z}{K_y}\right\}^{1/2}}. \tag{A8}$$

Notice if n and U both are zero, (A8) becomes:

$$A_0^*(x, \lambda_2) = \frac{q}{2\pi h\lambda_2\sqrt{K_xK_y}}\exp\left[-2\pi x\sqrt{\frac{K_y}{K_x}}\lambda_2\right] \tag{A9}$$

Inversion of $A_0^*(x, \lambda_2)$ is obtained as

$$A_0(x,y) = \frac{q}{2\pi h \sqrt{K_x K_y}} \int\limits_{-\infty}^{+\infty} \frac{1}{\lambda_2} \exp\left[-2\pi x \sqrt{\frac{K_y}{K_x}} \lambda_2\right] \exp(-2\pi i \lambda_2 y) d\lambda_2. \qquad (A10)$$

The improper integral appearing in (A10) diverges (PISKUNOV, 1969), hence the solution of equation (A3) with $U = 0$ does not exist. This is in conformity with the solution of such elliptic partial differential equations described by KRASNOV et al., (1990) and ZAUDERER (1983). This feature is also observed in finding the Green function for the ordinary differential equation in an unbounded domain. We are still in the process of showing that the GREEN function for such a differential equation in an unbounded domain does not exist.

Thus inverting (A8) with respect to λ_2 using ∮917 (Campbell and Foster, 1957, pp, 125) for $U \neq 0$, we find:

$$A_n(x,y) = \frac{q \cos\left(\frac{n\pi}{h} H\right)}{\pi h \sqrt{K_x K_y}} \exp\left(\frac{Ux}{2K_x}\right) K_0 \left\{ \pi \left(\frac{U^2}{4\pi^2 K_x} + \frac{n^2}{h^2} K_z\right)^{1/2} \left(\frac{x^2}{K_x} + \frac{y^2}{K_y}\right)^{1/2}\right\}. \qquad (A11)$$

The solution of equation (2) for nonzero wind speed is given by

$$c(x,y,z) = \frac{q}{2\pi h \sqrt{K_x K_y}} \exp\left(\frac{Ux}{2K_x}\right) \left[K_0 \left\{ \frac{U}{2\sqrt{K_x}} \left(\frac{x^2}{K_x} + \frac{y^2}{K_y}\right)^{1/2}\right\} \right.$$
$$+ \sum_{n=1}^{\infty} K_0 \left\{ \left(\frac{U^2}{4K_x} + \frac{n^2 \pi^2}{h^2} K_z\right)^{1/2} \left(\frac{x^2}{K_x} + \frac{y^2}{K_y}\right)^{1/2}\right\}$$
$$\left. \times \left\{ \cos\left(\frac{n\pi}{h}(z+H)\right) + \cos\left(\frac{n\pi}{h}(z-H)\right)\right\}\right]. \qquad (A12)$$

Appendix B

The equation (6) can be rewritten as:

$$c(x,y,z) = T_1 + T_2 \qquad (B1)$$

where

$$T_1 = \frac{q}{2\pi h \sqrt{K_x K_y}} \exp\left(\frac{Ux}{2K_x}\right) K_0 \left\{ \frac{U}{2\sqrt{K_x}} \left(\frac{x^2}{K_x} + \frac{y^2}{K_y}\right)^{1/2}\right\}$$

and

$$T_2 = \frac{q}{2\pi h \sqrt{K_x K_y}} \exp\left(\frac{Ux}{2K_x}\right) \sum_{n=1}^{\infty} K_0 \left\{ \left(\frac{U^2}{4K_x} + \rho_n^2 K_z\right)^{1/2} \left(\frac{x^2}{K_x} + \frac{y^2}{K_y}\right)^{1/2} \right\}$$

$$\times \left\{ \cos(\rho_n(z+H)) + \cos(\rho_n(z-H)) \right\}$$

in which $\rho_n = \frac{n\pi}{h}$. As h becomes very large (i.e., tends to infinity), T_1 becomes zero because of the presence of h in the denominator. In the expression for T_2, h appears in the denominator as well as in the argument of the modified Bessel function and cosine function. The limiting value of T_2 is found by converting the summation series into an integral using the property of Riemann integral (GRADSHTEYN and RYZHIK, 1980) and accordingly, the equation (B1) becomes:

$$c(x,y,z) = \frac{q}{2\pi^2 \sqrt{K_x K_y}} \exp\left(\frac{Ux}{2K_x}\right) \int_0^{\infty} K_0 \left\{ \left(\frac{U^2}{4K_x} + \rho^2 K_z\right)^{1/2} \left(\frac{x^2}{K_x} + \frac{y^2}{K_y}\right)^{1/2} \right\} \quad \text{(B2)}$$

$$\times \left\{ \cos(\rho(z+H)) + \cos(\rho(z-H)) \right\} d\rho$$

The integral in (B2) is evaluated using $\oint 43$ (BATEMAN, 1954) and accordingly equation (B2) becomes:

$$c(x,y,z) = \frac{q}{4\pi \sqrt{K_x K_y k_z}} \exp\left(\frac{U_x}{2K_x}\right)$$

$$\times \left[\frac{\exp\left\{-\frac{U}{2\sqrt{K_x}} \left(\frac{x^2}{K_x} + \frac{y^2}{K_y} + \frac{(z+H)^2}{K_z}\right)^{1/2}\right\}}{\left(\frac{x^2}{K_x} + \frac{y^2}{K_y} + \frac{(z+H)^2}{K_z}\right)^{1/2}} + \frac{\exp\left\{-\frac{U}{2\sqrt{K_x}} \left(\frac{x^2}{K_x} + \frac{y^2}{K_y} + \frac{(z-H)^2}{K_z}\right)^{1/2}\right\}}{\left(\frac{x^2}{K_x} + \frac{y^2}{K_y} + \frac{(z-H)^2}{K_z}\right)^{1/2}} \right]$$

$$\text{(B3)}$$

REFERENCES

ARYA, S.P. (1995), *Modeling and Parameterization of Near-source Diffusion in Weak, Winds*, J. Appl. Meteorol. *34*, 1112—1122.

ARYA , S. P., *Air Pollution Meteorology and Dispersion* (Oxford University Press, New York, 1999).

BATEMAN, H., *Table of Integral Transforms* (Mc-Graw Hill, Inc. New York, 1954).

BEYRICH, F. (1997), *Mixing Height Estimation from Sodar Data. A Critical Discussion*, Atmos. Environ. *31*, 3941–3953.

BRIGGS, G.A., *Diffusion Estimation for Small Emissions*, U.S. National Oceanic and Atmos. Admin. (E.R.L. Report, ATDL-106, 1973).

BATCHELOR, G. K. (1949), *Diffusion in a Field of Homogeneous Turbulence. I: Eulerian Analysis*, Aust. J. Sci. Res. *2*, 437–450.

CAMPBELL, G. A. and FOSTER, R. M., *Fourier Integrals for Practical Applications* (Van Nostrand, New York, 1957)

CIRILLIO, M.C. and POLI, A. A. (1992), *An Intercomparison of Semi-empirical Diffusion Models under Low Wind Speed, Stable Conditions*, Atmos. Environ. 26A, 765–774.

DEGRAZIA, G. A., ANFOSSI, D., CARVALHO, J. C., MANGIA, C., TIRABASSI, T., and CAMPOS VELHO, H. F. (2000), *Turbulence Parameterisation for PBL Dispersion Models in all Stability Conditions*, Atmos. Environ. *34*, 3575–3583.

DOBBINS, R. A., *Atmospheric Motion and Air Pollution* (John Wiley and Sons, New York, 1979).

DORAN, J.C., ABBEY, O.B., BUCK, J.W., GLOVER, D. W., HORST, T.W., *Field Validation of Exposure Assessment Models*, vol. 1 (Data Environ. Sci. Res. Lab, Research Triangle Park, NC. , 1984).

DORAN, J.C. and HORST, T. W. (1985), *An Evaluation of Gaussian Plume-depletion Models with Dual-tracer Field Measurements*, Atmos. Environ. *19*, 939–951.

DU, S. (2002), *On the Inter-dependency between Lateral Diffusion and Vertical Diffusion in the Atmospheric Surface Layer*, Atmos. Environ. *36*, 3049–3054.

ERBRINK, J.J. (1991), *A Practical Model for the Calculation of σ_y and σ_z for Use in an on-line Gaussian Dispersion Model for Tall Stacks Based on Wind Fluctuations*, Atmos. Environ. *25A*, 277–283.

GRADSHTEYN, I. S. and RYZHIK, I. M., *Table of Integrals* (Series and Products, Academic Press, New York, 1980).

GREEN, A.E.S., SINGHAL, R.P., and VENKATESWAR, R. (1980), *Analytical Extensions of the Gaussian Plume Model*, JAPCA, *30*, 773–776.

GRYNING, S. E., *Elevated Source SF_6-tracer Dispersion Experiments in the Copenhagen Area*, (Risφ Nat. Lab. Report No. R-446, Roskilde, Denmark, 1981).

GRYNING, S.E., and LYCK, E. (1984), *Atmospheric Dispersion from Elevated Sources in an Urban Area: Comparison between Tracer Experiments and Model Calculations*, J. Climate and Appl. Meteorol. *23*, 651–660.

GRYNING, S.E., HOLTSLAG, A. A. M., IRWIN, J., and SIVERTSEN, B. (1987), *Applied Dispersion Modeling Based on Meteorological Scaling Parameters*, Atmos. Environ. *21*, 79–89.

HANNA, S. R, BRITTER R., and FRANZESE, P. (2003), *A Baseline Urban Dispersion Model Evaluated with Salt Lake City and Los Angeles Tracer Data*, Atmos. Environ. *27*, 5069–5082.

HANNA, S.R., BRIGGS, G.A., and HOSKER, R.P., Jr., *Handbook on Atmos. Diffusion* (U.S. Dept. of Energy Report COE/TIC-11223, Washington, DC, 1982).

HANNA, S.R., STRIMAITIS, D.G., and CHANG, J.C., *Hazard Response Modeling Uncertainty* (*A Quantitative Method*), vol. II: *Evaluation of Commonly-used Hazardous Gas Dispersion Models* (AFESC Contract No. FO8635-89-C-0136, H.Q. AFESC/RDVS, Tyndall AFB, FL 32403, 1991).

IRWIN, J. S. (1983), *Estimating Plume Dispersion–A Comparison of Several Sigma Schemes*, J. Climate and Appl. Meteorol. *22*, 92–114.

KRASNOV, M., KISELEV, A., MAKARENKO, G., and SHIKIN, E., *Mathematical Analysis for Engineers* vol. I (Mir Publishers, Moscow, 1990).

LIN, J. S. and HILDEMANN, L. M. (1996), *Analytical Solutions of the Atmospheric Diffusion Equation with Multiple Sources and Height-dependent Wind Speed and Eddy Diffusivities*, Atmos. Environ. *30*, 239–254.

LLEWELYN, R. P. (1983), *An Analytical Model for the Transport, Dispersion and Elimination of Air Pollutants Emitted from a Point Source*, Atmos. Environ. *17*, 249–256.

NIEUWSTADT, F.T.M. (1979), *Application of Mixed-layer Similarity to the Observed Dispersion from a Ground-level Source*, J. Appl. Meteorol. *8*, 157–161.

OETTL, D., ALMBAUER, R. A., and STURM, P. J. (2001), *A New Method to Estimate Diffusion in Stable, Low-wind Conditions*, J. Appl. Meteorol. *40*, 259–268.

OLESEN, H. R. (1995), *The Model Validation Exercise at Mol: Overview of Results*, Internat. J. Environ. Pollution *5*, 761–784.

PHILIP, J.R. (1997), *Windward Diffusion*, J. Appl. Meteorol. *36*, 974–979.

RANGOTHAMAN, S., NARASIMHA, R., and VASUDEVA MURTHY, A.S. (2003), *Evolution of Nocturnal Temperature Inversions: A Numerical Study*, Il Nuovo Cimento, *25*, 147–163.

ROBSON, R. E. (1983), *On the Theory of Plume Trapping by an Elevated Inversion*, Atmos. Environ. *17*, 1923–1930.

SAGENDORF, J.F. and DICKSON, C.R. (1974), *Diffusion under Low Wind-speed, Inversion Conditions*, U.S. Nat. Oceanic and Atmos. Admin. Technical Memorandum ERL ARL-52.

SEINFIELD, J.H., *Atmospheric Chemistry and Physics of Air Pollution* (John Wiley and Sons, New York. 1986).

SHARAN, M., MCNIDER, R.T., GOPALALKRISHNAN, S.G., and SINGH, M.P. (1995a), *Bhopal Gas Leak: A Numerical Simulation of Episodic Dispersion*, Atmos. Environ. *29*, 2061–74.

SHARAN, M., YADAV, A,K., and SINGH, M.P. (1995b), *Comparison of Various Sigma Schemes for Estimating Dispersion of Air Pollutants in Low Winds*, Atmos. Environ. *29*, 2051–59.

SHARAN, M., SINGH, M.P., YADAV, A.K., Agarwal, P. and Nigam, S. (1996a), *A Mathematical Model for Dispersion of Air Pollutants in Low Wind Conditions*, Atmos. Environ. *30*, 1209–1220.

SHARAN, M., SINGH, M.P., and YADAV, A.K. (1996b), *A Mathematical Model for the Atmospheric Dispersion in Low Winds with Eddy Diffusivities as Linear Functions of Downwind Distance*, Atmos. Environ. *30*, 1137–1145.

SHARAN, M., YADAV, A.K., and SINGH, M.P. (1996c), *A Time-dependent Mathematical Model Using Coupled Puff and Segmented Approaches*, J. Appl. Meteorol. *35*, 1625–1631.

SHARAN, M. and YADAV, A.K. (1998), *Simulation of Diffusion Experiments under Light Wind, Stable Conditions by a Variable K-theory Model*, Atmos. Environ. *32*, 3481–3492.

SHARAN, M. and GUPTA, S. (2002), *Two-dimensional Analytical Model for Estimating Crosswind-integrated Concentration in a Capping Inversion: Eddy Diffusivity as a Function of Downwind Distance from the Source*, Atmos. Environ. *36, 97*–105.

SHARAN, M., YADAV, A.K., and MODANI, M. (2002), *Simulation of Short-range Diffusion Experiment in Low-wind Convective Conditions*, Atmos. Environ. *36*, 1901–1906.

SHARAN, M. and GOPALAKRISHNAN, S.G. (2003), *Mathematical Modeling of Diffusion and Transport of Pollutants in the Atmospheric Boundary Layer*, Pure Appli Geophys. *160*, 357–394.

SHARAN, M., MODANI, M., and YADAV, A. K. (2003), *Atmospheric Dispersion: An Overview of Mathematical Modeling Framework*, Proc. Indian Nat. Sci. Academy *69A*, 725–744.

SINGH, M.P., AGARWAL, P., NIGAM, S., and GULATI, A., *Tracer Experiments—A Report*. (Tech. Report, CAS, IIT Delhi, 1991).

SIRAKOV, D. E. and DJOLOV, G. D. (1979), *Atmospheric Diffusion of Admixtures in Calm Conditions*, Geophysique *32*, 891–892.

SMITH, F.B. (1968), *Conditioned Particle Motion in a Homoparticles in the Atmosphere*, Quart. J. Roy. Meteorol. Soc. *87*, 82–101.

TAYLOR, G.I. (1921), *Diffusion by Continuous Movements*, Proc. London Math. Soc., Ser 2, XX, 196–212.

TIKHONOV, A.N. and SAMARSKII, A. A., *Equation of Mathematical Physics*. (Pergamon Press. Oxford) (1963), 781 pp.

TIRABASSI, T. and RIZZA, U. (1997), *Boundary Layer Parameterization for a non-Gaussian Puff Model*, J. Appl. Meteorol. *36*, 1031–1037.

TURNER, D.B., *Workbook of Atmospheric dispersion Estimates* (Office of Air Programs Pub. No. AP-*26*, U.S. Environmental Protection Agency, Research Triangle Park, NC, 1970).

VAN DER HOVEN, I. (1976), *A Survey of Field Measurements of Atmospheric Diffusion under Low-wind-speed Inversion Conditions*, Nuclear Safety *17*, 223–230.

VAN ULDEN, A.P. (1978), *Simple Estimates for Vertical Diffusion from Sources near the Ground*, Atmos. Environ., *12*, 2125–2129.

VASUDEVA MURTHY, A.S., SRINIVASAN, J., and NARASIMHA, R. (1993), *A Theory of the Lifted Temperature Minimum on Calm Clear Nights*, Phil. Trans. R. Soc. London, *A 344*, 183–206.

VENKATRAM, A. (1980), The Relationship between the Convective Boundary Layer and Dispersion from Tall Stacks, Atmos. Environ. *14*, 763–767.

VENKATRAM, A. (1996), *An Examination of the Pasquill-Gifford-Turner Dispersion Scheme*, Atmos. Environ. *30*, 1283–1290.

VENKATARAM, A. (1999), *Applying a Framework for Evaluation the Performance of Air Quality Models*, Sixth Internat. Conf. Harmonisation within Atmospheric Dispersion Modelling for Regulatory Applications, Rouen, France, October 11–14, 1999.

WEBER, R. O. (1998), *Estimators for the Standard Deviations of Lateral, Longitudinal and Vertical Wind Components*, Atmos. Environ. *32*, 3639–3546.

WEIL, J.C., SYKES, R. I., and VENKATRAM, A. (1992), *Evaluating Air-quality Models: Review and Outlook*, J. Appl. Meteorol. *31*, 1121–1145.

YADAV, A. K. and SHARAN, M. (1996), *Statistical Evaluation of Sigma Schemes for Estimating Dispersion in Low Wind Conditions*, Atmos. Environ. *30*, 2595–2606.

YOKOHAMA, O. (1979), *Study on Atmospheric Dispersion of Pollutants*, Rep. National Res. Inst. For Pollution and Resources, MITI, 15, 408.

ZANNETTI, P., *Air Pollution Modelling* (Computational Mechanics Publications. Southampton, 1990).

ZAUDERER, E., *Partial Differential Equations of Applied Mathematics* (John Wiley and Sons, New York 1983) 779 pp.

(Received March 3, 2004, accepted August 19, 2004)
Published Online First: June 8, 2005

 To access this journal online:
http://www.birkhauser.ch

Pure appl. geophys. 162 (2005) 1893–1917
0033–4553/05/101893–25
DOI 10.1007/s00024-005-2697-4

© Birkhäuser Verlag, Basel, 2005

| Pure and Applied Geophysics

Uncertainty Analysis in Atmospheric Dispersion Modeling

K. Shankar Rao

Abstract—The concentration of a pollutant in the atmosphere is a random variable that cannot be predicted accurately, but can be described using quantities such as ensemble mean, variance, and probability distribution. There is growing recognition that the modeled concentrations of hazardous contaminants in the atmosphere should be described in a probabilistic framework. This paper discusses the various types of uncertainties in atmospheric dispersion models, and reviews sensitivity/uncertainty analysis methods to characterize and/or reduce them. Evaluation and quantification of the range of uncertainties in predictions yield a deeper insight into the capabilities and limitations of atmospheric dispersion models, and increase our confidence in decision-making based on models.

Key words: Atmospheric dispersion models, Concentration prediction, Uncertainty analysis, Stochastic uncertainty, Probabilistic framework, Regulatory modeling.

1. Introduction

Atmospheric dispersion models are routinely used to assess the impact of emission sources on air quality for varying meteorological conditions, source configuration, and topography. In addition, these models are also used at nuclear and chemical plants to estimate exposure to hazardous contaminants accidentally released into the atmosphere. A risk assessment is undertaken to quantify the potential hazard to exposed populations and to evaluate response or remediation measures. An uncertainty analysis is recommended as an integral part of any risk assessment to quantify the degree of confidence in the estimate of risk (IAEA, 1989).

Atmospheric dispersion is a stochastic phenomenon and, in general, the concentration observed at a given time and location downwind of a source cannot be predicted precisely (CHATWIN, 1982). Since concentration is a random variable, it should be described statistically, using quantities such as ensemble mean, variance, and probability distribution (CSANADY, 1973; LEWELLEN and SYKES, 1989). It was long recognized that uncertainties in atmospheric dispersion models must be studied

Atmospheric Turbulence and Diffusion Division, Air Resources Laboratory, National Oceanic and Atmospheric Administration, P. O. Box 2456, Oak Ridge, Tennessee, 37831, U.S.A.
E-mail: Shankar.Rao@noaa.gov

as part of any comprehensive model performance evaluation (e.g., FOX, 1984). There is a growing trend to move away from the simple deterministic predictions of ensemble mean concentrations and move towards development of probabilistic results which seek to describe a range of likely events and their associated probabilities (DABBERDT and MILLER, 2000; HOGREFE and RAO, 2001). This review paper addresses the analysis and quantification of various types of uncertainties associated with the prediction of concentrations from atmospheric dispersion models.

2. Uncertainty in Dispersion Models

A concentration estimate from an atmospheric dispersion model typically represents an ensemble-average of numerous repetitions of the same event at a specific site. The event is characterized by measured values or "known" parameters that are input to the model, e.g., wind speed, mixing layer depth, atmospheric stability, source conditions, etc. However, in addition to the known parameters, there are unmeasured or unknown variations in the conditions of this event, e.g., unresolved details of the atmospheric flow or the subgrid-scale atmospheric processes. Thus, even with a "perfect" model that predicts the correct ensemble-mean, there are likely to be deviations from the observed concentrations in individual realizations of the event, because of unknown variations in the conditions. In general, both observations and model predictions are uncertain. Therefore, meaningful model verification requires not only the average values but also the probability distributions of target variables.

Uncertainty in atmospheric dispersion model predictions is associated with: (a) "data" or "parameter" uncertainty resulting from errors in the data used to execute and evaluate the model, uncertainties in empirical model parameters, and initial and boundary conditions; (b) "model" or "structural" uncertainty arising from inaccurate treatment of dynamical and chemical processes, approximate numerical solutions, and internal model errors; and (c) "stochastic" uncertainty, which results from the turbulent nature of the atmosphere as well as from unpredictability of human activities related to emissions. The uncertainties associated with (a) and (b) can be minimized by making better (more accurate and more representative) measurements and improving model dynamics, parameterization schemes, and numerical methods, but it may not be feasible to reduce them beyond a certain level. The stochastic uncertainty term (c), arising from the natural variability of the atmosphere, can be expressed by the mean square concentration fluctuations:

$$\sigma_c^2 = \overline{(C^{(r)} - \overline{C})^2}, \qquad (1a)$$

where σ_c is the standard deviation and the overbar denotes the ensemble average. $C^{(r)}(\vec{x}, t)$ is the concentration in one (r-th) realization of the experiment at spatial location $\vec{x}$ and time t, and $\overline{C}(\vec{x}, t)$ is the ensemble-average concentration defined as

$$\bar{C}(\vec{x}, t) = \lim_{n \to \infty} \left[\frac{1}{n} \sum_{r=1}^{n} C^{(r)}(\vec{x}, t) \right]. \tag{1b}$$

The stochastic uncertainty in modeled and observed concentrations cannot be eliminated, but it can be quantified in a statistical sense. Understanding the various uncertainties and their causes is required to correctly interpret monitoring data and modeling results.

From the differences (residuals) between the observed and predicted concentrations, C_o and C_p, the bias (model error), $\bar{d}$, and the variance (total model uncertainty), $\overline{(C_o - C_p)^2}$, can be expressed (RAO and HOSKER, 1993) as

$$\bar{d} = \overline{C_{oa}} - \overline{C_{pa}}, \tag{2}$$

$$\overline{(C_o - C_p)^2} = \underbrace{\overline{(\delta C_p)^2}}_{(a-1)} + \underbrace{\overline{(\delta C_o)^2}}_{(a-2)} + \underbrace{(\bar{d})^2}_{(b)} + \underbrace{\sigma_c^2}_{(c)}, \tag{3}$$

where $\bar{C}_{oa}$ is the actual ensemble-average observed (without instrument errors) concentration for a given set of external conditions, $\bar{C}_{pa}$ is the corresponding prediction based on error-free input data, δC_o is the error in the observation of C_o, δC_p is the error in C_p due to input data errors, and σ_c^2 is the stochastic uncertainty, defined in Eq. (1), resulting from atmospheric turbulence.

The various components of the total model uncertainty, identified by (a-1), (a-2), (b), and (c) on the RHS of Eq. (3), are discussed above. The sum of the first two terms, (a-1) and (a-2), represents the data errors term (a). The third term on the right in Eq. (3), which is the square of model bias defined in Eq. (2), represents the model errors (e.g., in parameterizations, physics, coding) term (b). The last term (c) is the stochastic uncertainty.

The likely variations of the uncertainty terms in Eq. (3) versus the number of meteorological parameters in the model are depicted in Figure 1. We hope that errors in the model physics can be reduced (as shown) by increasing the number(N) of meteorological parameters in the model. This is the primary justification for using complex prognostic 3-D mesoscale meteorological models. The stochastic uncertainty should also decrease as N increases, as more and more of the atmospheric variability is "explained" by the model. On the other hand, the data errors are likely to increase monotonically with N. It can be seen in Figure 1 that the data errors term is a major contributor to total model uncertainty. Note that the finite value of this error at $N = 0$ is due to instrument errors in the observed concentrations. Though the variations depicted in Figure 1 are not directly based on data, they are qualitatively supported by available work in the literature. For example, LEWELLEN and SYKES (1989) showed that the meteorological input data uncertainty accounts for more than half of the total uncertainty in predicting the observed 1-hr ground level concentrations (GLCs) using a complex second-order closure plume model, which has improved model physics compared to the standard Gaussian plume model. It is

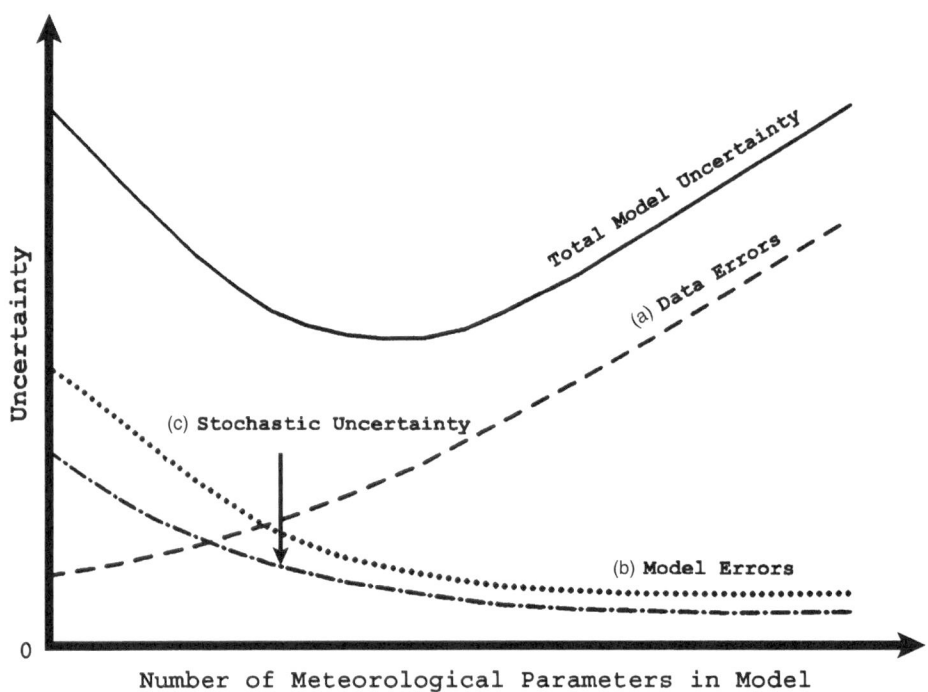

Figure 1
Variations of total model uncertainty and its components versus the number of meteorological input parameters in the model. The various uncertainty components are identified in Eq. (3) and discussed in the text. These conceptual variations are approximate and not drawn to scale.

clear that, for any given application, there is an optimum model complexity (defined here by the value of N) that minimizes the total uncertainty. This optimum level of complexity cannot be defined *a priori*, but can be determined by trial and error based on experience.

The data errors curve in Figue 1 represents uncertainties in all measurements such as emissions, meteorology, concentrations, and other input parameters. Even if an instrument is accurate, there can be large uncertainties in the data due to its unrepresentative siting. For example, data from a single meteorological tower cannot represent the flow over a large area or over a moderately complex terrain. Even in research-grade field measurements, there are significant uncertainties in the various measured parameters (HANNA, 1988). The total data error is not a simple sum of the errors in individual data components. Uncertainties in input data such as wind speed, dispersion parameters, and plume rise can be propagated through the model and the output concentration distributions can be analyzed through sensitivity or uncertainty analyses using analytical or numerical methods discussed in the next section.

It is important to distinguish between uncertainty and variability. The latter refers to the true heterogeneity that is observed in nature, whereas uncertainty characterizes our lack of complete knowledge of a specific quantity of interest. Variability is a property of the system being studied, whereas uncertainty is considered a property of the analyst. Different analysts, with different states of knowledge or access to different data sets, may have different levels of uncertainty in the predictions they make for a quantity of interest. MORGAN and HENRION (1990) suggest that variability is usually described by frequency distributions, whereas uncertainty is described by probability distributions. In exposure assessments, common sources of variability are due to differences in characteristics such as intake rates and activity patterns between individuals. However, there may also be uncertainty in the characteristics of specific individuals in the population, due to lack of data or measurement errors. CULLEN and FREY (1999) discuss the differences between uncertainty and variability, and suggest that it is desirable to characterize them separately for problems where these distinctions are important. Under specific conditions, it is possible for variability to be interpreted as uncertainty. For example, natural variability of the atmosphere leads to stochastic uncertainty, as discussed above, in the prediction of short-term air pollutant exposure for a given individual. This is sometimes referred to as *Type A uncertainty* to distinguish it from *Type B uncertainty* resulting from lack of knowledge about the appropriate mathematical models or parameter values (IAEA, 1989).

3. Data Uncertainty Analysis Methods

The relative importance of the uncertainties in input parameters to model outputs can be determined using sensitivity or uncertainty analyses. The objectives of these analyses are to: (1) study the robustness of the model with respect to potential changes in inputs and parameters; and (2) provide quantitative estimates of the overall uncertainty incorporated in model predictions.

Sensitivity analysis is the systematic study of the behavior of a model over ranges in variation of inputs and parameters. Often, the response of the model output to very small changes in a given uncertain parameter are studied, while all the other parameters are fixed. This gives only the local gradient of the model response surface with respect to that parameter. Uncertainty analysis is the quantitative assessment of how the uncertainties in model physics and input data, as well as the random variability in input parameters, propagate through the model to give a single measure of uncertainty in the model results. The collective uncertainty in all the parameters is studied, yielding more complete information on uncertainty propagation with respect to the multidimensional model response surface. The uncertainty in an atmospheric dispersion model is generally presented as a 90% or 95% confidence interval for the predicted concentrations. While sensitivity analysis is somewhat widely used in the atmospheric dispersion model evaluation, parametric uncertainty studies are much less common.

Both sensitivity analysis and uncertainty analysis attempt to rank the order of the parameters according to their contribution to overall model error. Sensitivity analysis approaches the problem by taking the partial derivatives of model equations with respect to individual parameters. Results are usually stated as the changes in model predictions to be expected from small changes in parameter values. A differential analysis provides good local information about the inputs, but does not extend well to global interpretation. Sensitivity analysis assumes that errors in model predictions can be approximated by examining small perturbations in the parameters, and the error contribution of each parameter can be examined separately. Thus, higher-order effects resulting from simultaneous errors in many parameters are often ignored.

In contrast, uncertainty analysis considers each parameter as a random variable. Assumptions in uncertainty analysis involve the statistical distributions of parameters and their means, variances and covariances. Uncertainty in predictions is estimated by Monte Carlo simulation with each parameter selected either from independent distributions or from multivariate distributions specified by a covariance matrix. The contribution of each parameter to model uncertainty is determined by statistical analysis of simulation results. The calculation of correlation coefficients (between each input parameter and the model output) from Monte Carlo simulations is the most relevant approach for ranking the parameters (GARDNER et al., 1981).

3.1. Sensitivity Analysis

A sensitivity study examines the way a particular model responds to variations in values of input variables or internal parameters. The results of the study are not directly related to model accuracy and may not correspond to physical reality, but can identify the relative importance of the input variables. This is helpful in indicating how accurately the input variables must be measured. Sensitivity studies are also useful for checking if the model is performing as expected. If it is not, then additional verification and/or development may be necessary.

Sensitivity analysis quantitatively estimates the relationships between changes in the input variables and the resulting changes in the model output. This relationship, described by sensitivity coefficients (SC), will be fairly accurate in a weak response regime where the output is linearly related to the inputs for a base case with a specific value for each input variable. For a model with the j-th output given by $C_j(t_i)$, where $t_i(i = 1, 2, - - -, n)$ are independent input variables, the SC (S_{ij}) are defined as

$$S_{ij} = \partial C_j / \partial(\ln\ t_i), \tag{4}$$

and the derivatives are evaluated for the base-case values of the inputs. The change in the j-th model output, ΔC_j, due to a newly modified input variable set can be calculated by summing over all input variable changes from the base case:

$$\Delta C_j = \sum_{i=1}^{n} [S_{ij} \cdot \Delta(\ln\ t_i)]. \tag{5}$$

For small finite changes in variables, $\Delta(\ln\ t_i) = \ln\ t_i^* - \ln\ t_i = \ln\ (t_i^*/t_i)$, where t_i is the base case value and t_i^* is the new value of the i-th input variable. Equation (5) assumes that the relationship between the changes in the input variable set, $\Delta(\ln\ t_i)$, and the resulting change in the model output, ΔC_j, is linear. The new estimated model output is

$$C_j^* = C_j + \Delta C_j. \tag{6}$$

The standard deviation σ_j of the j-th model output can be estimated as

$$\sigma_j = \left[\sum_{i=1}^{n} \{S_{ij} \cdot \ln(R_i)\}^2 \right]^{1/2}, \tag{7}$$

where $R_i = (t_i + \sigma_i)/t_i$ and σ_i is the standard deviation of the i-th input variable. This equation provides an efficient method to estimate the model output uncertainties when the input variable uncertainties are small.

Many of the commonly used sensitivity methods have been developed for use on simple models and are impractical for large models. However, sophisticated sensitivity analysis techniques, which can be applied to the governing equations of multidimensional models describing dynamical and chemical behavior in the atmosphere, are described by DUNKER (1981), RABITZ et al. (1983), and others.

3.2. Uncertainty Analysis

Once the input parameters to be studied are identified, uncertainty analysis is a two-step sequential procedure. The first step involves assigning probability distribution to each key input parameter either by its cumulative distribution function (CDF), or by its probability density function (PDF), which is the derivative of the CDF with respect to the parameter. These distributions should include measurement errors as well as random variations, and should also account for dependencies and correlations among various parameters. When sufficient data are available, these are used to generate the distribution for each parameter. Uniform, triangular, and normal distributions are among those widely used, which are shown in Figure 2. As long as the mean and variance are held constant, the exact shape of the distribution of a parameter has minimal effect on the distribution of the model prediction (HOFFMAN and HAMMONDS, 1992). When the range of a parameter extends over one or more orders of magnitude, logarithmic form of the distribution is used. For some problems, discrete or custom probability distributions may be needed to describe the inputs, as shown in Figure 2.

The second step in the uncertainty analysis involves propagating the joint probability distribution of the uncertain input parameters through the model to

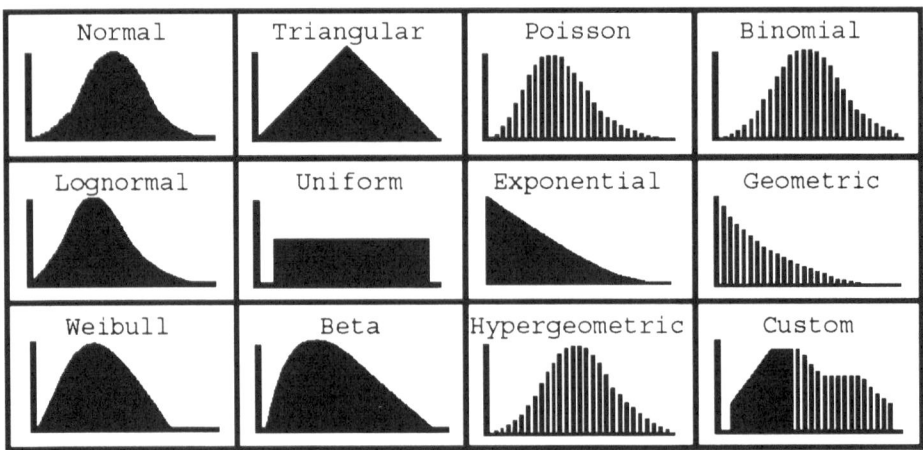

Figure 2
Probability distributions of input parameters used in uncertainty analysis. Continuous probability
distributions are shown by solid figures. Discrete probability distributions are shown by vertical bars.

generate a probability distribution of model prediction. Based on this distribution,
one can formulate a quantitative description in the form of confidence intervals (CIs)
in which the unknown *true* prediction should lie. These CIs are referred to as
subjective when the input probability distributions of the uncertain model parameters are derived using subjective judgment in the absence of data.

3.2.1. Expert Judgment

An uncertain input parameter should not be treated as a constant simply because
data are unavailable to define a range and a distribution. Professional judgment is
often employed in the absence of data. There has been remarkable progress over the
past thirty years in the understanding of how both experts and laypersons make
judgments on variables and events in the face of uncertainty, and in the development
of techniques for elicitation of professional judgment and encoding of probabilities
(MORGAN and HENRION, 1990).

The assessment of the range and statistical distribution for each uncertain
parameter is the most difficult task in quantitative uncertainty analysis, and requires
a high level of expertise. The dependence on expert opinion in this exercise can be
reduced in proportion to the amount of theoretical and experimental information
directly relevant to the parameter that exists. When there is a large body of
information about a parameter, the experts should not differ substantially in their
opinions. In situations where information is scarce, experts must rely on experience,
theoretical insight and ability to extrapolate information from one situation to
another (IAEA, 1989). In this case, the opinions of different experts are expected to
differ considerably. Therefore, it is necessary to have a survey of expert opinion when
model predictions provide input to important decisions.

The knowledge that the expert (assessor) has about the parameter to be assessed is referred to as *substantive expertise*. A substantive expert should on the average assign high probabilities to those events that turn out to occur, and low ones to those that do not. The skill of the expert in expressing his or her beliefs in probabilistic form is referred to as *normative expertise*. An expert is said to be well calibrated (reliable) if the assessed probability of events corresponds with their empirical frequency of occurrence. The expert's power to discriminate between different levels of probability is referred as *resolution*.

The assumptions and sources of information used by the expert must be documented in the elicitation process. When multiple experts are used, the probability distributions elicited from each expert are combined to form an aggregate distribution for each elicitation variable. HORA and IMAN (1989) outlined a formal approach for eliciting information from experts. HARPER *et al.* (1994) and Cook *et al.* (1994) discussed this approach, application, and results for the dispersion and deposition uncertainty assessment in the probabilistic uncertainty analysis of radiological accident consequences. HANNA *et al.* (1998) utilized an informal elicitation method based on responses from ten experts to specify the uncertainties in 109 input parameters, including those related to emissions, meteorology and boundary conditions for a photochemical grid model.

3.3. Uncertainty Analysis Methods

There are two main classes of uncertainty analysis methods: analytical and numerical. The choice of a specific method will depend on the complexity of the model, amount of information desired from uncertainty analysis, effort and time required, and costs for obtaining this information. In general, numerical methods are necessary for analysis of complex nonlinear physical systems. Monte Carlo methods (MCMs) involve simple random sampling (e.g., IMAN and CONOVER, 1980) from the probability distributions of input parameters and successive model runs with each parameter set until a statistically significant distribution of the output is obtained for uncertainty analysis. A partial rank correlation analysis between the output variable and the input parameters is performed, with the underlying premise that the greater the correlation for a parameter, the more influence that parameter has in dictating the model response. MCMs have been used for many years in diverse fields such as finance, engineering, and physical sciences.

3.3.1. Analytical Methods
For relatively simple models such as the Gaussian plume dispersion model, when the number of input parameters to be varied is small and their relationship to the model prediction can be expressed as an algebraic equation, uncertainty analysis can be performed using analytical methods. Among the analytical methods are moment matching (IAEA, 1989) and variance propagation (HOFFMAN and HAMMONDS,

1992). Moment matching permits the derivation of subjective confidence intervals by identifying a distribution function with the same mean, variance, and third and fourth moments as the subjective PDF of the model prediction. Thus, moment matching requires these four moments of the unknown output PDF to be obtained from known moments of the input parameters. Variance propagation can be easily applied to simple additive models, where the mean and the variance of the output distribution are equal to the sum of the means and variances, respectively, of the input parameters (e.g., MORGAN and HENRION, 1990).

In the case of a simple multiplicative model of the form:

$$X = p \times q \times r, \tag{8}$$

a logarithmic transformation is applied to put it in an additive form:

$$Y = \ln(X) = \ln(p) + \ln(q) + \ln(r). \tag{9}$$

The distribution of X will tend to be approximately lognormal even when the parameters p, q, and r are assigned distribution shapes other than lognormal (HOFFMAN and HAMMONDS, 1992). The median value (or geometric mean, X_g) of the lognormal distribution can be obtained by adding the means of the logarithmic terms on the RHS of Eq. (9) and exponentiating the sum ($\bar{Y}$):

$$X_g = e^{\bar{Y}}. \tag{10}$$

The standard deviation S_x of the distribution is obtained by adding the variances of the terms on the RHS of Eq. (9) and taking the exponential of the square root of the sum (S_y^2):

$$S_x = e^{S_y}. \tag{11}$$

S_x is often referred to as the geometric standard deviation of the lognormal distribution. The upper and lower limits for a 90% subjective CI are then calculated as

$$X_{95} = X_g S_x^{1.65} = \exp(\bar{Y} + 1.65\ S_y), \tag{12a}$$

$$X_5 = X_g / S_x^{1.65} = \exp(\bar{Y} - 1.65\ S_y). \tag{12b}$$

The upper and lower confidence limits for a 95% subjective CI are calculated as

$$X_{97.5} = X_g\ S_x^{1.96} = \exp(\bar{Y} + 1.96\ S_y), \tag{13a}$$

$$X_{2.5} = X_g / S_x^{1.96} = \exp(\bar{Y} - 1.96\ S_y). \tag{13b}$$

FREEMAN et al. (1986) used an error propagation formula, which involved expanding the atmospheric concentration C in terms of independent input variables in a Taylor series and retaining only terms of second order or less. This can be written as follows:

$$C = f(t_1, t_2, \ldots, t_n), \tag{14}$$

$$S_c^2 = \sum_{i=1}^{n} \left(\frac{\partial f}{\partial t_i} \right)^2 S_{t_i}^2 + \frac{1}{2} \sum_{i=1}^{n} \left(\frac{\partial^2 f}{\partial t_i^2} \right)^2 S_{t_i}^4 + \sum_{i=1}^{n} \sum_{j=1}^{n} \left(\frac{\partial^2 f}{\partial t_i \partial t_j} \right)^2 S_{t_i}^2 S_{t_j}^2. \tag{15}$$

This equation expresses S_c, the uncertainty in the predicted value of C, as a function of the uncertainties S_{t_i} in the input variables. For a Gaussian dispersion model, FREEMAN et al. (1986) showed that S_c calculated from this approach will be generally within 25% of the uncertainty of model-predicted C values calculated from Monte Carlo simulation of a randomly perturbed input data set for all stability cases and distances up to 15 km from the source.

For a Gaussian plume dispersion model, the concentration C can be expressed as

$$C/Q = p(y, \sigma_y) q(z, \sigma_z; H) r(U), \tag{16}$$

where Q is emission rate, U is mean wind speed, H is effective plume height, y and z are horizontal crosswind and vertical distances, and σ_y and σ_z are plume dispersion parameters, $r = 1/U$, and p and q are the horizontal and vertical probability density functions given by:

$$p(y, \sigma_y) = \frac{1}{\sqrt{2\pi}\sigma_y} \exp\left\{ \frac{-y^2}{2\sigma_y^2} \right\}, \tag{17}$$

$$q(z, \sigma_z; H) = \frac{1}{\sqrt{2\pi}\sigma_z} \left[\exp\left\{ -\frac{(z-H)^2}{2\sigma_z^2} \right\} + \exp\left\{ -\frac{(z+H)^2}{2\sigma_z^2} \right\} \right]. \tag{18}$$

Using a logarithmic transformation of Eq. (16), one can express it in a form identical to Eq. (9), with $X = C/Q$. From the above equations, we can determine the standard deviation S_y for the function $Y = f(U, \sigma_y, \sigma_z, H)$, for specified probability distributions in input parameters: U, σ_y, σ_z, and H. The partial derivatives required in Eq. (15) are derived by differentiating the functions p, q, and r with respect to these variables.

The analytical method described above has been applied by the present author to assess the effects of uncertainties in the input parameters for off-site radiological consequence estimates from hypothetical nuclear power plant accidents (see HARPER et al., 1994, vol. 2). Typical uncertainties in wind speed U range from 0.1 m/s (for research grade data) to 1 m/s (for routine monitoring data); these errors arise due to poor calibration and maintenance of anemometers, and use of wind data unrepresentative of the plume transport, especially at night, because of mesoscale or terrain variability and wind shear. Uncertainties in σ_y and σ_z are difficult to estimate, but probably range from 10% to 40%; in addition to measurement uncertainty, they result from differences between the experiment site and the application site, and errors in the stability classification. Uncertainty in the effective plume height H, arising primarily due to errors in plume rise estimation, usually ranges from 10% to

30%. For simplicity, it is assumed that U, σ_y, σ_z, H are uncorrelated and normally distributed, with means given by the input values and standard deviations $(S_u, S_{\sigma_y}, S_{\sigma_z}, S_H)$ by the magnitudes of their respective uncertainties. The model-calculated value of the relative concentration $\overline{C/Q}$ is taken to be the median or 50th percentile value; this determines $\bar{Y} = \ln(\overline{C/Q})$. Once $\bar{Y}$ and S_y are calculated, Eqs. (12) and (13) can be used to estimate the 90% and 95% subjective CIs, respectively. Note that these CIs are based only on uncertainties in input parameters, and do not include the model errors and stochastic uncertainty.

3.3.2. Numerical Methods

Given the PDFs of input parameters, MCMs can be used to obtain the probability distributions of model response (output) variables. MCMs have been applied to atmospheric dispersion models ranging from simple Gaussian plume models (KOCHER et al., 1987; IRWIN et al., 1987) to complex photochemical models (e.g., HANNA et al., 1998).

In practice, the large dimensionality of multiple uncertain parameters in complex models is a major problem in uncertainty analysis. The regular MCM, based on a random sampling of the entire input parameter space, is computationally expensive for nonlinear, coupled models with a large number of uncertain input parameters. In modified Monte Carlo methods such as Latin hypercube sampling (LHS; MCKAY et al., 1979; IMAN and HELTON, 1988) and the Fourier Amplitude Sensitivity Test (FAST; Cuckier et al., 1978; MCRAE et al., 1982), sampling from the probability distribution is stratified across the range of each input variable and accomplished in an efficient manner so that the number of necessary solutions (model runs) is greatly reduced, compared to the simple MCM. The main idea of the secondary model techniques such as the "response surface" method (RSM; DOWNING et al., 1985; IMAN and HELTON, 1988) is the construction of a comparatively simple approxima-tion (secondary model) of the original (primary) model; the latter is then replaced by the simpler model in subsequent analysis of relations between input and output uncertainties.

LHS is a stratified MCM, designed mainly for reducing the variance of some statistics of the response variables. In this method, the number of model runs is equivalent to the number of intervals M for each input parameter range between its high and low values. The M intervals are formed such that each interval contains an equal area under the parameter distribution. In standard LHS, a single value is then randomly sampled from each interval, which is equivalent to assuming a uniform distribution over that interval. In *mid-point* LHS, the median value of each interval is selected. This is repeated for all the uncertain input parameters. A scenario is generated by selecting one value at random for each of the input parameters, but without replacement, from the M sample values for each parameter. This finally yields M scenarios, with each value for each input parameter being used only once.

The mean and variance of the sample in LHS represent the parameter distribution more accurately than in unstratified random sampling. Using the CDF of the model output resulting from the LH sample of the inputs, one can obtain estimates of the percentile points of the output. LHS can reduce the number of runs required to obtain a given variance typically by a factor of 10 compared to the simple MCM. IMAN and SHORTENCARIER (1984) described a computer program for the generation of LH samples. IMAN and HELTON (1988) recommended the use of LHS method, because it is easy to use and is applicable to many different modeling situations and gives reliable results. However, it is harder to compute the statistics for the LHS method, using standard tests for estimating the precision of the results, than for the MCM (MORGAN and HENRION, 1990, HANNA et al., 1998).

FAST uses continuous probability distributions of model input parameters as data, and determines the relative contributions of each input parameter to the variances of model outputs. All input parameters are varied simultaneously through the ranges of possible values following their given PDFs. Each input parameter is assigned a different frequency, which determines the number of times that the complete range of the parameter is traversed. With each input parameter oscillating at a different characteristic frequency, each model run has a different set of input parameter values such that every value is used only once. The mean and variance for model output parameters, characterizing the uncertainty due to the variability of the input parameters, are then calculated. Fourier analysis of each output for all model runs is used to separate the response of the model to the oscillation of particular input parameters. Summation of those Fourier coefficients corresponding to a particular input parameter frequency and its harmonics determines the contribution of that parameter to the model output variances.

The FAST method provides information on the model sensitivity to particular input parameters. ULIASZ (1988) and COLLINS and AVISSAR (1994) applied FAST to atmospheric problems. They concluded that the efficiency of FAST is comparable to LHS, with the number of sampling points equivalent to the number of values chosen for each input parameter. Nevertheless, FAST may not generate sufficiently accurate joint PDF of the response variables without requiring a large number of statistical points (TATANG et al., 1997).

In RSM, the original (primary) model is replaced by a comparatively simple approximation (secondary model). The latter is then used as a surrogate for the primary model in the subsequent analysis of relations between the input and output uncertainties. The method consists of screening the original model to determine the subset of important input parameters, fitting a response-surface (usually a polynomial) to the model in terms of these inputs, obtaining moments of the response surface (secondary) model, and fitting a Pearson or Johnson distribution to the moments to obtain a statistical model of the proxy to the output distribution. The output CDFs, the variances of the output variables, and ranking of parameters by their contributions to variances are then derived. There are several pitfalls in this approach. The selection of the most important variables, especially when interactions among parameters cannot be neglected, is a major problem. The fitting of a RSM is not straightforward. In general, highly nonlinear mathematical models cannot be adequately approximated with a response surface except over a very limited range, and this yields misleading results from uncertainty analysis.

TATANG et al. (1997) described the probabilistic collocation method (PCM), which approximates model response surfaces using orthogonal polynomials, whose weighting functions are the PDFs of the input parameters. This PCM method was shown to be potentially a factor of 25 to 60 times faster than the MCM for parametric uncertainty analysis, converging exponentially with increasing orders of polynomial expansions. ISUKAPALLI et al. (1998) described the application of stochastic response surface methods (SRSM) for uncertainty characterization in environmental and biological systems. These methods, which can be thought of as "models of a model," are computationally efficient when the number of degrees of freedom is not very large.

3.3.3. Other Methods for Uncertainty Assessment

FISHER (2003) discussed the application of fuzzy set theory to the air pollution problem. Methods based on fuzzy sets usually rely on a judgment of the degree of uncertainty, or fuzziness, associated with the output of an air quality model. The comparison between the output from the model and an air quality objective is similar to the comparison between two fuzzy numbers. Fuzzy aggregation provides a way of combining or comparing fuzzy quantities. These techniques, similar to MCM, rely on expert judgment to set the range of uncertainty, and their predictions have some measure of uncertainty attached to them. FISHER suggested that the fuzzy method is more economical in terms of computing, since functions of fuzzy numbers can be used to express the uncertainty in models without the need for the large number of replications required by MCM.

In Bayesian methods (BOX and TIAO, 1973), which differ from the classical or frequentist methods discussed above, one starts with approximate *prior* distributions for a set of uncertain parameters, representing the available information. The likelihood of the data, given the model, is then used to derive the *posterior* distributions of the set of parameters by applying Bayes theorem. Since the posterior distributions are guided by the data, they should have less uncertainty than the prior distributions. The Real-time Online Decision Support system (RODOS; EHRHARDT et al., 1993) of the Commission of European Communities (CEC), which deals with off-site management of nuclear accidents, is based on Bayesian methods (FRENCH, 1997). Bayesian Monte Carlo analysis (BERGIN and MILFORD, 2000) provides a means of combining subjective prior distributions developed by MCM with the information about the agreement between model outputs and field observations. The resulting posterior uncertainty estimates reflect the model's performance, as well as subjective judgments about uncertainties in the model inputs and parameters.

4. Model Physics Uncertainty

A scientific assessment of the basis and physics of the model formulations should be the first step in a model evaluation. The next step is an evaluation of model performance by comparing its predictions with suitable data from laboratory and/or field experiments. A good model should be based on sound physical principles and give "good" predictions for the "right" reasons. This gives the model user faith in model predictions beyond the range of available data, and confidence in modeling new situations with different dispersion climatologies (WEIL et al., 1992). However, it is important to limit model applications and evaluations only to situations for which the model was designed (RAO and HOSKER, 1993).

The agreement between the observed and predicted concentrations, C_o and C_p, is studied using statistics and scatter plots. The statistics include performance measures of bias and error, such as fractional bias (FB) and normalized mean square error (NMSE), defined (HANNA, 1988) as

$$FB = 2(\bar{C}_o - \bar{C}_p)/(\bar{C}_o + \bar{C}_p), \quad NMSE = \overline{(C_p - C_o)^2}/(\bar{C}_o\bar{C}_p). \tag{19}$$

The figure of merit in space (FMS) is a statistical coefficient of space analysis used for evaluating the plume footprint predicted by the model. FMS is calculated at a fixed time for a fixed concentration level, and is defined (e.g., MOSCA et al., 1998) as the percentage of overlap of the observed (A_o) and predicted (A_p) plume contour areas divided by their sum:

$$FMS = [(A_o \cup A_p)/(A_o \cap A_p)] \cdot 100. \tag{20}$$

A high value of FMS indicates a good model performance, but a low value need not necessarily correspond to a bad model performance. Two areas very similar in shape but shifted in space (due to errors in mean wind direction, for example) may have a low FMS. In this case, a simple rotation of the predicted area may increase the FMS value significantly. For this reason, MOSCA et al. (1998) suggest evaluation of the FMS value together with a graphical representation of the observed and predicted contour areas. However, based on the experience of the present author, simple rotation of the predicted contours may not be possible or useful sometimes in improving the FMS values, especially in cases of complicated plume behavior (e.g., for significant wind shear, with plume bifurcation or direction change).

For hazard prediction models, it may be useful to delineate the overlap, over- and under-prediction regions of plume footprint for a given threshold concentration value. A false negative area (A_{fn}) is where hazard is observed but not predicted, and a

false positive area (A_{fp}) is where hazard is predicted but not observed. WARNER *et al.*
(2001) plot a measure of effectiveness (*MOE*) for which the *x*-axis corresponds to the
ratio of overlap area to observed area, calculated as $(1 - A_{fn}/A_o)$, and the *y*-axis
corresponds to the ratio of overlap area to predicted area, calculated as $(1 - A_{fp}/A_p)$.
As *x* increases from 0 to 1, the false negative area fraction (A_{fn}/A_o) decreases from 1
to 0. Similarly, as *y* increases from 0 to 1, the false positive area fraction (A_{fp}/A_p)
decreases from 1 to 0. The *MOE* plot is an alternate form of *FMS* representation for
model evaluation.

   The scatter plots of predicted versus observed concentrations are widely used for
model evaluation. If *R* is the correlation coefficient of the regression fit between C_o
and C_p, then R^2 indicates the fraction of the variance that is explained by the model.
An effective method of assessing model physics uncertainty term (b) in Eq. (3) is to
plot residuals $d = C_o - C_p$ against each of the key model variables separately.
Ideally, the points should be symmetrically distributed about $d = 0$ and show no
trend (VENKATRAM, 1982). If a clear trend is evident in the plot, then there is a need
for modifying the model's physics or parameterizations. Often, it is useful to divide
the points into a finite number of intervals of each key variable and calculate the
mean $\bar{d}$ and its uncertainty (estimated as 95% confidence limits of a normal
distribution using the standard deviation S_d for the points in each interval,
$\bar{d} \pm 1.96\,S_d$) and plot them. If the uncertainty limits of $\bar{d}$ in each interval include zero,
then the model error is not statistically significant. Otherwise, the model physics and
parameterizations should be improved, and the new results should be plotted until
this condition is satisfied. This method can also be used with residual plots of the
form $\ln(C_p/C_o)$ versus each of the key variables, where the 95% confidence limits in
each interval are determined using the geometric standard deviation. This was
illustrated by WEIL *et al.* (1992) and RAO and HOSKER (1993). The model physics
error analysis based on scatter and residual plots is an iterative process in which the
identified model deficiencies are corrected, and the modified model is re-evaluated
until the uncertainty is reduced to acceptable limits.

5. *Stochastic Uncertainty*

   The stochastic uncertainty, σ_c^2 in Eq. (3), is caused by atmospheric turbulence. It
arises because of the variability in the details of the velocity field in each realization
of the turbulent flow, and the finite averaging-time of the measured concentration. σ_c
decreases with an increase in the averaging time. In addition to varying in space and
time, σ_c also depends on the meteorological and source conditions defining the
ensemble (CHATWIN, 1982). It is difficult and expensive to conduct an ensemble of
dispersion experiments in the atmosphere. The conditions in the ensemble do not
repeat with sufficient frequency because of the inherent variability of the atmosphere.
Hence, concentration fluctuations are often determined from laboratory experiments

(e.g., FACKRELL and ROBINS, 1982; DEARDORFF and WILLIS, 1984) or models (e.g., PANWAR et al., 1994). However, field studies of concentration fluctuations in full-scale atmospheric experiments facilitate proper representation of large-scale eddy motions, which contribute to plume meander, and the full range of atmospheric stability conditions. Among such field studies are those by SAWFORD (1987), MYLNE and MASON (1991), and YEE et al. (1993).

For many dispersion problems in the atmosphere, the concentration fluctuation intensity $\sigma_c/\bar{C}_o \sim 1$ for averaging times of about 1 h and distances less than 30 km. Laboratory data on tall stack plumes show that $\sigma_c/\bar{C}_o$ can be as large as 6 near the surface for averaging times of a few minutes. Nevertheless, most atmospheric dispersion models presently used in regulation and safety assessment predict only the ensemble-mean concentration, and generally ignore the concentration fluctuations because of the difficulty in making reliable estimates of σ_c for general atmospheric conditions and arbitrary sources.

The concentration fluctuations include contributions both from in-plume fluctuations (relative diffusion) due to small-scale "inertial subrange" turbulence, and from plume meandering due to large-scale "energy containing" eddies in a turbulent flow (GIFFORD, 1959; CSANADY, 1967). Meandering causes plume intermittency, i.e., the concentration measured at a fixed sampler essentially varies between "in-plume" peaks and a zero value in the environment. Meandering dominates the fluctuations from small sources for downwind travel times $t < \tau_L$, the Lagrangian integral time scale. The in-plume component dominates the fluctuations from small sources at large travel times. For a surface release, vertical meandering is small, but lateral meandering can lead to significant concentration fluctuations.

Models for the probability distributions of C or σ_c are useful in predicting the expected exceedance of threshold concentration value (TCV) for a toxic contaminant dispersing in the atmosphere. Such methods transform the deterministic model result into a probabilistic form. The PDF, $p(\theta; \vec{x}, t)$, of the concentration $C(\vec{x}, t)$ of a hazardous pollutant dispersing in the atmosphere can be defined (CHATWIN, 1982) as

$$p(\theta; \vec{x}, t)\delta\theta = \text{prob} \left[\theta \leq C(\vec{x}, t) < \theta + \delta\theta \right]. \tag{21}$$

The ensemble mean concentration and variance are then given by

$$\bar{C} = \int_0^\infty \theta p \, d\theta, \tag{22}$$

$$\sigma_c^2 = \int_0^\infty (\theta - \bar{C})^2 p \, d\theta. \tag{23}$$

Measurements of the frequency distribution of fluctuating plume concentrations have revealed that the PDF is strongly skewed to the right with an upper tail that is heavier than that of the Gaussian form (YEE and CHAN, 1997). Several models are proposed for the PDF of concentration in the literature over the past two decades. Among them are the lognormal distribution (CSANADY, 1973) and the exponential distribution (HANNA, 1984). LEWIS and CHATWIN (1997) proposed a three-parameter model based on a weighted sum of exponential and generalized Pareto distributions, which was shown to fit the concentration data over different atmospheric conditions. SCHOPFLOCHER and SULLIVAN (2002) proposed a double Beta distribution for the PDF of a scalar diffusing in a turbulent flow, which emphasizes the representation of the underlying physical structure of the concentration field, as well as accurate modeling of the high-concentration tails of the distribution.

LEWELLEN and SYKES (1989) described an advanced modeling approach, which generates confidence limits for the very high but rare concentrations, which are important for the atmospheric dispersion of hazardous substances. This method involves predicting the variance of concentration fluctuations from a second-order closure model, and estimating the probability distribution from the mean and the variance using a truncated Gaussian (referred to as "clipped normal") distribution, which replaces any unphysical negative tail (i.e., negative concentrations) in the Gaussian with a delta function at zero concentration. YEE and CHAN (1997) showed that a clipped gamma PDF, derived from the same two model-predicted parameters, fits the concentration data better and is more flexible in fitting the full range of observed PDF behavior in a dispersing plume.

The SCIPUFF model (SYKES *et al.*, 1989), which uses this probabilistic approach, is one of the few atmospheric dispersion models that can predict a concentration probability density function as a function of time and space. This PDF can be used to specify probabilities of interest, such as the probability of exceeding a TCV for a given time period at a given location. The expected CDF and associated confidence bounds can be estimated directly by random sampling of the predicted PDF of concentration at each location and time. Model evaluation then consists of checking if the observed concentration samples are consistent with what could be expected from a single realization of the range of CDFs estimated from the concentration PDF; see LEWELLEN and SYKES (1989) and SYKES *et al.* (1989) for details and examples. Though physically realistic, this method has not been widely used since very few dispersion models are capable of predicting the concentration PDF. In addition, the probabilistic evaluation approach is difficult to interpret.

HANNA and DAVIS (2002) applied this method for the evaluation of a photochemical grid model by estimating the PDF of predicted hourly concentrations at 66 ozone monitors from 100 MCM runs, assuming that the PDF was completely determined only by the uncertainties in input variables. Using the set of observed (C_o) and predicted ensemble-mean ($\bar{C}_p$) concentrations and the estimated PDF, the CDF of observed model residuals ($C_o - \bar{C}_p$) at each location was calculated (by integrating the PDF from 0 to the concentration of interest) and plotted, as shown in Figure 3. Since this curve fell completely within the 95%

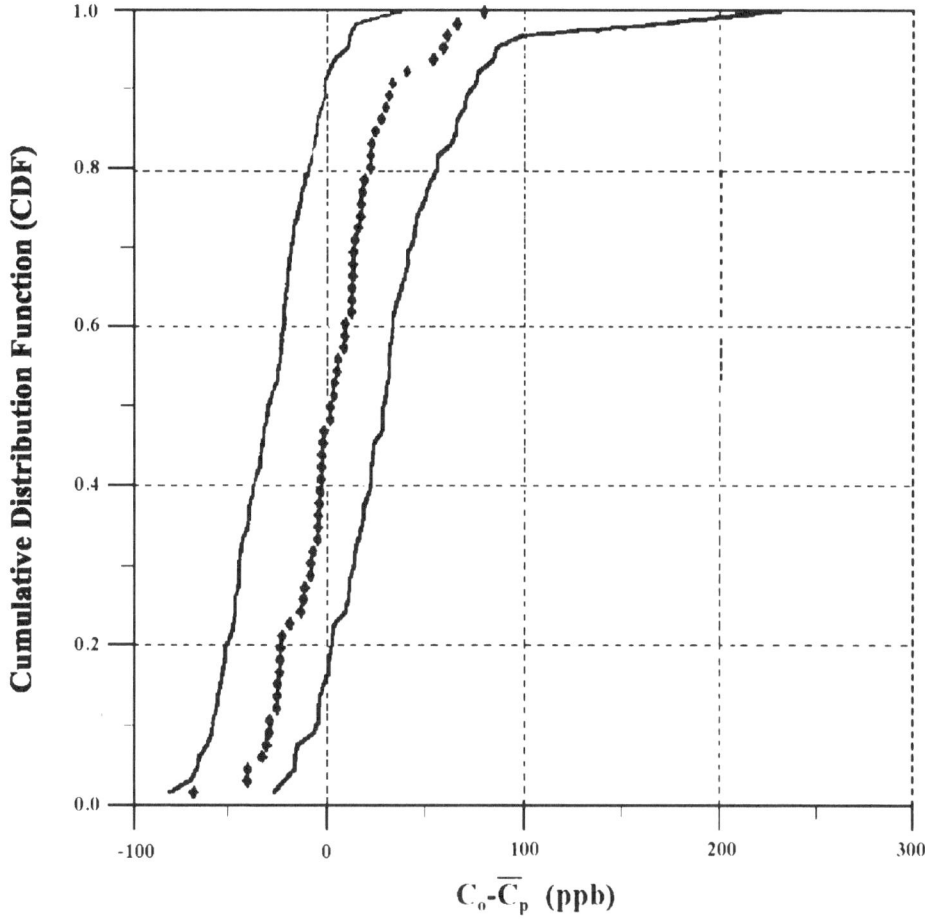

Figure 3

CDF for model ozone concentration residuals, $C_o - \bar{C}_p$, for 66 monitoring sites in the model domain for 10 July, 1995. C_o is observed concentration and $\bar{C}_p$ is predicted mean hourly averaged ozone concentration (ppb), and represent maximum values for that day at that monitoring site. Diamonds represent the actual observations. Solid lines denote the 95% confidence bounds, defined as the range of 100 alternate CDFs determined from predictions of 100 Monte Carlo runs (from HANNA and DAVIS, 2002).

confidence bounds of the CDFs of predicted model residuals determined from the 100 MC runs (each MC run gives a new CDF), it was concluded that the model was performing as well as expected. Though this evaluation accounts for the uncertainties in the input variables, it did not account for the model physics errors, stochastic fluctuations in concentrations, and errors in observed concentrations. Therefore, the actual concentration PDF would likely have a larger variance than the PDF used in this work.

5.1. Meteorological Data Errors

Atmospheric dispersion models such as SCIPUFF are intended for use over a wide range of distances from short-range (30 km) boundary layer scale to long-range (3000 km) mesoscale applications, and are generally driven by diagnostic (mass-consistent) wind field models for emergency response applications. The output quality of a diagnostic wind-field model depends directly on the quantity and quality of the input wind data. The uncertainty in the predicted concentration field also critically depends on the detail of the available wind field.

The SCIPUFF model accounts only for the stochastic uncertainty arising from the atmospheric turbulence. However, the model results will also depend on the uncertainties in input data and parameters, as well as on model physics errors. The errors in meteorological inputs result from the differences among various types of data (e.g., tower, aircraft, rawinsonde, wind profiler, etc.), averaging times, and representativeness. LEWELLEN and SYKES (1989) showed that the meteorological input uncertainty accounts for more than 50% of the total uncertainty in predicting the observed 1-hr GLCs with a second-order closure plume model. They found that the meteorological input error was dominated by the horizontal wind variance. In general, uncertainty in the transport wind direction (WEIL et al., 1992) is a major contributor to the total model error even at short distances over flat terrain (HANNA, 1988). Simulation of the effective path of mean transport of pollutants is even more difficult and uncertain in situations involving complex terrain (e.g., BANTA et al., 1996) and regional or long-range transport (KAHL and SAMSON, 1986; STOHL, 1998).

6. Uncertainty in Regulatory Air Quality Modeling

Air quality models used for regulation are generally designed to estimate the ensemble average concentrations, while the air quality standards focus on the extreme values of the distribution of observed pollutant concentrations, i.e., the rare event is more significant than the common event. Thus, there is an inherent uncertainty in using the modeling results in an absolute sense for attainment demonstrations. HOGREFE and RAO (2001) described a probabilistic approach for use in decision-making related to compliance with National Ambient Air Quality Standards (NAAQS) for ozone pollution. They utilized extreme value statistics (ROBERTS, 1979) and bootstrap resampling techniques (EFRON, 1982) to estimate the probability of exceeding the NAAQS for both 1-hr and 8-hr ozone concentrations. Exact theory of extreme values can be applied to calculate the CDF of a specified order statistic (GUMBEL, 1958), e.g., fourth-highest value in a sample of 1-hr ozone concentrations over a consecutive 3-year period. Instead of thinking of attainment process in a pass/fail mode in the regulatory framework, such an approach helps policy-makers in assessing the probability that a certain emission control strategy would lead to compliance with the NAAQS.

Much of the uncertainty work in the literature has dealt with the characterization of the effects of uncertainties in the data input to regulatory models based on Gaussian plume or puff dispersion. FREEMAN et al. (1986) used the analytical method, Eq. (15), to propagate the input data errors in a plume model. Uncertainties in emission rate and height, wind speed and direction, dispersion parameters, and mixing depth were considered for several stability classes and downwind distances. IRWIN et al. (1987) used MCM to relate the error bounds of meteorological data input to a Gaussian plume dispersion model to the uncertainty in the estimates of the maximum concentration and its downwind distance from the source.

RAO et al. (1985) applied a probabilistic approach to air quality model performance evaluation by determining the model's ability to simulate the tails of the CDF of the observed concentrations. Extreme value statistics and bootstrap resampling techniques were applied to develop confidence intervals for each percentile value of the CDF of observed tail concentrations, and then the CDF of predicted concentrations was superimposed over it. Air quality model performance evaluation consisted of checking whether the model predictions lie within the confidence intervals of observed concentrations at each probability point. The resampling technique was also applied to the differences between the observed and predicted concentrations to estimate the model uncertainty.

7. Conclusions

Atmospheric Dispersion is a stochastic phenomenon, and it is necessary to consider the concentration as a random variable that must be described in terms of probabilistic quantities such as ensemble mean, variance and probability distribution. The predicted distribution can be used to specify the probabilities of interest to the user, such as the probability of exceeding a TCV for a particular time period at a given location. It is now widely recognized that uncertainty analysis should be an essential part of modeling the dispersion of flammable and/ or hazardous substances in the atmosphere. This paper discussed the various uncertainties in atmospheric dispersion model predictions and outlined available techniques to quantify or reduce them. While this review does not claim to be thorough, an attempt is made to provide a comprehensive discussion of key issues and methods. Though space constraints did not permit detailed discussions, relevant references are provided for obtaining more information.

We have to make decisions in the face of uncertainty. Identification and estimation of uncertainties in predictions yield a deeper insight into the capabilities and limitations of atmospheric dispersion models, and increase our confidence in decision-making based on these models. They provide more credibility to the environmental risk assessment process, and indicate a proper direction for action or further investigation. Uncertainty analysis will enable the assessor to rank the atmospheric contaminants (and their pathways) more accurately, and to derive a subjective probability distribution about which CIs can be formed to represent the uncertainty in the risk. This information can be used to guide decisions. For example, if a 5% (lower) CI concentration value is above a regulatory standard or threshold value of concern, then appropriate remedial measures (e.g., emission controls, evacuations, etc.) are probably needed. If the 95% (upper) CI is below the standard, remedial action is probably not required. If the 95% upper CI is above the standard but the 50th percentile is below the standard, further study should be recommended on the key parameters that dominate the overall uncertainty. If the 50th percentile is also above the standard, further study might still be recommended, but one might also proceed with cost-effective remedial measures for risk reduction. Thus, incorporation of uncertainty analysis into atmospheric dispersion modeling provides a valuable tool for decision-making and optimal use of resources.

REFERENCES

BANTA, R.M., OLIVIER, L.D., GUDIKSEN, P.H., and LANGE, R. (1996), *Implications of Small-Scale Flow Features to Modeling Dispersion over Complex Terrain*, J. Appl. Meteor. 35, 330–342.
BERGIN, M.S. and MILFORD, J.B. (2000), *Application of Bayesian Monte Carlo Analysis to a Lagrangian Photochemical Air Quality Model*, Atmos. Environ. 34, 781–792.
BOX, G.E.P. and TIAO, G.C., *Bayesian Inference in Statistical Analysis* (Wiley-Interscience, New York 1973).
CHATWIN, P.C. (1982), *The Use of Statistics in Describing and Predicting the Effects of Dispersing Gas Clouds*, J. Hazard. Mater. 6, 213–230.
COLLINS, D.C. and AVISSAR, R. (1994), *An Evaluation with the Fourier Amplitude Sensitivity Test (FAST) of which Land-Surface Parameters are of Greatest Importance in Atmospheric Modeling*, J. Climate 7, 681–703.
COOK, R.M., GOOSSENS, L.J.H., and KRAAN, B.C.P. (1994), *Methods for CEC/USNRC Accident Consequence Uncertainty Analysis of Dispersion and Deposition*, EUR-15856EN, Department of Mathematics, Delft University of Technology, The Netherlands, 209 pp.
CSANADY, G.T. (1967), *Concentration Fluctuations in Turbulent Diffusion*, J. Atmos. Sci. 24, 21–28.
CSANADY, G.T., *Turbulent Diffusion in the Environment* (Reidel Publishing Co., Boston 1973).
CUCKIER, R.I., LEVINE, H.B., and SHULER, K.E. (1978), *Nonlinear Sensitivity Analysis of Multiparameter Model Systems*, J. Comput. Phys. 26, 1–42.
CULLEN, A.C. and FREY, H.C., *Probabilistic Techniques in Exposure Assessment: A Handbook for Dealing with Variability and Uncertainty in Models and Inputs* (Plenum Press, New York 1999).
DABBBERDT, W.F. and MILLER, E. (2000), *Uncertainty, Ensembles and Air Quality Dispersion Modeling: Applications and Challenges*, Atmos. Environ. 34, 4667–4673.
DEARDORFF, J.W. and WILLIS, G.E. (1984), *Ground Level Concentration Fluctuations from a Buoyant and a Non-buoyant Source within a Laboratory Convectively Mixed Layer*, Atmos. Environ. 18, 1297–1309.
DOWNING, D. J., GARDNER, R.H., and HOFFMAN, F.O. (1985), *Response Surface Methodologies for Uncertainty Analysis in Assessment Models*, Technometrics 27, 151–163.
DUNKER, A. M. (1981), *Efficient Calculation of Sensitivity Coefficients for Complex Atmospheric Models*, Atmos. Environ. 15, 1155–1161.

EFRON, B. *The Jackknife, the Bootstrap, and Other Resampling Plans*, CBMMS-NSF-38 (Soc. for Indust. and Appl. Math., Philadelphia 1982).

EHRHARDT, J., PASLER-SAUER, J., SCHULE, O., BENZ, G., RAFAT, M., and RICHTER, J. (1993), *Development of RODOS, a Comprehensive Decision Support System for Nuclear Emergencies in Europe–An Overview*, Radiation Protection Dosimetry *50*, 195–203.

FACKRELL, J.E. and ROBINS, A.G. (1982), *Concentration Fluctuations and Fluxes in Plumes from Point Sources in a Turbulent Boundary Layer*, J. Fluid Mech. *117*, 1–26.

FISHER, B. (2003), *Fuzzy Environmental Decision-Making: Applications to Air Pollution*, Atmos. Environ. *37*, 1865–1877.

FOX, D.G. (1984), *Uncertainty in Air Quality Modeling*, Bull. Amer. Meteor. Soc. *65*, 27–36.

FREEMAN, D.L., EGAMI, R.T., ROBINSON, N.F., and WATSON, J.G. (1986), *A Method for Propagating Measurement Uncertainties through Dispersion Models*, J. Air Poll. Cont. Assoc. *36*, 246–253.

FRENCH, S. (1997), *Source Term Estimation, Data Assimilation and Uncertainties, Sixth Topical Meeting on Emergency Preparedness and Response*, San Francisco, Proceedings, Vol. II, American Nuclear Society, 427–430.

GARDNER, R.H., O'NEILL, R.V., MANKIN, J.B., and CARNEY, J.H. (1981), *A Comparison of Sensitivity Analysis and Error Analysis Based on a Stream Ecosystem Model*, Ecological Modeling *12*, 173–190.

GIFFORD, F.A. (1959), *Statistical Properties of a Fluctuating Plume Dispersion Model*, Adv. Geophys. *6*, 117–138.

GUMBEL, E.J., *Statistics of Extremes* (Columbia University Press, New York 1958)

HANNA, S.R. (1984), *The Exponential Probability Density Function and Concentration Fluctuations in Smoke Plumes*, Bound.-Layer Meteor. *29*, 361–375.

HANNA, S.R. (1988), *Air Quality Model Evaluation and Uncertainty*, J. Air Poll. Cont. Assoc. *38*, 406–412.

HANNA, S.R., CHANG, J.C., and FERNAU, M.E. (1998), *Monte Carlo Estimates of Uncertainties in Predictions by a Photochemical Grid Model (UAM-IV) due to Uncertainties in Input Variables*, Atmos. Environ. *32*, 3619–3628.

HANNA, S.R., and DAVIS (2002), *Evaluation of a Photochemical Grid Model using Estimates of Concentration Probability Density Functions*, Atmos. Environ. *36*, 1793–1798.

HARPER, F.T., GOOSSENS, L.H., COOKE, R.M., HELTON, J.D., HORA, S.C., JONES, J.A., KRAAN, B., LUI, C., MCKAY, M.D., MILLER, L.A., PASLER-SAUER, J., and YOUNG, M.L. (1994), *Probabilistic Accident Consequence Uncertainty Analysis: Dispersion and Deposition Uncertainty Assessment*, Volumes 1, 2, 3, NUREG/CR-6244, U.S. Nuclear Regulatory Commission, Washington, DC.

HOFFMAN, F.O. and HAMMONDS, J.S. (1992), *An Introductory Guide to Uncertainty Analysis in Environmental and Health Risk Assessment*, ES/ER/TM-35, Environmental Sciences Div., Oak Ridge Natl. Lab., Oak Ridge, TN.

HOGREFE, C. and RAO, S.T. (2001), *Demonstrating Attainment of the Air Quality Standards: Integration of Observations and Model Predictions into the Probabilistic Framework*, J. Air and Waste Managem. Assoc. 51, 1060–1072.

HORA, S.C. and IMAN, R.L. (1989), *Expert Opinion in Risk Analysis: the NUREG-1150 Methodology*, Nuclear Sci. Engin. *102*, 323–331.

IAEA, *Evaluating the Reliability of Predictions Made using Environmental Transfer Models* (Safety Series No. 100, Intern. Atomic Energy Agency, Vienna 1989) 106 pp.

IMAN, R.L. and CONOVER, W.J. (1980), *Small Sample Sensitivity Analysis Techniques for Computer Models, with an Application to Risk Assessment*, Communications in Statistics. Part A. Theory and Methods 17, 1749–1842.

IMAN, R.L. and HELTON, J.C. (1988), *An Investigation of Uncertainty and Sensitivity Analysis Techniques for Computer Models*, Risk Analysis 8, 71–90.

IMAN, R.L. and SHORTENCARIER, M.J. (1984), *A FORTRAN 77 Program and User's Guide for the Generation of Latin Hypercube and Random Samples for Use with Computer Models*, NUREG/CR-3624, SAND83-2365, Sandia National Labs., Albuquerque, NM.

IRWIN, J.S., RAO, S.T., PETERSEN, W.B., and TURNER, D.B. (1987), *Relating Error Bounds for Maximum Concentration Estimates to Diffusion Meteorology Uncertainty*, Atmos. Environ. *21*, 1927–1937.

ISUKAPALLI, S.S., ROY, A., and GEORGEOPOULOS, P.G. (1998), *Stochastic Response Surface Methods (SRSMs) for Uncertainty Characterization and Propagation: Application to Environmental and Biological Systems*, Risk Analysis *18*, 351–363.

KAHL, J.D. and SAMSON, P.J. (1986) *Uncertainty in Trajectory Calculations due to Low Resolution Meteorological Data*, J. Clim. Appl. Meteor. *25*, 1816–1831.

KOCHER, D.C., WARD, R.C., KILLOUGH, G.G., DUNNING, D.E., HICKS, B.B., HOSKER, R.P., KU, J.-Y., and RAO, K.S. (1987), *Sensitivity and Uncertainty Studies of the CRAC2 Computer Code*, Risk Analysis *7*, 497–507.

LEWELLEN, W.S. and SYKES, R.I. (1989), *Meteorological Data Needs for Modeling Air Quality Uncertainties*, J. Atmos. Ocean. Tech. *6*, 759–768.

LEWIS, D.M. and CHATWIN, P.C. (1997), *A Three-Parameter PDF for the Concentration of an Atmospheric Pollutant*, J. Appl. Meteor. *36*, 1064–1075.

MCKAY, M.D., BECKMAN, R.J., and CONOVER, W.J. (1979), *A Comparison of Three Methods for Selecting Values of Input Variables in the Analysis of Output from a Computer Code*, Technometrics *21*, 239–245.

MCRAE, G.J., TILDEN, J.W., and SEINFELD, J.H. (1982), *Global Sensitivity Analysis—A Computational Implementation of the Fourier Amplitude Sensitivity Test (FAST)*, Comp. Chem. Eng. *6*, 15–25.

MORGAN, M.G. and HENRION, M., Uncertainty: *A Guide to Dealing with Uncertainty in Quantitative Risk and Policy Analysis* (Cambridge Univ. Press, New York 1990).

MOSCA, S., GRAZIANI, G., KLUG, W., BELLASIO, R., and BIANCONI, W. (1998), *A Statistical Methodology for the Evaluation of Long-Range Dispersion Models: An Application to the ETEX Exercise*, Atmos. Environ. *32*, 4307–4324.

MYLNE, K.R. and MASON, P.J. (1991), *Concentration Fluctuation Measurements in a Dispersing Plume at a Range of up to 1000 m*, Quart. J. Roy. Met. Soc. *117*, 117–206.

PANWAR, T.S., COWAN, I.R., and BRITTER, R.E. (1994), *Concentration Fluctuation Models: A Comparative Study*, Technical Report CUED/A-AERO/TR *24*, Engineering Dept., Cambridge Univ., 42 pp.

RABITZ, H., KRAMER, M., and DACOL, D. (1983), *Sensitivity Analysis in Chemical Kinetics*, Ann. Rev. Phys. Chem. *34*, 419–461.

RAO, S.T., SISTLA, G., PAGNOTTI, V., PETERSEN, W.B., IRWIN, J.R., and TURNER, D.B. (1985), *Resampling and Extreme Value Statistics in Air Quality Model Performance Evaluation*, Atmos. Environ. *19*, 1503–1518.

RAO, K.S. and HOSKER, R.P. (1993), *Uncertainty in the Assessment of Atmospheric Concentrations of Toxic Contaminants from an Accidental Release*, Radiation Protection Dosimetry *50*, 281–288.

ROBERTS, E.M. (1979), *Review of Statistics of Extreme Values with Applications to Air Quality Data*, Part I. Review, J. Air Poll. Cont. Assoc. *29*, 632–637.

SAWFORD, B.L. (1987), *Conditional Concentration Statistics for Surface Plumes in the Atmospheric Boundary Layer*, J. Clim. Appl. Meteor. *24*, 1152–1166.

SCHOPFLOCHER, T.P. and SULLIVAN, P.J. (2002), *A Mixture Model for the PDF of a Diffusing Scalar in a Turbulent Flow*, Atmos. Environ. *36*, 4405–4417.

STOHL, A. (1998), *Computation, Accuracy and Application of Trajectories—A Review and Bibliography*, Atmos. Environ. *32*, 947–966.

SYKES, R.I., LEWELLEN, W.S., PARKER, S.F., and HENN, D.S. (1989), *A Hierarchy of Dynamic Plume Models Incorporating Uncertainty Volume 4: Second-Order Closure Integrated Puff*, EPRI EA-6095 Volume 4, Project 1616-28, Electric Power Research Institute, Palo Alto, CA.

TATANG, M.A., PAN, W., PRINN, R,G., and MCRAE, G.J. (1997), *An Efficient Method for Parametric Uncertainty Analysis of Numerical Geophysical Models*, J. Geophys. Res. *102* (D18), 21,925–21,932.

ULIASZ, M. (1988), *Application of the FAST Method to Analyze the Sensitivity–Uncertainty of a Lagrangian Model of Sulphur Transport in Europe*, Water, Air and Soil Pollut. *40*, 33–49.

VENKATRAM, A. (1982), *A Framework for Evaluating Air Quality Models*. Bound.-Layer Meteor. *24*, 371–385.

YEE, E., KOSTENIUK, P.R., CHANDLER, G.M., BILTOFT, C.A., and BOWERS, J.F. (1993), *Statistical Characteristics of Concentration Fluctuations in Dispersing Plumes in the Atmospheric Surface Layer*, Bound.-Layer Meteor. *65*, 69–109.

YEE, E. and CHAN, R. (1997), *A Simple Model for the Probability Density Function of Concentration Fluctuations in Atmospheric Plumes*, Atmos. Environ. *31*, 991–1002.

WEIL, J. C., SYKES, R.I., and VENKATRAM, A. (1992), *Evaluating Air Quality Models: Review and Outlook*, J. Appl. Meteor. *31*, 1121–1145.

WARNER, S., PLATT, N., and HEAGY, J.F. (2001), *User-Oriented Measures of Effectiveness for the Evaluation of Transport and Dispersion Models*, Proc. of Seventh Intern. Conf. on Harmonisation within Atmospheric Dispersion Modelling for Regulatory Purposes, Belgirate, Italy, 24–29.

(Received November 15, 2003, accepted April 14, 2004)
Published Online First June 21, 2005

 To access this journal online:
http://www.birkhauser.ch

Pure appl. geophys. 162 (2005) 1919–1939
0033–4553/05/101919–21
DOI 10.1007/s00024-005-2698-3

© Birkhäuser Verlag, Basel, 2005

❘ Pure and Applied Geophysics

Assessing the Comparability of Ammonium, Nitrate and Sulfate Concentrations Measured by Three Air Quality Monitoring Networks

EDITH L. GEGO,[1] P. STEVEN PORTER,[2] JOHN S. IRWIN,[3] CHRISTIAN HOGREFE,[4]
S. TRIVIKRAMA RAO[3]

Abstract—Airborne fine particulate matter across the United States is monitored by different networks, the three prevalent ones presently being the Clean Air Status and Trend Network (CASTNet), the Interagency Monitoring of PROtected Visual Environment Network (IMPROVE) and the Speciation and Trend Network (STN). If combined, these three networks provide speciated fine particulate data at several hundred locations throughout the United States. Yet, differences in sampling protocols and samples handling may not allow their joint use. With these concerns in mind, the objective of this study is to assess the spatial and temporal comparability of the sulfate, nitrate and ammonium concentrations reported by each of these networks. One of the major differences between networks is the sampling frequency they adopted. While CASTNet measures pollution levels on seven-day integrated samples, STN and IMPROVE data pertain to 24-hour samples collected every three days. STN and IMPROVE data therefore exhibit considerably more short-term variability than their CASTNet counterpart. We show that, despite their apparent incongruity, averaging the data with a window size of four to six weeks is sufficient to remove the effects of differences in sampling frequency and duration and allow meaningful comparison of the signals reported by the three networks of concern. After averaging, all the sulfate and, to a lesser degree, ammonium concentrations reported are fairly similar. Nitrate concentrations, on the other hand, are still divergent. We speculate that this divergence originates from the different types of filters used to collect particulate nitrate. Finally, using a rotated principal component technique (RPCA), we determined the number and the geographical organization of the significant temporal modes of variation (clusters) detected by each network for the three pollutants of interest. For sulfate and ammonium, the clusters' geographical boundaries established for each network and the modes of variations within each cluster seem to correspond. RPCA performed on nitrate concentrations revealed that, for the CASTNet and IMPROVE networks, the modes of variation do not correspond to unified geographical regions but are found more sporadically. For STN, the clustered areas are unified and easily delineable. We conclude that the possibility of jointly using the data collected by CASTNet, IMPROVE and STN has to be weighed pollutant by pollutant. While sulfate and ammonium data show some potential for joint use, at this point, combining the nitrate data from these monitoring networks may not be a judicious choice.

Key words: Air quality, particulate matter, monitoring networks, moving average, principal component analysis.

[1]University Corporation for Atmospheric Research, Idaho Falls, ID, 84401, USA. E-mail: e.gego@onewest.net

[2]Department of Civil Engineering, University of Idaho, Idaho Falls, ID, 83401, USA

[3]NOAA Atmospheric Sciences Modeling Division, On Assignment to the U.S. Environmental Protection Agency, Research Triangle Park, NC, 27711, USA

[4]Atmospheric Sciences Research Center, University at Albany, Albany, NY, 12222, USA

1. Introduction

Depending on its size, air-borne particulate material is commonly divided into two classes: fine and coarse. The fine particles are those whose diameter is less than or equal to 2.5 micrometers (μm), justifying the acronym '$PM_{2.5}$'. The particles exceeding this threshold but whose diameter is less than or equal to 10 μm constitute the coarse class size ('$PM_{coarse} = PM_{10} - PM_{2.5}$'). $PM_{2.5}$ is mostly composed of secondary particles, i.e., particles that are not directly emitted in the atmosphere but are formed from primary gaseous emissions. Among the important $PM_{2.5}$ constituents are sulfates, nitrates and ammonium. Sulfate has its origins in SO_2 emissions from power plants and industrial facilities; nitrates are formed by the oxidation of the nitrous oxides (NOx) emitted from power plants, automobiles and other types of combustion sources, and ammonium predominantly originates from human and animal wastes and agriculture (fertilizer application). New research shows that, while air-borne particulate sulfate and ammonium are indeed mostly present in the fine particle class ($PM_{2.5}$), a substantial portion of nitrate can be found in the coarse class size (PM_{10}), especially in coastal areas (ZHUANG *et al.*, 1999; CAMPBELL *et al.*, 2002). Growing concerns about the adverse effects of particulate matter on human health and the visual quality of the environment justify the recently promulgated particulate matter National Ambient Air Quality Standards (NAAQS) for $PM_{2.5}$ and PM_{10} (see http://www.epa.gov/ttn/naaqs). Should a NAAQS be violated, it is important to know the chemical composition of particulate material so as to identify its origin and develop meaningful emission control strategies to attain the NAAQS for $PM_{2.5}$.

Currently, speciation information is provided by several networks; the three prominent ones being the Clean Air Status and Trend Network (CASTNet), the Interagency Monitoring of PROtected Visual Environment Network (IMPROVE) and, more recently, the Speciation and Trend Network (STN). In addition to these three networks, there are other smaller networks, such as the Atmospheric Integrated Research Monitoring Network developed by the National Oceanic and Atmospheric administration (see http://www.arl.noaa.gov/research/programs/airmon.html) that collect the same type of information. Combining as many air quality data as possible to obtain the most accurate and aerially extensive representation of the atmosphere would be helpful for evaluation of the performance of regional-scale air quality models, primary tools used for designing control strategies aimed at meeting and maintaining the above-mentioned NAAQS. If combined, all networks provide speciated air quality data at several hundred locations throughout the United States. However, differences in network sampling protocols and samples handling may not allow their joint use. With these concerns in mind, the objective of this study is to assess the spatial and temporal comparability of the sulfate, nitrate and ammonium concentrations reported by the three prominent networks (CASTNet, IMPROVE and STN). Although some authors have investigated differences between CASTNet and IMPROVE observations (AMES and MALM, 2001), to our knowledge, no study of the comparability of these three networks (CASTNet, IMPROVE and STN) is available yet. Our objective is to uncover similarities and differences among the reported observations. By doing so, we hope to inform about the difficulties ahead when blending data from multiple networks with different sampling protocols. It is not our goal to provide a universal recipe for performing this task.

While the concentrations reported by IMPROVE and STN reflect one-in-three days 24-hour air samples, CASTNet observations describe one-week integrated samples. Since a shorter sampling interval leads to a higher amount of short-term variability in the time series of measurements and *vice versa*, the data from the different networks need to be somehow harmonized before they can meaningfully be combined. One means of harmonizing the data is by calculating temporal averages so that the averaged data from all networks contain similar information regarding temporal changes. We intend to determine the shortest averaging time interval after which fluctuations of data gathered by the three networks may become comparable. Second, a rotated principal component analysis (RPCA) will be used to summarize the regional organization of each contaminant, as can be assessed with each network. Finally, the temporal patterns recorded by each network in corresponding geographical areas will be compared.

2. Brief Description of Monitoring Networks and their Sampling Protocols

Created in 1990 to measure dry deposition fluxes, CASTNet now comprises over 70 monitoring sites in the United States, located mostly in rural areas. The air sampler at a CASTNet site is a non size-selective three-stage filter pack located 10 m above ground level, continuously supplied with a 1.5 l/min flow rate in the eastern United States (U.S.) and 3 l/min in the western U.S. Unlike the IMPROVE protocols, filters are not equipped with any particle size limiting device. Yet, the flow rate utilized and the height of the instrument are thought adapted to limit the entrance of coarse particles into the filter (FINKELSTEIN, 2003, personal communication). Filters are changed every week; measured concentrations, therefore, are seven-day average estimates. The nitrate, sulfate and ammonium ions, collected on the first of the three consecutive filters, composed of teflon, are interpreted as particulate species. Nitrate and sulfate are quantified by ion chromatography and ammonium concentrations are estimated by the indophenol method. Presently, all CASTNet concentrations are standardized to a temperature of 25°C and a pressure of 1013 mb before being reported. A new database providing CASTNet concentrations in ambient conditions is currently being built (FINKELSTEIN, 2003, personal communication).

Initiated in 1985, the IMPROVE network essentially aims at monitoring air quality conditions in Class I areas, i.e., in national parks and wilderness areas that receive special protection from adverse air quality impacts through the U.S. Environmental Protection Agency's Prevention of Significant Deterioration (PSD) program (U.S. EPA, 1980). The air sampler at IMPROVE sites consists of 4 modules located 3 m above ground level and equipped with a device that stops particles larger than 2.6 μm. Sulfate concentration is calculated by stoichiometry from the mass of sulfur extracted from a teflon filter and analyzed for by X-Ray fluorescence. Nitrate is determined from particles extracted from a nylon filter preceded by an acidic vapor diffusion denuder which eliminates nitric acid vapor (non-particulate nitrate). Nitrates are determined by ion chromatography. Ammonium concentrations at IMPROVE sites are not determined directly but are calculated by stoichiometry, assuming that all the sulfates and nitrates in the particulate phase have been neutralized by ammonium. A 24-hour integrated air sample is collected every three days. Measured concentrations are reported at ambient temperature and pressure conditions, in contrast to CASTNet.

Established by the USEPA to supplement $PM_{2.5}$ mass estimates provided by the Federal Reference Method (FRM) network, the STN network began operation in late 1999. Contrary to CASTNet and IMPROVE, STN sites are located in urban, suburban, and rural environments. The data they provide will allow, among other things, assessment of trends in fine particles in urban areas across the country. Eventually, the number of STN sites will surpass the combined number of IMPROVE and CASTNet sites. At this early stage of its development, the network is not as simply describable as CASTnet and IMPROVE networks. A variety of air samplers and sampling protocols has been approved for use in the STN while the other networks use identical equipment at all their sites and standard analytical techniques. It appears, though, that the STN sampling methodology resembles that of IMPROVE. As with IMPROVE, a one-in-three days sampling schedule has been adopted. Nitrates are extracted from a nylon (or quartz filter), as is the case at IMPROVE sites, rather than the teflon filter used at CASTNet sites. Nitrates are determined by ion chromatography. Depending on the sampling equipment, sulfates are extracted from a teflon, nylon or quartz filter and analyzed by ion chromatography. Unlike IMPROVE sites, ammonium is determined directly via ion chromatography from the same filter used to determine sulfates. Measured concentrations are reported at ambient temperature and pressure conditions.

Further details about the sampling protocols utilized by each network, as well as the data they provide, are available at http://vista.cira.colostate.edu/IMPROVE/, http://www.epa.gov/CASTNet and http://www.epa.gov/oar/oaqps/pm25 for the IMPROVE, CASTNet, and STN, respectively.

3. Methods

3.1. Evaluation of an Appropriate Averaging Time Interval for Removal of the Effects of Distinct Sampling Frequency and Duration

As stated earlier, the frequency and duration of air sample collection at CASTNet sites are different from those used at IMPROVE and STN sites. While the latter two provide non-consecutive (one-in-three days) 24-hour samples, CASTNET data represent seven-day averages. Consequently, the corresponding contaminant time series are very distinct, with the IMPROVE and STN data exhibiting more short-term variability than these of CASTNet. Using an interative moving average filter such as described in RAO *et al.* (1997) and HOGREFE *et al.* (2000) to separate variation at frequencies less than 2.5 months^{-1} from those at frequencies greater than 2.5 months^{-1}, GEGO *et al.* (2003) showed that, despite their apparent incongruity, the low frequency signals embedded in the sulfate time series reported by CASTNet and IMPROVE are comparable.

We propose to extend the results of GEGO *et al.* (2003) by identifying the shortest temporal averaging interval that minimizes the effects of different sampling durations and frequency. After identification of the shortest averaging interval and construction of the temporally averaged signals, we believe that the remaining differences between the three networks are no longer attributable to differences in the sampling frequency, but result from differences in site locations, instrumentation used or analytical techniques employed. There is no indicated method for precisely determining that shortest averaging time interval. In this study, we calculated, for each contaminant and network, the variance of the measurement averages corresponding to window sizes of 1, 2, 4, 6 and 8 weeks. Averaging may be considered sufficient when the variance of the IMPROVE and STN average signals no longer exceed that of CASTNET average signals. We also calculated, for each pair of networks and each window size, the correlation coefficient, the slope of the regression line, and the average and maximum difference between signals. The slope of the regression line between a pair of networks is not expected to vary as the window size increases, as it indicates systematic differences caused by differences in sampling location or instrumentation. On the other hand, increasing the averaging time is expected to increase the strength of the correlation between the respective signals and reduce the amplitude of their differences by reducing the differences in the amount of short-term variability introduced by their respective sampling frequencies.

3.2. Rotated Principal Component Analysis

Principal component analysis (PCA) is a multivariate technique designed to facilitate interpretation of large data sets involving numerous mutually dependent variables. By summarizing the correlations (i.e., identifying the redundancies) between all variables, PCA allows determination of the 'true' dimensionality of a data set. It also allows building of a new data set (the principal components data set) whose dimensions reflect the true dimensionality of the original data set and whose variables are mutually orthogonal. EDER (1989) provided insights on how to use PCA to analyze and summarize the temporal correlation of time series of a given air contaminant measured at numerous monitored sites. In EDER's (1989) approach, a sample individual corresponds to a sampling event (date) and a variable is a monitoring site. PCA used in this framework allows classification of all monitoring sites into a limited number of categories, each of which corresponds to a specific contaminant's temporal evolution (specific succession of rises, falls and plateaus), i.e., a specific mode of variation.

Practically speaking, PCA begins with the construction of the correlation (or covariance) matrix summarizing the site-to-site correlation between all pairs of sites. For this study, all observations at a given site were standardized to zero mean and unit variance before evaluation of the correlation matrix. This procedure is thought to limit the impact of heteroscedasticity (inequality of variances) and facilitate interpretation of results (PALM, 1999).

After determination of the eigenvectors and eigenvalues of the correlation matrix (KENDALL *et al.*, 1983), the principal components were calculated. The first principal component (PC) is obtained by multiplying the original set of variables by the first eigenvector of the correlation matrix. Its orientation is such that it maximizes the part of the variance of the original data that can be explained by a single variable. The second and higher order PCs are defined in similar fashion. The second PC, for instance, is obtained by multiplying the original variables by the second eigenvector of the correlation matrix. The variance of the second PC is the second eigenvalue of the correlation matrix, and so on. Since higher order eigenvalues are progressively smaller, successive PCs explain less and less of the variance of the original data.

One may consider that the information included in the original data set can be reasonably described by a limited number of PCs. The number of PCs retained is representative of the true dimensionality of the original data set. In our case, it also represents the number of 'distinct modes of variations' or the number of clusters we wish to differentiate in the data set. There are several methods for deciding the number of PCs to retain, among them the "Rule N" method (OVERLAND and PREISENDORFER, 1982), and the Scree test (CATTELL, 1966; WILKS, 1995). No one approach is thought superior to the others. In this study, the number of clusters retained for each air pollutant and network is the number of eigenvalues greater than 1 (EDER, 1989).

Orthogonally rotating the PCs retained so as to increase their correlation with the original data, a procedure often referred to as varimax (KAISER, 1958), has been shown to facilitate interpretation of the principal components (HOREL, 1981). We, therefore, chose to use it as well. The successive use of PCA to determine the number of PCs to retain and of varimax to better segregate the variables (in this case, the monitoring sites) is often referred to as rotated principal component analysis

(RPCA). Monitoring stations where nitrate, sulfate or ammonium concentrations fluctuate in a similar manner (i.e., that are grouped in the same cluster) are those that are more correlated with a given rotated principal component than with the others.

4. Data

This study utilizes the nitrate, sulfate and ammonium concentrations reported by IMPROVE, CASTNet and STN at sites located east of 100° longitude west (eastern U.S.), from July 1st, 2001 to July 31st, 2002. Only those sites with less than 20% missing values were retained. Because RPCA cannot handle missing data, missing data at a given site were substituted for using a temporal linear interpolation scheme. Although not exceeding the number of missing data threshold, several clustered STN sites in the urban corridor between Pennsylvania to Massachusetts were excluded from the analysis to reduce overrepresentation of the urban corridor in the principal component analysis. Also, due to the unavailability of pertinent meteorological information (pressure), the concentrations reported by CASTnet were not converted into 'ambient conditions', a condition that would simplify their comparison with IMPROVE and STN data. A total of 51 CASTNet sites, 39 IMPROVE sites and 26 STN sites were utilized in the RPCA. Sites used in the study are presented in Figure 1.

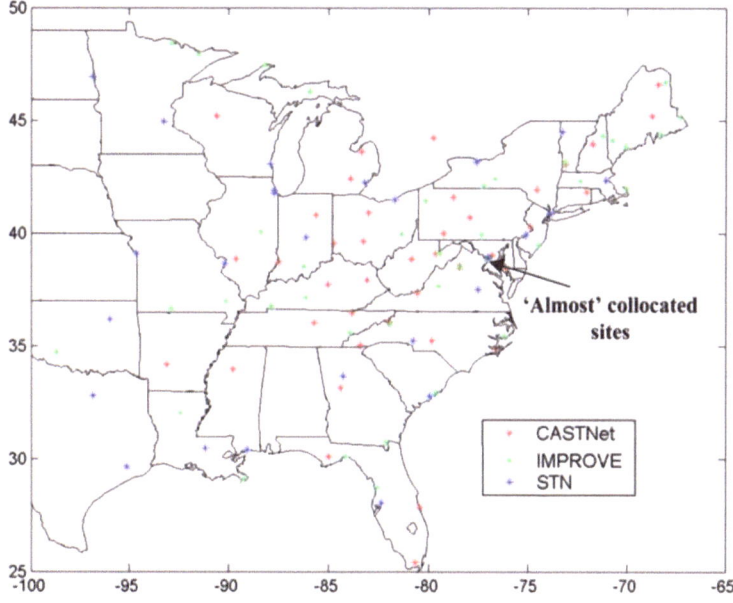

Figure 1
Location of the monitoring sites.

5. Results

5.1. Evaluation of an Appropriate Averaging Time Interval

Ideally, the appropriate averaging time interval for the removal of short-term variability linked to their specific sampling frequencies should be determined by analysis of signals recorded by the three networks at collocated sites. In this case, indeed, the differences between the signals are only attributable to dissimilarities in sampling frequency and sampling equipment, and not to spatial variability. While there are instances of collocated IMPROVE and CASTNet sites, presently there are no sites where all three networks are represented. As the best alternative, for this study, we identified the set of the three closest CASTNet, IMPROVE and STN stations and used the data collected at these stations to evaluate an appropriate averaging time. See Figure 1 for the location of these sites and Table 1 for specification of their respective names and coordinates.

Justifying the need for averaging the information reported by all networks, Figure 2 depicts a scatter plot of weekly nitrate, sulfate and ammonium concentrations at the almost collocated CASTNet, IMPROVE and STN sites during the time period studied. Weekly IMPROVE and STN estimates were obtained by averaging

Table 1

Names and locations of the 'quasi-collocated' sites

Network	Station name or number	Location	Longitude	Latitude
CASTNet	BEL116	Beltsville, Maryland	−76.8172	39.0284
IMPROVE	WASN1	District of Columbia	−77.0343	38.8761
STN	110010043	District of Columbia	−77.0125	38.9188

The distance between the CASTNet and IMPROVE sites is 25 km; the distance between the STN and IMPROVE sites is five km.

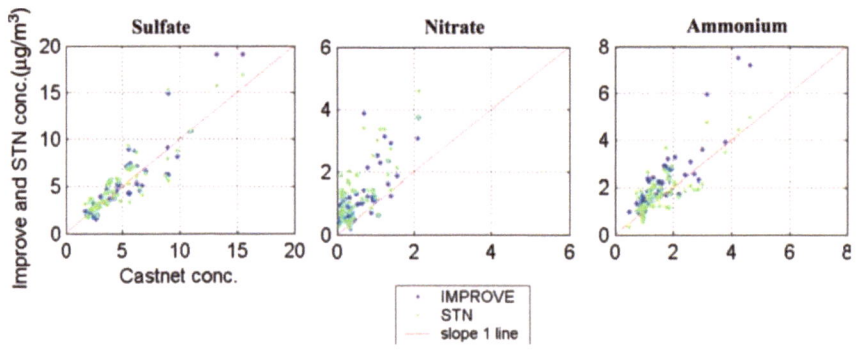

Figure 2

Scatter plots of IMPROVE and STN weekly average concentrations ($\mu g/m^3$) versus CASTnet concentrations ($\mu g/m^3$).

Table 2

Evolution of the variance $(\mu g/m^3)^2$ of the moving averaged signals for different averaging intervals

Network	Averaging interval				
	1 wk	2 wk	4 wk	6 wk	8 wk
Sulfate					
Castnet	8.25	6.23	5.37	4.42	3.47
Improve	12.81	10.27	8.07	6.26	4.82
STN	10.05	7.76	6.00	4.76	3.64
Nitrate					
Castnet	0.27	0.19	0.17	0.16	0.15
Improve	0.62	0.51	0.44	0.38	0.35
STN	1.45	1.17	0.87	0.38	0.63
Ammonium					
Castnet	0.73	0.52	0.43	0.34	0.25
Improve	1.65	1.27	0.95	0.69	0.50
STN	0.80	0.55	0.37	0.26	0.18

the 24-hour observations that fall within a given CASTNet week. The significant dispersion observable in Figure 2 proves that, at the time scale of a week, the three networks provide different information and that this information needs to be harmonized if the networks are to be jointly used. Table 2 shows the effects of averaging on the variances of the signals reported by the three networks. It is obvious that averaging tends to homogenize the variability in the data from all networks.

In the case of sulfate (Tables 3a and 3b), the CASTNet signals tend to be slightly less than those of STN, which are themselves slightly less than IMPROVE (Table 3a). These differences may be the result of different local environments. Indeed, although close in space, the local settings of these monitors are quite different. While the CASTNet site is in range-land, the IMPROVE site is in Washington D.C. and the STN station is classified as 'Urban and Center City'. Differences between recorded signals are therefore to be expected. Despite the various local environments, for averaging intervals of 4 weeks and longer, the mean relative difference between all observations is less than 15% and the correlation (R) more than 95% (Table 3b). Since our estimation is supported by examination of only

Table 3a

Slope of the regression lines of sulfate concentrations recorded by each pair of networks

Networks	Slope of the regression line
Improve vs Castnet	1.10
Stn vs Castnet	1.05
STN vs Improve	0.95

Table 3b

Absolute values of the mean and the largest relative differences (%) between sulfate concentrations recorded by each pair of networks, as a function of the averaging time interval

Network	Averaging interval				
	1 wk	2 wk	4 wk	6 wk	8 wk
Absolute value of the mean relative difference (%)					
Improve vs Castnet	22	17	15	11	12
Stn vs Castnet	23	17	13	11	11
STN vs Improve	15	13	12	8	5
Absolute value of the largest relative difference (%)					
Improve vs Castnet	68	46	31	22	20
Stn vs Castnet	80	51	30	28	25
STN vs Improve	86	45	29	20	10
Correlation coefficient					
Improve vs Castnet	.90	.95	.96	.97	.96
Stn vs Castnet	.88	.98	.99	.99	.99
STN vs Improve	.97	.98	.98	.99	.99

a single set of 'quasi-collocated' sites, our results are quite uncertain and should be reconfirmed with additional sets of collocated sites, when available.

The encouraging sulfate results seen for 4-week averaging do not apply to nitrate. Although the correlation between signals is quite high after averaging the observations, CASTNet nitrate estimates are always much less (50%) than their STN and IMPROVE counterparts, regardless of the averaging interval (Tables 4a and 4b). We believe that such extreme differences do not solely result from the local settings of the monitors but are mostly due to the different collection devices utilized by the networks. In CASTNet, the nitrate interpreted as particulate material is collected on a teflon filter (nitrate is also collected from a nylon filter but the latter is assumed to represent nitric acid concentrations). It has been shown that volatilization of ammonium nitrate from teflon filters may cause substantial loss of nitrate and considerable underestimation of nitrate concentration (HERRING and CASS, 1999). At IMPROVE and STN sites, nitrates are collected on a nylon filter preceded by an acidic vapor diffusion denuder placed in the system to eliminate all nitric acid vapor. The nitrates present on the nylon filter are therefore interpreted as particulate matter.

Table 4a

Slope of the regression lines of nitrate concentrations recorded by each pair of networks

Networks	Slope of the regression line
Improve vs Castnet	1.68
Stn vs Castnet	2.12
STN vs Improve	1.29

Table 4b

Absolute values of the mean and the largest relative differences (%) between nitrate concentrations recorded by each pair of networks, as a function of the averaging time interval

Network	Averaging interval				
	1 wk	2 wk	4 wk	6 wk	8 wk
Absolute value of the mean relative difference (%)					
Improve vs Castnet	229	202	182	169	166
Stn vs Castnet	323	290	261	247	236
STN vs Improve	37	33	31	33	32
Absolute value of the largest relative difference (%)					
Improve vs Castnet	845	820	479	529	468
Stn vs Castnet	1804	1290	844	550	472
STN vs Improve	140	107	62	62	51
Correlation coefficient					
Improve vs Castnet	.74	.77	.84	.89	.94
Stn vs Castnet	.59	.68	.87	.88	.94
STN vs Improve	.89	.92	.97	.99	.99

In this kind of apparatus, the reliability of the vapor remover has occasionally been questioned. It has been shown that high humidity may cause some nitric acid vapor to return to the sample stream, causing an overestimation of the estimated particulate nitrate concentrations (HICKS, 2003, personal communication). AMES and MALM (2001) provide a rather thorough review of the respective strengths and weaknesses of teflon vs. nylon filters for determination of nitrate concentrations. Their article also refers to different studies aimed at quantifying the biases between techniques. Since our investigation focuses on a single set of nearby stations (not strictly collocated), we did not attempt to quantify these biases, although their existence seems unquestionable. As mentioned in the introduction of this paper, our only intention at this time is to alert the reader of the difficulties ahead when blending data from the three networks considered, however providing solutions to effectively deal with this problem will require further effort.

Examination of ammonium signals tends to show that the concentrations calculated by IMPROVE, under the assumption that ammonium is the only cation used for the neutralization of sulfate and nitrate, are overestimated by about 30% and 20% in comparison to CASTNet and STN signals, respectively (Tables 5a

Table 5a

Slope of the regression lines of ammonium concentrations recorded by each pair of networks

Networks	Slope of the regression line
Improve vs Castnet	1.37
Stn vs Castnet	1.07
STN vs Improve	0.78

Table 5b

Absolute values of the mean and the largest relative differences (%) between ammonium concentrations recorded by each pair of networks, as a function of the averaging time interval

Network	Averaging interval				
	1 wk	2 wk	4 wk	6 wk	8 wk
	Absolute value of the mean relative difference (%)				
Improve vs Castnet	41	40	39	36	39
Stn vs Castnet	34	28	25	25	24
STN vs Improve	22	19	17	13	14
	Absolute value of the largest relative difference (%)				
Improve vs Castnet	115	89	68	68	57
Stn vs Castnet	138	100	70	69	45
STN vs Improve	64	43	41	25	31
	Correlation coefficient				
Improve vs Castnet	.88	.90	.93	.94	.93
Stn vs Castnet	.78	.83	.82	.84	.90
STN vs Improve	.90	.90	.91	.94	.95

and 5b). These results suggest that the assumption used in the IMPROVE network to calculate ammonium concentrations (all the sulfates and nitrates in the particulate phase are neutralized by ammonium) may not always be valid.

Based on the preceding results, it appears that an averaging window size between 4 and 6 weeks might be appropriate for harmonization of the short-term variability of all signals and, therefore, removing the effect of the different sampling frequencies. Hence, for the rest of this study, we chose to present the results relative to a 5-week window size rather than the raw information when plotting a contaminant time series. Since that window size is large enough to eliminate the short-term effects of synoptic forcings, any time series constructed by applying a 5-week moving average filter is hereafter referred to as a 'longer-term' signal.

Note that in the case of sulfate, a window size as little as 2 or 3 weeks may be judged sufficient for network blending. A shorter window leads to slightly lesser correlation and larger mean relative difference between networks, but preserves more temporal details. Depending on the study at hand, shorter intervals may be found more pertinent. In this case, we chose to apply the same 5-week window to all three species considered.

Figure 3 depicts scatter plots of IMPROVE and STN versus CASTNet signals, averaged by blocks of successive 5-week intervals over the period considered (Panel A) and the long-term signals of sulfate, nitrate and ammonium observations (Panel B). The close agreement between sulfate estimates is rather obvious (Fig. 3, upper graphs). As obvious is the divergence between nitrates estimated by CASTNet and by the two other networks. Regularly but mostly in the high concentration season, CASTNet's concentration is less than a third that of IMPROVE or STN (Fig. 3, middle graphs). Ammonium concentrations correspond fairly well for all networks,

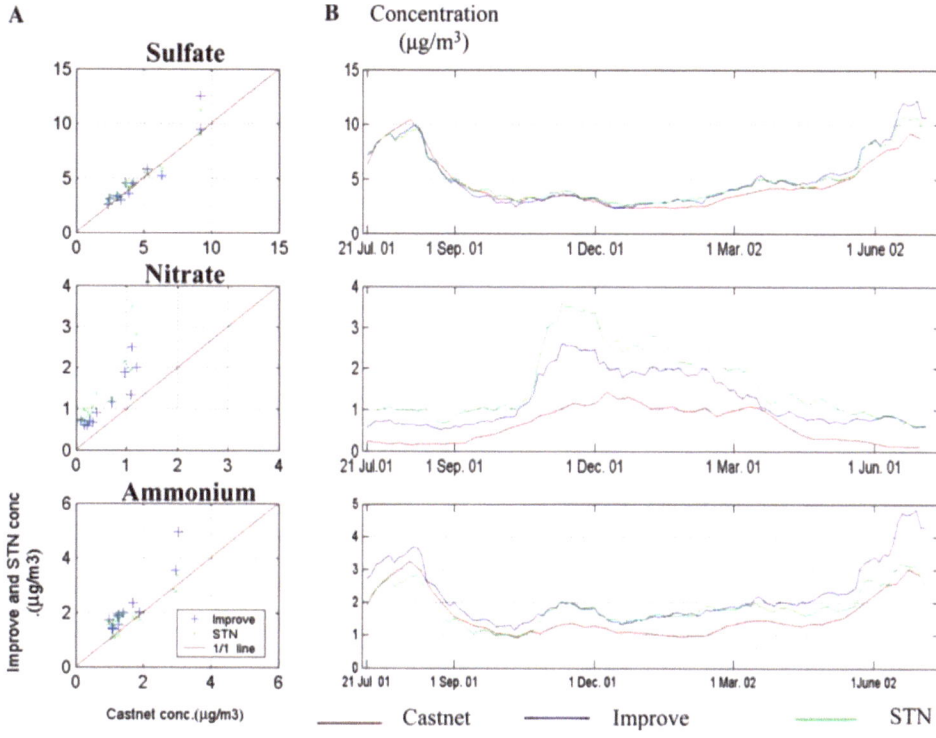

Figure 3
Panel A — Scatter plots of the block average concentrations calculated for STN and IMPROVE vs the block average concentrations calculated for CASTNet (window: 5 weeks), Panel B — time series of the longer-term signals (5-week moving average) of STN, IMPROVE and CASTNet.

with, again, a tendency for CASTNet estimates to be lowest. The differences between STN and IMPROVE estimates are minor at low concentrations but IMPROVE estimates significantly exceed that of STN at higher concentrations (greater than $2 \ \mu g/m^3$).

5.2. Rotated Principal Component Analysis

5.2.1. Organization of monitoring sites into distinct 'modes of variation'

RPCA allows regrouping in a single category (cluster) all sites in a given network responding to the same mode of variation, i.e., where undulations (changes) occur simultaneously and with a reasonably similar amplitude (in terms of standardized scores). It was performed on the raw data (not temporally averaged) collected by each network individually. Attempts to perform RPCA on all networks simultaneously (on the weekly averages, the longer-term signals or the differences between weekly averages and longer-term signals) proved unsuccessful.

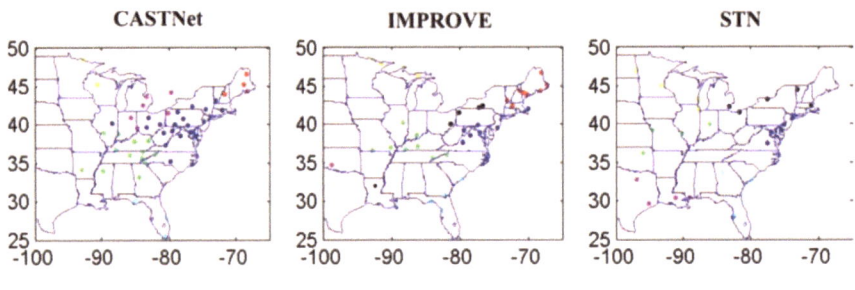

Figure 4

Identification of sites where sulfate concentrations present the same mode of variation (each mode of variation is represented by a different marker and color).

SULFATE

RPCA of the sulfate concentrations measured at the 51 CASTNet sites included in this study indicates the presence of seven distinguishable modes of variation (seven eigenvalues greater than 1). Performed on IMPROVE sites, PCA suggests the existence of eight groups, two of them constituted of a single site. Similarly, the correlation matrix of sulfate concentrations at STN sites has six eigenvalues greater than 1 (six modes of variations). Varimax rotation of the principal component axes and computation of the correlation coefficient between the original time series and the rotated principal component allows identification of sites (clusters) exhibiting the same mode of variation.

Figure 4 displays the location of sites corresponding to each one of the modes of variation identified for sulfate at CASTNet, IMPROVE and STN sites. In the three networks, each mode of variation corresponds to a distinct and unified geographical region. The number of IMPROVE and STN sites is too limited to clearly identify the geographical boundaries of each cluster. Yet, it appears that the limits delineated for CASTNet are somehow compatible with those of IMPROVE and STN. To facilitate that comparison, identical colors have been chosen to identify corresponding clusters in each network.

NITRATE

RPCA of the nitrate concentrations at the CASTNet sites indicates nine distinguishable modes of variation (nine eigenvalues greater than 1), suggesting formation of nine homogeneous clusters (or groups); IMPROVE and STN data suggest the existence of nine and five nitrate clusters, respectively. The clusters identifying location of the different nitrate modes of variations in the CASTNet, IMPROVE and STN networks are presented in Figure 5. Neither the CASTNet nor

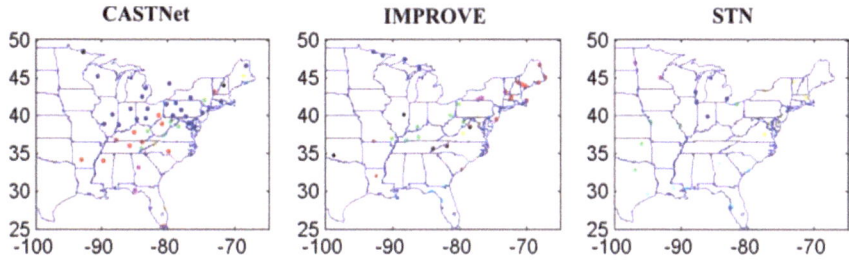

Figure 5

Identification of sites where nitrate concentrations present the same mode of variation (each mode of variation is represented by a different marker and color).

the IMPROVE network allows clear delineation of the limits of each cluster. Several modes of variation are occasionally observed in a limited geographical area, as seen by the classification of sites in New England and in the mid-Atlantic States, for CASTNet and IMPROVE, respectively. For STN , each cluster forms a unified geographical area.

AMMONIUM

RPCA performed on ammonium concentrations led to fairly sharp clustering in the three networks, with all modes of variation corresponding to unified areas (Fig. 6), despite some anomalies. Among the anomalies are the central Florida IMPROVE site that is grouped with one site in Vermont; and the Michigan, Pennsylvania and Vermont sites that are grouped together in CASTNet, although separated by two widespread modes of variation. There are strong similarities between the ammonium and sulfate clusters in the IMPROVE network (compare Figs. 6 and 4, middle panel), an obvious situation that was expected since

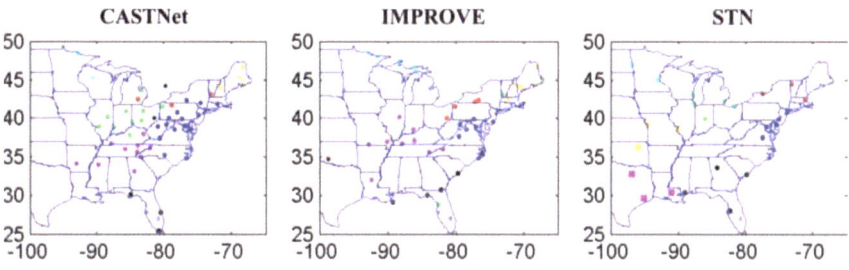

Figure 6

Identification of sites where ammonium concentrations present the same mode of variation (each mode of variation is represented by a different marker and color).

ammonium concentrations at IMPROVE sites are calculated (by stoichiometry) from sulfate and nitrate concentrations.

5.2.2. Assessment of the similarities between the average longer-term signals of each network in corresponding zones

RPCA of the sulfate data collected at CASTNet sites led to delineation of seven modes of variation corresponding to distinct regions whose limits seem compatible with IMPROVE and STN classifications. This finding suggests the possibility of jointly using the sulfate measurements reported by the three networks. Although the boundaries of some clusters seem to agree, it is possible that the mode of variation they bound are substantially different. For instance, the time series of the data collected by one network in a given area may hypothetically represent mostly seasonal variation while the data collected by another network in the same area may reflect synoptic processes. In such a case, combining data from the two networks is not advised.

To address this concern, we identified for each clustered area the spatially averaged longer-term signals of each network and simply compared these signals. Since monitoring sites are not collocated and the networks monitor different environments (CASTNet sites are located in rural areas, IMPROVE sites are predominantly in pristine areas, STN monitors are placed in urban and rural sites), the average concentrations they measure may be different. The most relevant comparison, therefore, is of the synchronism and direction (increase or decrease) of the change reported. Synchronous changes would indicate that the information reported by the different networks is correlated; another clue of their potential joint use.

RPCA and temporal characterization of each mode of variation are valuable tools for a global comparison of networks because they involve all monitoring sites. However, they only allow assessment of the correlation between long-term signals but not of other important parameters for joint use of all data, such as biases between networks, the latter being only identifiable through comparison of data at collocated sites.

Because of the numerous disagreements between the cluster boundaries assessed for nitrate with the three networks, there is insignificant purpose in providing comparisons of the longer-term nitrate signals from the three networks. Therefore, the proposed technique was only applied to sulfate and ammonium data, and then only for those areas where there is reasonable agreement in the geographic boundaries of the clusters by all three networks.

SULFATE

Figure 7 shows the average longer-term signals calculated for the three networks in five corresponding clusters. Each line on the figure represents the spatially

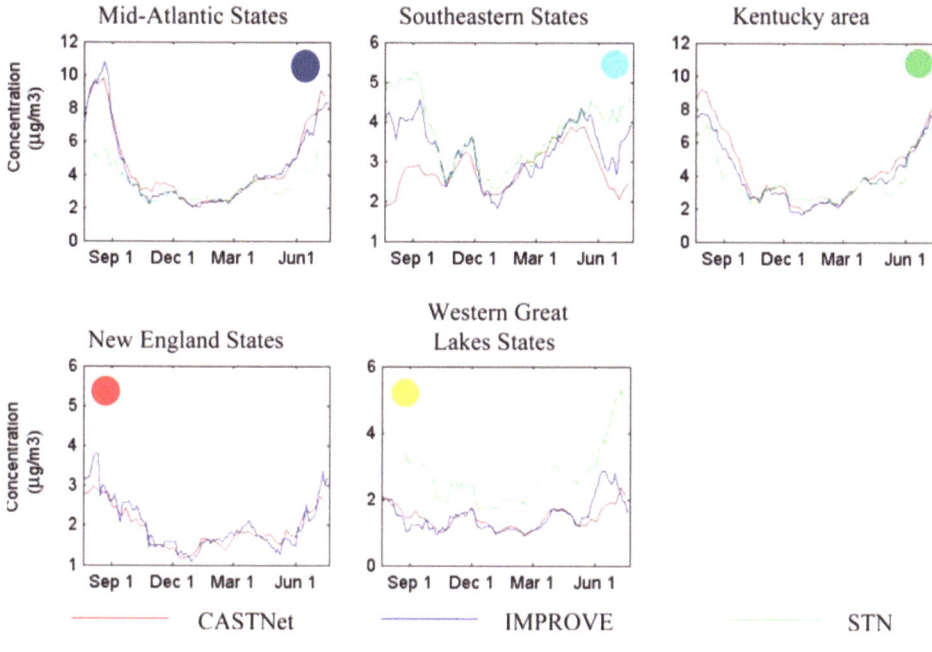

Figure 7

Longer-term signal (5-week moving average) of sulfate concentrations measured by CASTNet, IMPROVE and STN in 5 homogenous areas – average of all stations in each cluster.

averaged longer-term signals of all sites in a given network that are grouped in the same cluster, i.e., that are identified by the same marker. In the case of the cluster overlapping the mid-Atlantic States (dark blue cluster), for instance, the CASTNet line is the average of longer-term signals from all 22 CASTNet sites within the blue cluster. The corresponding IMPROVE line is the average of the eight IMPROVE sites within that cluster, and the STN line is the average of the five sites in the dark blue cluster.

The average longer-term signals in all three networks are very similar, not only in terms of the timing of the changes they report, but their amplitudes as well in the cluster overlapping the mid-Atlantic States (dark blue), the New England cluster (red) and that centered on Kentucky (green). Sulfate fluctuations in the Western Great Lakes States (yellow) are synchronous, but STN concentrations are higher than those reported by CASTNet and IMPROVE, probably because the former pertain to an urban environment while the other networks are located in rural areas or National parks. In the case of the Southeastern States (cyan), the timing of major breaks in the longer-term signals is synchronous. All networks report a local maximum around November 15, 2001, and again around May 1, 2002. Yet, the CASTNet signal is lower than those of STN and IMPROVE, perhaps because the former are exclusively located along the Florida shoreline and not further inland.

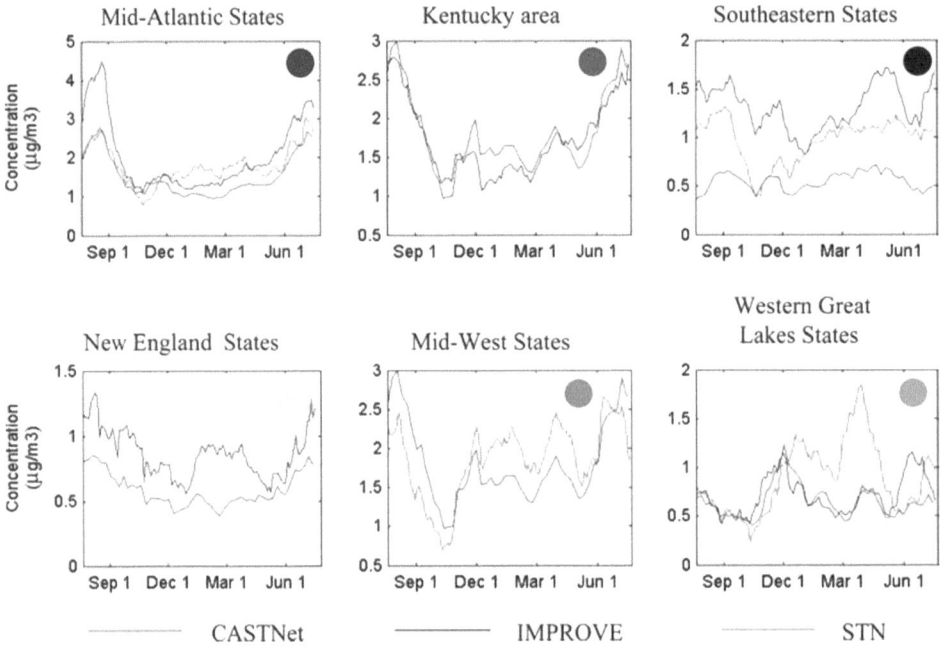

Figure 8

Longer-term signal (5-week moving average) of ammonium concentrations measured by CASTNet, IMPROVE and STN in 6 homogenous areas – average of all stations in a cluster.

AMMONIUM

As observed with sulfate, the spatially averaged longer-term signals of all three networks in the mid-Atlantic States (dark blue cluster) are very similar, both in terms of the timing of the changes and their amplitudes (Fig. 8). In New England (yellow) the longer-term signal in IMPROVE always exceeds that of CASTNnet, suggesting that the calculated IMPROVE concentrations may be overestimating reality. Yet, this discrepancy between CASTNet and IMPROVE signals is not visible in the cluster centered on Kentucky (magenta), nor that centered on the Western Great Lakes States (cyan). These findings illustrate once again a limitation of the RPCA technique, as it is not sensitive to systematic biases between networks. Notice that, presumably because they monitor urban environments, the STN signals indicate higher concentrations in the Western Great Lakes States (cyan) than those of CASTNet and IMPROVE. Still, the timing of major changes is reproduced. The same observation applies to the comparison of the CASTNet and STN signals in the cluster centered in the Midwest States of Illinois, Indiana and Ohio (green). The signals reported in the Southeastern States (black) show few similarities. If indulgent, one may see some synchronism in the CASTNet and STN signals, although the IMPROVE signal is clearly not correlated with the other two.

Since similar modes of variation are recorded by all networks in several corresponding zones, both for sulfate and ammonium concentrations, it appears that, for these contaminants at least, jointly using different networks may be possible. However, one needs to keep in mind that the actual observations may need some adjustment for biases, an issue not resolvable using RPCA.

6. Summary

The objective of this study is to compare some spatio-temporal characteristics of the particulate nitrate, sulfate and ammonium concentrations reported by the CASTNet, IMPROVE and STN networks. While CASTNet collects weekly-integrated samples, IMPROVE and STN have opted for one-in-three day sampling frequency and produce non-continuous results pertaining to 24-hour air samples. Due to these differences in sampling, the weekly-average concentrations calculated for STN and IMPROVE show manifest variability than their CASTNet counterpart.

Averaging the observations (in time) reduces the variability of STN and IMPROVE signals and enhances inter-network compatibility. Using a set of nearly collocated sites, we estimated the shortest moving average interval needed for comparing the three networks to be between four and six weeks. After such averaging, the inter-network correlations of sulfate, nitrate and ammonium time series become high. In addition, the longer-term sulfate concentrations reported by the three networks become very similar, possibly indicating that no systematic biases were introduced by the different sampling locations nor the instrumentation used at the sites studied. This interpretation has to be considered with caution because it relies on a single set of three nearly co-located sites. On the other hand, even after averaging the data within eight- or ten-week intervals, the nitrate concentrations reported by CASTNet are still very different and substantially less than those reported by the IMPROVE and STN. The differences were thought attributable to the distinct air sampling equipment used in the CASTNet program. Nitrate estimates are determined from the material collected on a teflon filter in the CASTNet protocol and from a nylon filter in the IMPROVE and STN protocols. Ammonium estimates seem rather consistent at low concentrations. At high concentrations, though, the IMPROVE signal exceeds those of CASTNet and STN. We speculated that the assumption used in the IMPROVE network to calculate ammonium concentrations (all the sulfates and nitrates in the particulate phase are neutralized by ammonium) may not always be valid.

RPCA was used to identify the distinct modes of variations (clusters) identified for each pollutant in a given network and their boundaries. It was shown that, for sulfate, the clusters formed for all three networks have clear geographical boundaries that seem to correspond. It was also shown that the average longer-term signals defined by each network within corresponding zones are very similar. In the case of nitrate, the clusters formed for the CASTNet and the IMPROVE networks cannot be clearly delineated but appear more mingled. Both networks, for instance, indicated that several modes of variation were occasionally coexisting in a limited geographical area. RPCA performed on ammonium concentrations led to fairly sharp clustering for the three networks, with the spatial limits of each cluster reasonably coinciding for the three networks. As for sulfate, we demonstrated that the modes of variation within a given cluster are very similar between the networks.

In summary, the results of this study indicate that sulfate and ammonium present some potential for networks integration. However, before blending the data from the three networks, basic issues such as that of the biases between networks first need to be resolved. A direct means to resolve network biases is by comparing data at collocated sites. Since there are presently no truly collocated sites, we were precluded from further investigating this issue. In the perplexing case of nitrate, severe incompatibilities between the observations reported by CASTNet and the other two networks prevent us from recommending any type of joint type use of the data from the three networks.

Whether for model evaluation purposes or for the generation of spatial maps depicting the most accurate and aerially extensive representation of the atmosphere, blending data from different sources is theoretically a beneficial option. Yet, as this study shows, that operation needs to be considered cautiously. While apparently possible for some species, the prospect of merging data sets is very uncertain for others. In fact, the extreme divergences between networks for some contaminants are bound to foster debate on the pertinence of the diverse sampling protocols utilized, and the representativeness of the measurements performed.

Disclaimer

The United States Environmental Protection Agency through its Office of Research and Development partially funded and collaborated in the research described here under Interagency Agreements (DW 13938634 and DW 13938483) with the Department of Commerce. The Department of Commerce partially funded and collaborated in the research described here under contracts with Dr. E. Gego (EA133R-03-SE-0710), with the University of Idaho to Dr. P. S. Porter (EA133R-03-SE-0372), and with the State University of New York to Dr. C. HOGREFE (EA133R-03-SE-0650). It has been subjected to Agency review and approved for publication.

REFERENCES

AMES, R.B. and MALM, W.C. (2001), *Comparison of Sulfate and Nitrate Particle Mass Concentrations Measured by IMPROVE and the CDN*, Atmos. Environ. *35*, 905–916.

CATTELL R.B. (1966), *The Scree Test for the Number of Factors*, Multivariate Behavioral Res. *1*, 245–276.

CAMPBELL, S.W., EVANS, M.C., and POOR, N.D. (2002), *Predictions of Size-resolved Aerosol Concentrations of Ammonium, Chloride And Nitrate at a Bayside Site Using EQUISOLV II*, Atmos. Envir. *36*, 4299–4307.

EDER, B.K. (1989), *A Principal Component Analysis of SO_4^{2-} Precipitation Concentrations over the Eastern United States*, Atmos. Environ. *23* (12), 2739–1750.

GEGO, E., HOGREFE, C., PORTER, P.S., IRWIN, J.S., and TRIVIKRAMA, RAO, S. (2003), *Comparison of the Space-time Signatures of Air Quality Data from Different Monitoring Networks*, Proc. 26th NATO/ CCMS International Technical Meeting on Air Pollution Modeling and Its Application, May 26 – 30, 2003, Istanbul, Turkey.

HERRING, S. and Cass, G. (1999), *The Magnitude of Bias in Measurement of $PM_{2.5}$ Arising from Volatilization of Particulate Nitrate from Teflon Filters*, J. Air and Waste Managem. Assoc. *49* (6), 725–733.

HOGREFE, C., RAO, S.T., ZURBENKO, I.G., and PORTER, P.S. (2000), *Information in Time Series of Ozone Observations and Model Predictions Relevant To Regulatory Policies in the Eastern United States*, Bull. Amer. Met. Soc. *81*, 2083–2106.

HOREL, J.D. (1981), *A Rotated Principal Component Analysis of the Interannual Variability of the Northern Hemisphere 500 mb Height Field*, Mon. Wea. Rev. *109*, 2080–2092.

KAISER, H.F. (1958), *The Varimax Criterion for Analytical Rotation in Factor Analysis*, Psychometrika *23*, 187–201.

KENDALL, M., STUART, A., and ORD, J.K., *The Advanced Theory of Statistics Volume 3: Design and Analysis and Time-Series*, 4th ed. (Macmillan , New York 1983).

OVERLAND, J.E. and PREISENDORFER, R.W. (1982), *A Significance Test for Principal Components Applied to a Cyclone Climatology*, Mon. Wea. Rev. *110*, 1–4.

PALM, R., *L'analyse des composantes principales: principes et applications*. Notes de statistiques et d'informatique (Faculté Universitaire des Sciences Agronomiques, Gembloux, Belgique 1999).

RAO, S.T., ZURBENKO, I.G., NEAGU, R., PORTER, P.S., KU, J.Y., and HENRY, R.F. (1997), *Space and Time Scales in Ambient Ozone data*, Bull. Amer. Met. Soc. *78*, 2153–2166.

U.S. ENVIRONMENTAL PROTECTION AGENCY, *Prevention of Significant Deterioration Workshop Manual*. *EPA-450/2-80-081*-NTIS PB-81-201410 (U.S. Environmental Protection Agency, Research Triangle Park, NC 1980).

WILKS, D.S., *Statistical Methods in the Atmospheric Sciences: An Introduction*, International Geophysic Series, Vol. 59 (Academic Press San Diego California 1995).

ZHUANG, H., CHAN, C.K., FANG, M., and WEXLER, A.S. (1999), Size Distributions of Particulate Sulfate, Nitrate and Ammonium at a Coastal Site in Hong Kong, Atmos. Environm. *33*, 843–853.

(Received October 7, 2003; accepted March 25, 2004)

 To access this journal online:
http://www.birkhauser.ch

Pure appl. geophys. 162 (2005) 1941–1954
0033–4553/05/101941–14
DOI 10.1007/s00024-005-2699-2

© Birkhäuser Verlag, Basel, 2005

❚Pure and Applied Geophysics

A Simple Model for the Movement of Fire Smoke in a Confined Tunnel

M. K.-A. Neophytou[1] and R. E. Britter[1]

Abstract—Fires in tunnels are unfortunately frequent occurrences often with tragic outcomes. A recent example is the fire on the funicular train at the ski resort in Kaprun (Austria), which caused nearly 160 deaths. Design engineers and risk analysts require knowledge of the fluid dynamics of the fire and smoke movement to answer questions such as how much oxygen can access and feed the fire, and what concentration of smoke will the people be exposed to. As an example in the Austrian accident the geometry was a long tunnel with fire doors closed at one end, and with a fire initiated near the closed (lower) end. The hot smoke from the fire is a source of buoyancy; the smoke reaches the ceiling of the tunnel, and then develops along the ceiling as a wall-bounded plume. The motion of the smoke is driven by a buoyancy force, but at the same time, mechanisms of turbulent heat and mass transfer act as a brake to this motion. In this paper we present how a generic model describing a semi-enclosed buoyancy-driven flow can be interpreted and used in the modelling of fire smoke movement in a confined tunnel. A consideration of the net pollutant volume flux through the tunnel leads to predictions for the variation of concentrations along the tunnel. The smoke concentrations near the fire smoke source scale linearly with the length of the tunnel, with higher concentrations at the lower section of the tunnel, as could be expected. Similarly the concentration of oxygen making its way through to the fire source decreases linearly with the length of the tunnel. A lower bound estimate of the smoke residence time can be obtained based on smoke concentration predictions from the model.

Key words: Buoyancy, semi-enclosed, dispersion, CFD, experiments.

1. Introduction

Many important engineering and geophysical flow applications are driven by sources of buoyancy; the dispersion of buoyant effluents from industrial plants, the snow avalanches, propagation of smoke from fires and the oil-water flow occurring during the production phase of an oil well are a few examples. Applications most often involve the flow interacting with boundaries. Using the situation with no confinement as a reference (e.g., the case of a free vertical buoyant plume), we could categorise the confined flows into three types based on the confinement-flow interaction Neophytou and Britter (2004): (a) *the wall-type confinement,* where the

---

[1] Department of Engineering, University of Cambridge, Trumpington Street, Cambridge, CB2 1PZ, UK. E-mail: neophytou@ucy.ac.cy

wall just restricts the direction of the fluid motion; (b) *the room-type confinement*, in which the confined space leads to a modification of the environment in which the buoyant flow takes place, and; (c) *the tunnel-type confinement,* where in addition to (a) and (b) above, the actual mixing mechanism is altered.

This paper relates to the latter of these three classes, the tunnel-type confinement, the importance of which is illustrated in a recent tragic example — a tunnel fire accident at a ski resort in Kaprun-Austria in 2000 (BRITISH BROADCASTING CORPORATION, 2001). On one trip of the funicular train a fire started at the lower end of the tunnel with the nearby fire doors closed. Fresh air was initially thought to have fed the fire from the top end of the tunnel, and smoke generation and movement led to the death of many people. Investigations into such a disaster, as well as the design and implementation of safety measures and procedures (such as fire-door installation and ventilation strategy) require knowledge about the fluid dynamics of the hot smoke movement. This knowledge is required in order to answer questions such as: how much oxygen can access and feed the fire, and what concentration of smoke will the people in the tunnel be exposed to. Of course we can invert the problem and consider the release of a material that is denser than the environment within a tunnel confinement. This is also an area of concern for risk analysts and consequence modellers, as many accidents involving stored chemicals produce dense materials (BRITTER, 1989; MOHAN *et al.*, 1995).

The behaviour of buoyant plumes is governed by the fundamental properties of the source: its volume flux, momentum flux and buoyancy flux, Q_0, M_0, and F_0 respectively, which correspond also to the initial values of the plume properties its volume flux (Q), its momentum flux (M), its buoyancy flux (F), as defined by (MORTON *et al.*, 1956). The source buoyancy flux produces a continuous increase in the momentum flux of the flow, and this will, at some position, dominate the momentum flux of the flow. In this paper we address the fire smoke movement in a confined tunnel as a semi-enclosed buoyancy-driven flow, inclined at an angle to the horizontal, with one end of the tunnel blocked. This paper builds on generic results produced by NEOPHYTOU and BRITTER (2004), in order to make deductions specifically applied to the area of fire smoke modelling.

2. Background: A Generic Model on Semi-enclosed Buoyancy-driven Flows

The results presented in this paper are based on an interpretation and application of a generic model developed in NEOPHYTOU (2001), and NEOPHYTOU and BRITTER (2004) in the context of fire smoke modelling. In this section we present the methodology adopted in the generic model development, so that the predictions made further on in the paper are more understandable. The approach in the generic model development entailed the complementary use of laboratory experiments, integral analysis, and computational fluid dynamics (CFD) simulations.

2.1. Laboratory Experiments

Laboratory experiments were carried out in order to investigate a buoyancy-driven flow in a semi-enclosed configuration; the geometrical configuration essentially mimicking that of a smoking fire at the lower end of an inclined tunnel, with closed fire doors at the lower end. Using water as working fluid, the experimental arrangement consisted of a rectangular duct immersed in a water tank; with the aspect (length-to-width) ratio of the duct being 15, and the duct width 0.05 m. The arrangement was interpreted as a two-dimensional flow. The input buoyancy flux was implemented by means of a discharge of salt water inside the top end of the duct; envisaging a negative buoyancy flux, input at the top end of the duct; the dynamics in either (positive or negative buoyancy) case is the same. For a two-dimensional flow the experimental arrangement produced an input volume flux per unit depth, Q_0, of 3.33×10^{-4} m^2/s and an inlet buoyancy flux per unit depth of $F_0 = 1.31 \times 10^{-4}$ m^3/s^3. A shadowgraph technique was used for flow visualisation and video frames were used to track the development of the flow. The purpose of the laboratory experiments was purely qualitative in order to discern the effect of flow confinement, and to support the adopted flow structure in the integral modelling, prior to any numerical simulations.

Two series of experiments were performed: The first had both ends of the duct immersed in the fresh water, while the second had the top end of the duct above the water surface so that the free surface could mimic a confined top, and hence a semi-enclosed flow. When both ends of the duct were immersed in fresh water (corresponding to both ends open, hence an unconfined tunnel), the discharge of salt water produced a bulk flow through the duct (NEOPHYTOU, 2001; NEOPHYTOU and BRITTER, 2004); this would correspond to a fire in a tunnel with all fire doors open. When the top end of the duct was above the water surface (the semi-enclosed buoyancy-driven flow) a totally different flow was produced: the flow adopted a symmetric pattern consisting of two counterflowing streams, with near-zero velocity at the middle, and high velocities at the edges. The width of the counterflowing streams was equal (to the half-width of the duct) and remained so along the entire length of the duct. The flow was more complex at both ends, but this occurred over a very short length of the duct and appeared to have no influence on the counterflow. The velocity of the outflowing stream was slightly higher than that of the inflowing stream; this was a result of the finite inlet volume flux at the source. This flow structure was retained at the angles of 30, 45 and 60 degrees. At 90 degrees, the flow had a different structure. The counterflow pattern is lost and it is replaced by a disorganised flow. With the vertical inclination, the counterflow cannot establish a preferred arrangement of the two streams (with the heavier below the lighter fluid).

2.2. Analysis: An Integral Model

Based on the experimental observations, a conceptual model was developed to describe the buoyancy-driven flow in a semi-enclosed region. The semi-enclosed buoyancy-driven flow is characterised by the input buoyancy flux into the region and takes the form, as observed experimentally, of two counterflowing streams that occupy most of the length of the duct. The flow was alternatively viewed as presented in Figure 1: The open end of the duct is replaced with a closed end and a buoyancy flux is removed through it. The input and output buoyancy flux must be equal for a steady state to exist. As the net volume flux into the region is zero, the counterflowing streams must carry equal volume fluxes (depicted in Fig. 1 as Q) at any position along the duct, and overall, the flow is seen then as equal counterflowing volume fluxes up

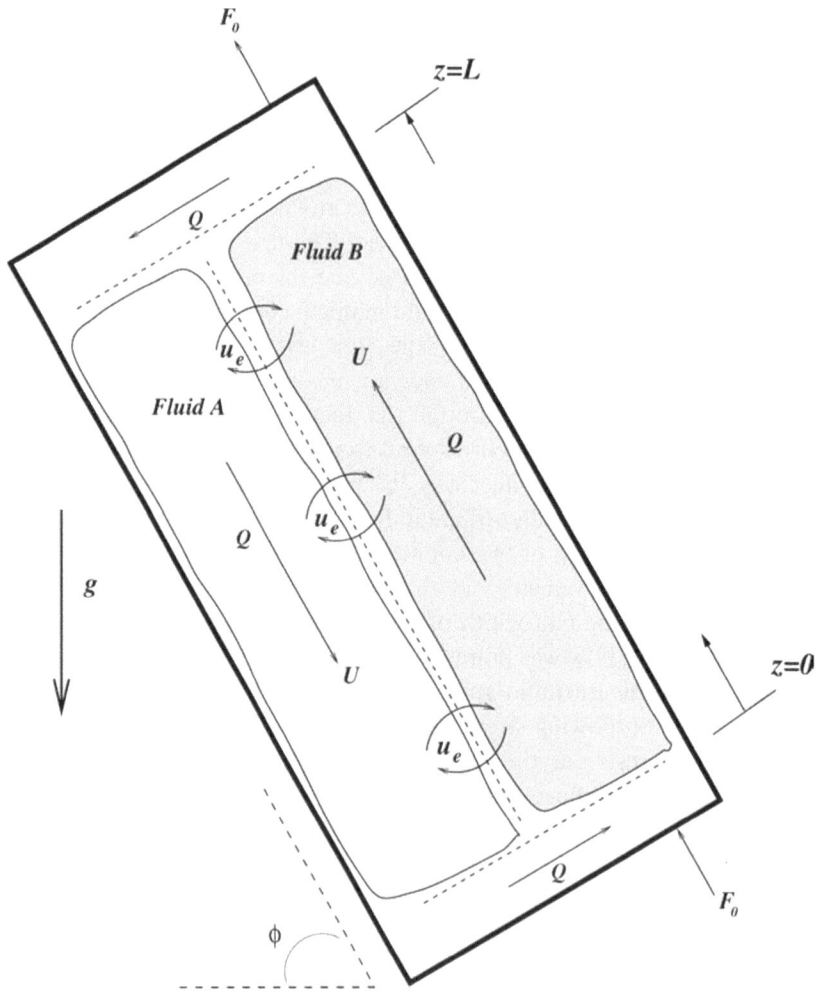

Figure 1
The fluid mechanical model of a buoyancy-driven counterflow in the inclined semi-enclosed configuration.

and down the duct with exchanges between the streams. The two end regions are modelled as local volume fluxes across the duct.

Although it is not an essential part of the integral analysis; the Boussinesq approximation is made, that is for small density differences, the densities are dynamically important only when they give rise to a buoyancy force. The analysis is also restricted to incompressible fluids; that is liquids or low-velocity gases. This restricts direct application of the model to real fire scenarios but the fluid dynamics remains the same. No particular form of the similarity profiles needed to be assumed; top-hat profiles were used to describe the velocity and density variations across the duct, only for reasons of simplicity. The use of other similarity profiles would lead to the introduction of shape factors into the integral analysis.

The turbulent counterflowing streams produce volume, mass and momentum exchanges across the interface between the two streams. The fluid density of each stream (ρ_A and ρ_B, respectively) varies along the duct due to the interfacial mass exchanges, and the density difference across the interface drives the flow in the duct. The interfacial exchange process is characterised by a turbulent entrainment velocity, u_e, which affects the rate at which the density of each layer varies along the duct. Overall, the density differences across the duct are such that they are always consistent with the buoyancy flux entering and leaving the duct, F_0.

By considering the equations of fluid motion (mass, volume and momentum equations) in integral form, it can be deduced that (a) the flows are of constant width equal to the duct half width, and that (b) the bulk counterflow velocity U is constant along the duct (NEOPHYTOU, 2001; NEOPHYTOU and BRITTER, 2004). The shear stress (τ) at the interface, which can also be interpreted as a momentum exchange between the streams, can be modelled by $\tau = 2\bar{\rho}u_eU$, where $\bar{\rho}$ is a reference density of the flow, e.g., that of the ambient environment. By assuming that $u_e = \alpha(2U)$ (where α is the entrainment coefficient and $2U$ is the velocity difference between the streams), it can be deduced that:

$$U = \left(\frac{F_0 \sin \phi}{8\alpha}\right)^{1/3}, \tag{1}$$

$$(\rho_A - \rho_B) = -\frac{2\bar{\rho}\alpha^{1/3}(F_0 \sin \phi)^{2/3}}{gb}, \tag{2}$$

$$\frac{d\rho_A}{dz} = \frac{4\bar{\rho}\alpha^{4/3}}{gb^2}(F_0 \sin \phi)^{2/3}, \tag{3}$$

where ϕ is the angle to horizontal at which the duct is inclined. Equations (1) to (3) are the main results of the integral model describing the buoyancy-driven counterflow of Figure 1. At small angles to the horizontal, there is a more stable density stratification, the turbulence is inhibited and consequently the entrainment coefficient must be small. As the inclination angle becomes steeper, the density stratification becomes less stable, the turbulence is enhanced and consequently the entrainment coefficient also increases. Thus the effect of the angle ϕ is not directly deducible as it depends on the competitive influences of $\sin \phi$ and the entrainment coefficient, α.

2.3. Computational Fluid Dynamics Simulations

In the context of a study for the fire smoke movement in a tunnel, it is interesting to consider the effect of some geometrical parameters such as the aspect ratio and the inclination angle of the tunnel on the produced flow. Computational Fluid Dynamics (CFD) is a tool in the understanding and modelling of fluid mechanical problems, complementary to analytical and experimental techniques, and often provides a cost-effective tool (both in resources and time) for further investigations such as parametric studies. For this purpose, a series of CFD simulations were performed in order to: (a) compare the qualitative results from the CFD modelling with the overall observations of the experiments, and (b) ensure consistency of the CFD results with the integral modelling.

To check the consistency with the integral modelling deductions, a reference test case was used. The geometry in the reference case involved a duct inclined at 45 degrees to horizontal, with an aspect ratio of 11.9 (the length L, and width $2b$ of the duct being 0.8 m and 0.13 m, respectively). The simulated flow was two-dimensional in all test cases. Gravitational acceleration in the reference flow case was set at 10 times g, in order to stress the buoyancy-driven effects. There is no other effect due to gravitational acceleration accounted for in the computational code, therefore the actual numerical value of g does not impact the applicability of results in real flow cases. The buoyancy source was implemented by means of temperature differences. The input buoyancy flux was envisaged as heated water being introduced into the duct immersed in cold water.

The turbulence was modelled in the CFD simulation using the standard $k - \varepsilon$ model, as the $k - \varepsilon$ family of models have proved to predict the growth rates as well as the mean flow quantities of jet and plume flows within engineering accuracy; WOODBURN (1995) reported differences in the induced mean velocities as well as the width of the flow between 0% and 10% (compared with Reynolds-Stress-Model results). However, the standard $k - \varepsilon$ model is less accurate if predictions of the actual turbulence quantities or transverse profiles of flow variables are desired (WOODBURN, 1995). In the context of this study, we essentially address the bulk flow properties as affected by the turbulence quantities, and not the turbulence quantities themselves. Therefore, the use of the standard $k - \varepsilon$ model within the context of this study is legitimate, with the benefit that the relatively low computational cost enables the use of the CFD as a tool for parametric studies. The employed CFD code was FLUENT, developed by Fluent Inc.

The results of the CFD simulations supported the general structure of the flow observed in experiments, and that adopted as well in the integral analysis. The results more importantly confirmed that the buoyancy flux (F_0) is the dominant independent variable of the flow problem. A test case with the buoyancy flux eight times larger than the one in the reference case yields a velocity field which doubles in magnitude. This verifies that the counterflow velocity U scales with $F_0^{1/3}$.

The CFD output data were processed in order to derive the averaged flow quantities that can provide more insight into the governing mechanisms of the flow. Two interesting aspects of the flow are the bulk counterflow velocity and the characterisation of mixing through an entrainment coefficient. The bulk counterflow velocity was obtained by averaging the mean velocity of each of the streams over the symmetric flow regime. The mean counterflow velocity U of the two streams was approximately found to be constant over the length of the duct, and was given by $U/F_0^{1/3} = 0.75$. By employing more detailed CFD data a representative entrainment velocity at all corresponding positions along the duct can be deduced and a local entrainment coefficient, α, (from $u_e = 2\alpha U$) can be derived. The computed average value of the entrainment coefficient (which could be now called the entrainment constant) for the reference test case is 0.12; a value which is typical of the values found experimentally for buoyant plumes (RICOU and SPALDING, 1961).

3. A Model for Fire Smoke Movement

Relevant to tunnel fire safety, there are a few interesting questions that can arise from a situation described by the flow configuration addressed in this paper; for example, given that fresh air is replenishing the tunnel from the open top end, an important issue is how much fresh air can reach, and therefore feed, the fire, and then how much smoke the people are exposed. These questions essentially relate to the question of how much smoke the fresh air will have mixed with by the time it reaches the "fire" end of the tunnel, and therefore what will the concentration of smoke be in the lower and upper half of the closed bottom end of the tunnel.

Experimental observations support the view that as the fresh air enters the top end of an inclined tunnel, it mixes with smoke-containing counterflowing fluid, originating from the smoke plume at the bottom end of the tunnel. Part of the resulting mixed fluid escapes from the tunnel as part of the outflowing fluid (over the upper half of the tunnel), while another part of the resulting mixed fluid carries on flowing into the tunnel as part of the inflowing stream over the lower half of the tunnel (Fig. 2).

The integral model derivations in NEOPHYTOU (2001) and NEOPHYTOU and BRITTER (2004) have included the Boussinesq approximation. It is likely that a fire smoke movement in fresh air will yield density differences that are not within the Boussinesq approximation. Nonetheless, studies on the differences between Boussinesq and non-Boussinesq plumes (GRÖBELBAUER et al., 1993; ROONEY and LINDEN, 1996) have shown that the behaviour of plumes (structure and overall shape of the

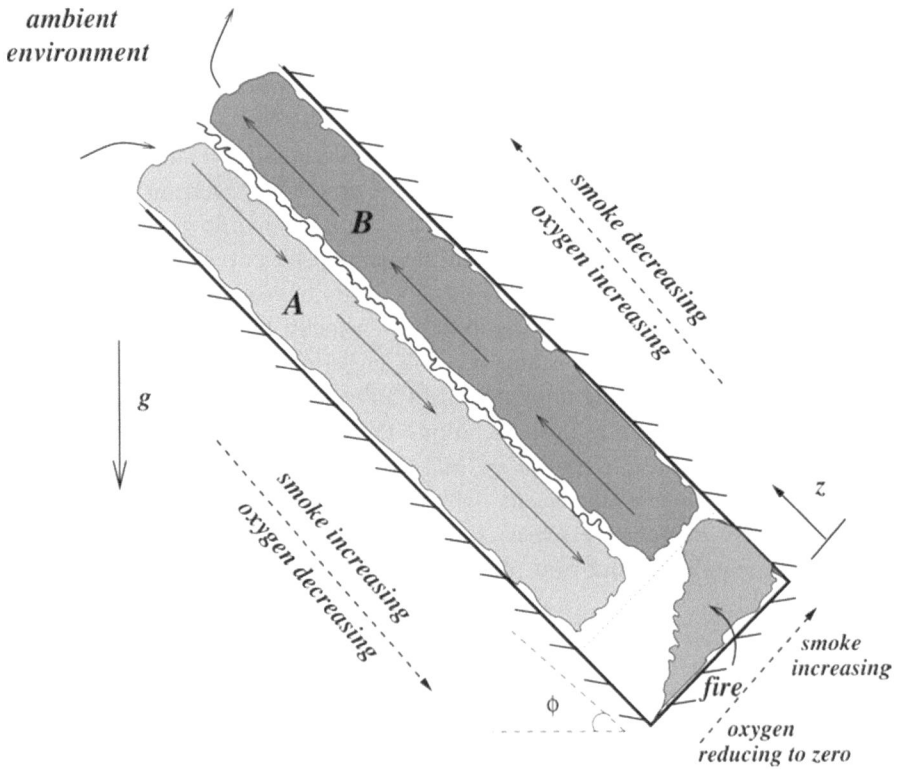

Figure 2
Schematic view of the resulting flow from fire-generated smoke inside a tunnel.

flow) remains unaffected, while quantitative predictions on the mass flux, momentum flux and buoyancy flux scale weakly on the ratio of the fluids densities. On this basis an extension of the Boussinesq model to the smoke movement in the configuration addressed in this paper will not be inappropriate; moreover, the observed density differences in the semi-enclosed tunnel will be less than those between freshly-generated smoke and fresh air, due to the continuous mixing along the entire length of the tunnel.

3.1. Prediction of Smoke Concentration Variation in the Tunnel

The buoyancy-driven flow can also be interpreted as a fluid flow driven by differences in mass (or volume) concentration of a pollutant in the fluid volume; the governing pollutant mass equations (for each counterflowing stream) being similar to those corresponding to counterflowing streams of different density—the pollutant concentration replacing density in the equations.

The input buoyancy flux addressed in the integral model (NEOPHYTOU, 2001; NEOPHYTOU and BRITTER, 2004) determines the density, or concentration, differences that result across the tunnel—not the actual value of concentrations; what will detemine that, is the reference pollutant conentration in the ambient environment at the exit of the tunnel, as well as the geometry of the tunnel, that is the region over which the fluid has travelled and mixed with the counterflowing stream.

We can calculate the smoke concentration variation of the inflowing and outflowing streams by considering the mass equations of the individual streams, as well as using the fact that at the top, open end of the tunnel (length L, width $2b$), the inflowing stream is always fresh fluid from the ambient environment, i.e., zero smoke concentration.

The mass equations for the pollutant in the outflowing (upper half) and the inflowing (lower half) streams give, respectively:

$$\frac{d}{dz}(C_B Ub) = -C_B u_e + C_A u_e \tag{4}$$

$$-\frac{d}{dz}(C_A Ub) = C_B u_e - C_A u_e, \tag{5}$$

which lead to

$$\frac{d}{dz}[(C_A - C_B)Ub] = 0, \tag{6}$$

a statement of that

$$(C_A - C_B) = \text{constant}, \quad \frac{dC_A}{dz} = \text{constant}, \quad \frac{dC_B}{dz} = \text{constant}. \tag{7}$$

The term $(C_A - C_B)$ can be rewritten in terms of the net pollutant volume flux (q) carried along the duct, as

$$q = (C_B - C_A)Ub, \tag{8}$$

which leads to

$$\frac{dC_B}{dz} = -u_e \frac{q}{(Ub)^2}, \quad \frac{dC_A}{dz} = -u_e \frac{q}{(Ub)^2}. \tag{9}$$

Integrating the above equations gives, respectively

$$C_B = -u_e \frac{q}{(Ub)^2} z + C_{B_0}, \quad C_A = -u_e \frac{q}{(Ub)^2} z + C_{A_0}. \tag{10}$$

where C_{A_0} and C_{B_0} are the integration constants and denote the smoke concentrations at the closed (bottom) end of the tunnel. For fluid A, that is the incoming fluid (occupying the lower half of the tunnel), the smoke concentration is zero at the open end, i.e., $C_A = 0$ at $z = L$; this, combined with equation (8), leads to

$$C_B = \frac{q}{Ub}\left(\frac{u_e}{Ub}(L-z)+1\right), C_A = \frac{q}{Ub}\left(\frac{u_e}{Ub}(L-z)\right). \quad (11)$$

The above equations are also interpreted as demonstrating that the outflowing stream contains a high smoke-concentration near the closed end $(C_B = \frac{q}{Ub}(u_e/UbL+1))$ at $z = 0)$. As it flows outwards along the tunnel mixing with fluid containing fresh air, its smoke concentration continuously reduces so that at the open end its smoke concentration becomes $C_B = q/Ub$ (at $z = L$). Similarly, the inflowing stream entering at the open end is fresh air ($C_A = 0$ at $z = L$) that mixes with smoke-containing fluid as it flows further inwards along the tunnel, thereby increasing continuously its smoke concentration so that at the fire end, its smoke concentration becomes $C_A = q/Ub(u_e/UbL)$.

3.2. Prediction of the Oxygen Concentration Feeding the Fire

The model predictions for the smoke concentration variation in the tunnel can be re-interpreted in terms of the oxygen concentration remaining in the induced inflow that reaches the fire. We introduce the reasonable assumption that the fire is ventilation- (and not diffusion-) controlled, i.e., that the fire will be controlled by how much air actually reaches from the open end up to the fire end. The oxygen removal at the fire, denoted by q', is defined as:

$$-q' = (C_B - C_A)Ub, \quad (12)$$

where C' denotes the oxygen concentration and subscripts A and B denote properties of the inflowing and outflowing streams, respectively. Mass flux considerations lead to

$$\frac{dC'_A}{dz} = -u_e\frac{(-q')}{(Ub)^2} = \frac{dC'_B}{dz}, \quad (13)$$

and by integrating, we deduce that

$$C'_A = u_e\frac{q'}{(Ub)^2}z + C'_{A_0}, C'_B = u_e\frac{q'}{(Ub)^2}z + C'_{B_0}, \quad (14)$$

where C'_{A_0} and C'_{B_0} are the integration constants and denote the oxygen concentrations at the closed (bottom) end of the tunnel. To calculate the integration constants we consider two boundary conditions: (a) that the fire is ventilation-controlled, so the oxygen concentration just after the fire in the outflowing stream is zero, i.e., that $C'_B = 0$ at $z = 0$; and (b) that the oxygen concentration of the inflowing stream at the open end of the duct is that of the ambient environment, i.e., that $C'_A = C_{ambient}$ at $z = L$. These two conditions conclude that the oxygen variation along the tunnel is given for each of the streams by

$$C'_A = u_e\frac{q'}{(Ub)^2}(z-L) + C'_{ambient}, \quad (15)$$

$$C'_B = u_e \frac{q'}{(Ub)^2} z. \tag{16}$$

An interesting question from the fire safety perspective is how much oxygen feeds the fire. For this we need to calculate C'_A at $z = 0$, that is

$$C'_A(z = 0) = -u_e \frac{q'}{(Ub)^2} L + C'_{\text{ambient}}. \tag{17}$$

Given equation (12) it follows that

$$C'_B(z = 0) = C'_A(z = 0) - \frac{q'}{Ub}. \tag{18}$$

We can derive a condition for q' in terms of the ambient oxygen concentration, based on the oxygen concentration just after the fire, $C'_B(z = 0)$, being 0

$$C'_{\text{ambient}} = \frac{q'}{Ub}\left(\frac{u_e L}{Ub} + 1\right). \tag{19}$$

Hence the fraction of oxygen concentration that feeds the fire (relative to the ambient) is given by

$$\frac{C'_A(z = 0)}{C'_{\text{ambient}}} = \frac{1}{1 + \frac{u_e L}{Ub}}. \tag{20}$$

3.3. Residence Times

Another issue of interest to tunnel designers is the smoke residence time in the tunnel once the fire has been extinguished; i.e., how long would it take for the smoke to clear the tunnel. A very similar question arises in risk analysis, when considering, for example, accidental releases of toxic gases in tunnels; an interesting issue would be how long does it take for the toxic material to escape from the tunnel into the surrounding environment.

A first-order estimate for the residence time could be obtained by relating the average mass of pollutant contained in the volume of the tunnel and the initial rate of loss of the pollutant. The residence time obtained from such an estimate is a lower bound as the loss rate of the pollutant will continuously decrease once the source has been switched off.

An indicative average concentration of pollutant in the tunnel is given by

$$\overline{C} = \frac{C_{A_{z=\frac{L}{2}}} + C_{B_{z=\frac{L}{2}}}}{2} \tag{21}$$

deducing an average mass of pollutant $\bar{M}$ contained in the volume of the tunnel V (equal to $2Lb$), of

$$\overline{M} = \overline{C} \times V = \frac{qL}{U}\left(\frac{u_e L}{U b} + 1\right). \tag{22}$$

The initial loss rate of the pollutant is q—the net flux of the pollutant through the tunnel. Therefore an indicative residence time (t_{res}) would be $\overline{M}/q$, yielding

$$\frac{L}{U}\left(\frac{u_e L}{U b} + 1\right). \tag{23}$$

4. An Example

In this section we illustrate the application and usefulness of such simple models and guide rules by means of an example. Let us assume that a 10 MW fire occurs in a tunnel that is 200 m long, has a 5 m-diameter, and is inclined at 20° to horizontal. The specific heat capacity of smoke c_p is taken to be 10^3 J/kgK, the ambient temperature T_∞ is 300 K and the ambient density of air ρ_∞ is 1.2 kg/m^3. As part of a safety analysis we would like to calculate (a) the velocity of the flow that is set-up in the tunnel keeping the fire-doors closed, (b) the temperature gradients along and across the tunnel, (c) the smoke concentration that the people will be exposed to, and (d) the smoke residence times.

The buoyancy flux of the fire smoke (B_0) given its thermal output (Q_c) can be calculated as (12) (Society of Fire Protection Engineers and National Fire Protection Association, 1988)

$$B_0 = g\left(c_p T_\infty \rho_\infty\right)^{-1} Q_c. \tag{24}$$

In the above example the buoyancy flux B_0 is 27.22 m^4/s^3, and corresponds for the two-dimensional model to $F_0 W$, where W would be the width into the two-dimensional plane in this example $W = 5$ m; therefore F_0 is 5.44 m^3/s^3. From the equations (4), (5), (6) we can calculate directly the induced counterflow velocity, and the density difference along and across the tunnel. The counterflow velocity U is determined to be as 1.325 m/s, the density difference across the tunnel $(\rho_A - \rho_B)$ to be 0.07 kg/m^3 ($b = 2.5$ m); and the density gradient along the tunnel to be 0.005 kg/m^3/m. These can be converted into temperature differences (given the width and length of tunnel) as 17.5 K across the tunnel and 250 K along the tunnel.

The prediction of large along-tunnel temperature difference does not negate the Boussinesq assumption, since the assumption refers to across-tunnel temperature differences, that are of the order of 17.5/300 ($\approx 6\%$) which is in the Boussinesq limit. However, the large along-tunnel temperatures will cause along-tunnel variations in the buoyancy flux which is not consistent with the model. Nonetheless, the variations in the induced bulk velocity due to the change in the buoyancy flux, F_0, would be much weaker, as they scale with $F_0^{1/3}$. In this example the maximum variation of F_0 due to the along-tunnel temperature changes is up to 1.8 F_0. This corresponds to a change in bulk velocity U of about 20% (or ± 10%), which we consider acceptable in the context of this simple model.

If an estimate of the smoke residence time was desired in this example, a first-order estimate from equation (23) would be t_{res} = 2566 s (approximately 45 minutes). This residence time estimate is going to be a lower-bound as explained in the previous section, and the observed is likely to be longer.

The results can also be interpreted for a three-dimensional buoyancy-driven counterflow inside a duct. This can be done if the buoyancy flux per unit depth, F_0, is converted appropriately into an inlet buoyancy flux passing through the cross section; for example by assuming a certain depth in the two-dimensional case, the corresponding buoyancy flux is calculated, and then for the equivalent inlet buoyancy flux, the inlet cross-sectional area is evaluated.

5. Conclusions

1. We have shown in this paper how a generic model describing a semi-enclosed buoyancy-driven flow can be interpreted and used in the modelling of fire smoke movement in a confined tunnel.

2. A consideration of the net pollutant flux through the region (tunnel) and the concentration variation along the tunnel leads to the prediction of the smoke concentrations for the lower and upper halves of the cross section of the tunnel. The smoke concentrations near the smoke source scale linearly with the length of the tunnel, with higher concentrations at the lower section of the tunnel, as could be expected. Similar arguments are used to determine the distribution of oxygen concentrations in the tunnel, particularly that just before the fire.

3. A lower bound estimate of the smoke residence time can be obtained based on smoke concentration predictions from the model; the lower bound resulting from the fact that the prediciton is based on the initial net pollutant volume fluxes in the tunnel, once the fire smoke source has been switched off.

Acknowledgements

MK—AN wishes to acknowledge the contribution of the Computational Fluid Dynamics Laboratory, Department of Engineering, University of Cambridge, in computational resources and expertise.

REFERENCES

BRITISH BROADCASTING CORPORATION (Sept. 2001), *Austria Tunnel Fire Blamed on Heater*, Official Website; http://news.bbc.co.uk/2/hi/europe/1528917.stm.

BRITTER, R.E. (1989), *Atmospheric Dispersion of Dense Gases*, Ann. Rev. Fluid Mech. *21*, 317–344.

GRÖBELBAUER, H.P., FANNELØP, T.K., and BRITTER, R.E. (1993), *The Propagation of Intrusion Fronts of High-density Ratios*, J. Fluid Mech. *250*, 669–687.

ROONEY, G.G. and LINDEN, P.F. (1996), *Similarity Considerations for non-Boussinesq Plumes in an Unstratified Environment*, J. Fluid Mech. *318*, 237–250.

LAUNDER, B.E., REECE, G.J., and RODI, W. (1975), *Progress in the Development of a Reynolds-stress Turbulence Closure*, J. Fluid Mech. *68*.

MOHAN, M., PANWAR, T.S., and SINGH, M.P. (1995), *Development of Dense Gas Dispersion Model for Emergency Preparedness*, Atmos. Environm. *29*(16), 2075–87.

MORTON, B.R., TAYLOR, G.I., and TURNER, G.S. (1956), *Turbulent gravitational convection from maintained and instantaneous sources*. In Proc. Roy. Soc. *A234*, pages 1–23.

NEOPHYTOU, M.K. (2001), *Modelling Buoyancy- and Momentum-driven Flows in Semi-enclosed Regions*, University of Cambridge, Depart. Engin., Ph.D. Thesis.

NEOPHYTOU, M.K. and BRITTER, R.E. (2004), *Modelling Buoyancy-driven Flows in Confined Regions: Experiments, Analysis and Computations*, J. Fluid Mech., submitted.

RICOU, F.P. and SPALDING, D.B. (1961), *Measurements of Entrainment by Axisymmetric Turbulent Jets*, J. Fluid Mech. *II*(21–32), 173.

RODI, W. (1985), *Calculation of Stably Stratified Shear Layer Flows with a Buoyancy-extended $k - \varepsilon$ Turbulence Model, Turbulent Diffusion in Stable Environment*, editor: J.C.R. Hunt.

Society of Fire Protection Engineers and National Fire Protection Association. *The SFPE Handbook of Fire Protection Engineering*, First Edition, 1988.

WOODBURN, P. (1995), *Computational Fluid Dynamics Simulation of Fire-generated Flows in Tunnels and Corridors*, Ph.D. Thesis, University of Cambridge, UK, January.

(Received October 10, 2003; accepted June 17, 2004)
Published Online First June 17, 2005

 To access this journal online:
http://www.birkhauser.ch

Pure appl. geophys. 162 (2005) 1955–1980
0033–4553/05/101955–26
DOI 10.1007/s00024-005-2700-0

© Birkhäuser Verlag, Basel, 2005

| Pure and Applied Geophysics

Observations and Numerical Simulations of Urban Heat Island and Sea Breeze Circulations over New York City

PETER P. CHILDS and SETHU RAMAN

Abstract—Observations from two SOund Detection And Ranging (SODAR) units, a 10 m micrometeorological tower and five Automated Surface Observing Stations (ASOS) were examined during several synoptic scale flow regimes over New York City after the World Trade Center disaster on September 11, 2001. An ARPS model numerical simulation was conducted to explore the complex mesoscale boundary layer structure over New York City. The numerical investigation examined the urban heat island, urban roughness effect and sea breeze structure over the New York City region. Estimated roughness lengths varied from 0.7 m with flow from the water to 4 m with flow through Manhattan. A nighttime mixed layer was observed over lower Manhattan, indicating the existence of an urban heat island. The ARPS model simulated a sea-breeze front moving through lower Manhattan during the study period consistent with the observations from the SODARs and the 10-m tower observations. Wind simulations showed a slowing and cyclonic turning of the 10-m air flow as the air moved over New York City from the ocean. Vertical profiles of simulated TKE and wind speeds showed a maximum in TKE over lower Manhattan during nighttime conditions. It appears that this TKE maximum is directly related to the influences of the urban heat island.

Key words: Urban heat island, sea breeze, TKE.

1. Introduction

Meteorological effects of urbanization are well documented throughout atmospheric literature. Most studies have focused on the urban heat island and its interactions with larger-scale atmospheric phenomena. The heat-island circulation (HIC) associated with an urban area can significantly alter lower tropospheric winds and low-level pollutant dispersion. When an urban area is located at the coast of a large body of water, complexities of the flow patterns increase because of additional circulations associated with sea and land breezes. Several studies have examined the urban heat island. SHREFFLER (1978, 1979) conducted observational studies of St. Louis, while ASAI (1970) analyzed observed data from Tokyo. Additional observational studies by TAKEUCHI and KIMURA (1976), BORNSTEIN (1975) and SAWAI (1978)

State Climate Office of North Carolina, North Carolina State University, Raleigh, NC, 27695–7236, U.S.A.

and physical modeling studies by Sethu Raman and Çermak (1974) and Mellor and Yamada (1974) have focused on the urban heat island and its effects on local circulations and flow alterations. Few numerical studies have attempted to simulate the very complex micro- and mesoscale meteorology associated with urban areas such as New York City; the focus region of this study.

Atmospheric boundary layer (ABL) inhomogeneity is most apparent over a dense urban center like NYC that lies adjacent to the ocean, especially when compared to the ABL over rural, inland areas. NYC's landuse is characterized as a highly developed urban core on Manhattan Island and a sprawling dense suburban area that covers northeastern New Jersey and western Long Island. Adding to this complex urban region is a highly variable coastline consisting of many small bays, rivers and sounds (Jamaica Bay, New York Harbor, Hudson River, East River and Long Island Sound). All these features and their influence on the lower atmosphere make attempts at modeling the region difficult (Michael et al., 1998).

Historical studies have focused both directly and indirectly on the unique small-scale variations of the ABL in and around NYC. The NYC urban blocking effect and urban heat island phenomena have been examined in detail (Bornstein and Johnson, 1977; Bornstein et al., 1994). The blocking effect can be described as the modification of the flow by an abnormally rough surface presented by Manhattan Island. The urban heat island develops because of both anthropogenic heating and heat-holding structures. This local heating further modifies the wind flow patterns over the city. Past research shows that wind speed along a streamline decreases below (increases above) those at sites outside of the city when synoptic scale winds speeds are above (below) 4 ms^{-1}. Above this critical value wind over the city is less than in rural areas and turns cyclonically as the air passes over the city due to increased frictional effect. Conversely, when the wind speed is below this value, the urban heat island tends to develop and the winds over the city are slightly stronger during both daytime and nighttime. This urban enhancement of the wind speed is a result of the increase in mesoscale baroclinicity and decrease in stability, which allows for efficient downward flux of momentum. This wind enhancement during lower wind speed regimes, especially at night, results in a more anticyclonic curvature of the wind trajectory as it passes over the city. Many studies have been performed on this topic (Angell et al., 1971; Wong and Dirks, 1978; Lee, 1979; Draxler, 1986) and most agree with this behavior of wind flow over a rough urban surface.

As a result of the urban heat island and the resulting heat island circulation, and their effect on surface-layer wind flow, numerical modeling over highly urbanized areas is complex. As the realization of potential bio-terrorism hazards in urbanized area continues to grow, reliable numerical modeling of urban areas is rapidly becoming an important research and operational issue. Studies involving high-resolution mesoscale models and pollutant dispersion models are underway (Arya, 1999). Historically, air pollution has been regarded as a serious problem only for large cities and commercial centers. As a result of the industrial revolution and the advent of the automobile, air quality in most of the large urban and industrialized areas has been suffering greatly. Various urban air quality models have been developed to facilitate the implementation of new strategies and techniques to help regulate pollutants being released from automobiles and industry.

The main objective of this investigation is to analyze observations and numerically simulate the mesoscale and microscale boundary layer structure over New York City. Observations from two SOund Detection And Ranging (SODAR) systems, a 10 m micrometeorological tower and five Automated Surface Observing Stations (ASOS) are examined during several synoptic scale flow regimes including the one that existed on September 11, 2001. Subsequently, a numerical simulation is conducted to explore the complex mesoscale boundary layer structure over New York City. The high resolution (1 km) numerical investigation examines the urban heat island, urban roughness effect and sea breeze structure over the region.

The National Exposure Research Laboratory, U.S. Environmental Protection Agency (EPA/NERL) has an instrumentation cluster that facilitates high-resolution temporal measurements near the surface. This ensemble consisted of three portable trailers that supported the Aerovironment Model 4000 miniSODAR, Aerovironment Model 2000 SODAR and a three-level 10 m micrometeorological tower. This cluster was deployed in lower Manhattan, New York in November 2001 to support the EPA and State Climate Office of North Carolina's study of pollutant exposure over lower Manhattan following the September 2001 disaster. To supplement these observations, data from five automated surface observation systems (ASOS) located in Newark, New Jersey, Teterboro, New Jersey, Central Park, New York, JFK Airport, New York and LaGuardia Airport, New York are also analyzed. The emphasis here is on diagnosing the synoptic scale flow regimes over the New York City during the autumn of 2001. Additionally, using the 10 m micrometeorological tower and the two SODARS, roughness length is estimated for the different flow regimes observed over the region. Observations from the ASOS and the 10 m tower are used to study the effect, if any, the urban heat island has on the temperature structure and wind fields.

A numerical simulation using the ARPS model was performed during a high ground-level pollution event observed between 13 November 2001 and 15 November 2001 over lower Manhattan (GILLIAM et al., 2003). This period was characterized by light and variable winds on November 13th, with a more southwesterly component developing on the 14th and 15th. The ARPS model is used to study the effects of the urban heat island and roughness length variations on the boundary layer structure and its diurnal evolution over New York City. This simulation employs the 1 km USGS surface characteristics and 30 second terrain information to define the lower boundary. A 48-hr case study for the 1 km domain is presented. The domain is initialized from the 32-km ARPS Data Assimilation System (ADAS). A 5 km intermediate domain is utilized to ensure that accurate lateral and upper boundary conditions are ingested into the model with 1 km grid spacing. Observations from the independent cluster and ASOS network are used to evaluate the model simulation. The 10 m tower data and the ASOS data are used to validate the model simulation of 10 m wind speed and wind direction associated with the sea breeze front and roughness induced deflections. The SODAR data are used to examine the vertical structure of the lower boundary layer and for validating the simulation of the sea breeze structure and urban heat island effect.

2. Instrumentation and Data

The miniSODAR used in this study is a high-resolution surface layer (15 to 200 m range at 5 m intervals, 10 min averaged) wind sampler. It transmits sound at a frequency of 4500 Hz, which facilitates mitigation of environmental noise interference (CRESCENTI, 1998) leading to a better representation of the surface layer wind distribution and variance. The miniSODAR has a wind speed uncertainty less than 0.50 ms^{-1} and a wind direction uncertainty of $\pm$ 5 deg. A previous study that evaluated the performance of ground-based instruments, including the miniSODAR, found a high correlation with tower measurements (CRESCENTI, 1999). SODAR systems, including the miniSODAR, use sound to sample the boundary layer emitting a pulse and receiving scatter from gradients of temperature and moisture. Turbulent mixing in the boundary layer often causes these gradients. Frequency shifts (Doppler effect) between the transmitted and returned signal are translated as moving parcels of air, and the velocity is directly related to the frequency shift. Algorithms extract other related parameters such as standard deviations of the wind components, vertical velocity, and return signal intensity (reflectivity).

The Aerovironment Model 2000 SODAR used in this study measured the same wind properties as the miniSODAR from 60 to 600 m at 30 m intervals, and averaged over a 10 min period. The Model 2000 SODAR has a wind speed error less than 0.50 ms^{-1} and a wind direction uncertainty of $\pm$ 5 deg. This unit provided important data for the convective mixed layer, and provided an independent source for comparison with the miniSODAR, while also yielding mixing height measurements below 600 m. This unit is also capable of assessing boundary layer structure and evolution after sunrise and before sunset.

The 10 m micrometeorological tower used in this study has instruments that measure wind (Young Model 05701 anemometer) at 2, 5 and 10 m along with temperature and relative humidity at 2 and 10 m. The wind direction is accurate to within $\pm$ 5 deg, while the wind speed is accurate to within 0.25 ms^{-1} at all levels. The temperature sensors are accurate to within 0.2 C at all levels. This "ground truth" instrumentation is important and valuable for evaluating the accuracy of the SODAR data, and provides the lower level observations that are not sampled by either SODAR. The temperature observations are also important, especially the difference between 2 m and 10 m providing valuable information regarding the static stability of the surface layer. All tower data used in this study are sampled each second, averaged and stored at 10 min intervals. A plan view showing the location of the instrumentation cluster in lower Manhattan is shown in Figure 1.

Hourly surface observations from five National Weather Service Automated Surface Observing System (ASOS) stations are also used in this study. The stations are Newark Airport, Teterboro Airport, Central Park, LaGuardia Airport, and John F. Kennedy Airport. Quality assured hourly ASOS data were acquired from the National Climatic Data Center (NCDC) for the study period. The ASOS 10 m wind and 2 m temperature data are recorded every minute and are representative of the previous 5 min average. The data are obtained at the bottom of the hour, approximately 51 minutes past each hour. Figure 2a shows the locations of the five ASOS sites used in this study. Figure 2b shows a high-resolution photograph of lower Manhattan on 12 September 2001. The smoke plume from the World Trade Center disaster site is evident on the photograph. The instrumentation cluster was located on Pier 25 in lower Manhattan and is labeled EPA/SCO on the photograph for reference.

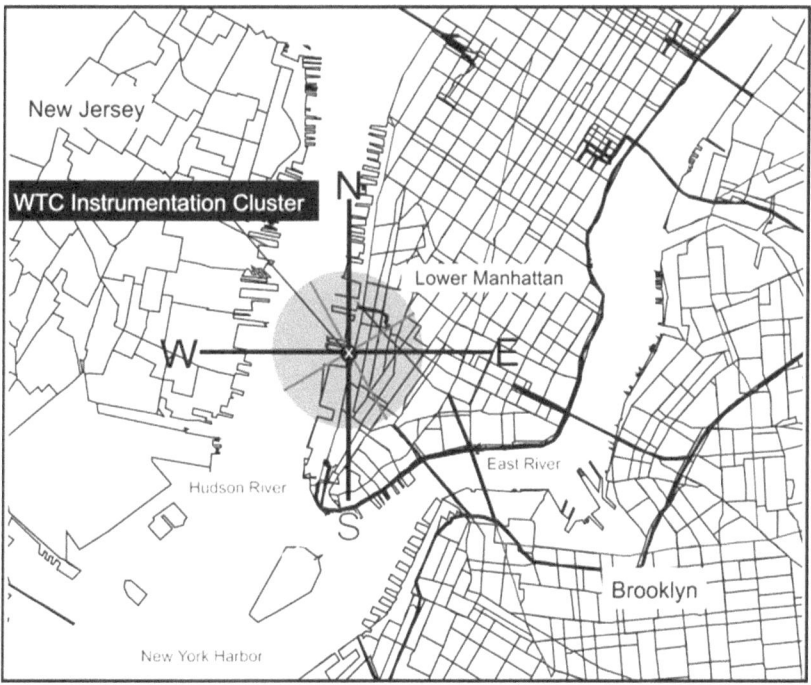

Figure 1
Plan view of the Lower Manhattan area, showing the location of NERL's 10 m Tower and two SODARs.

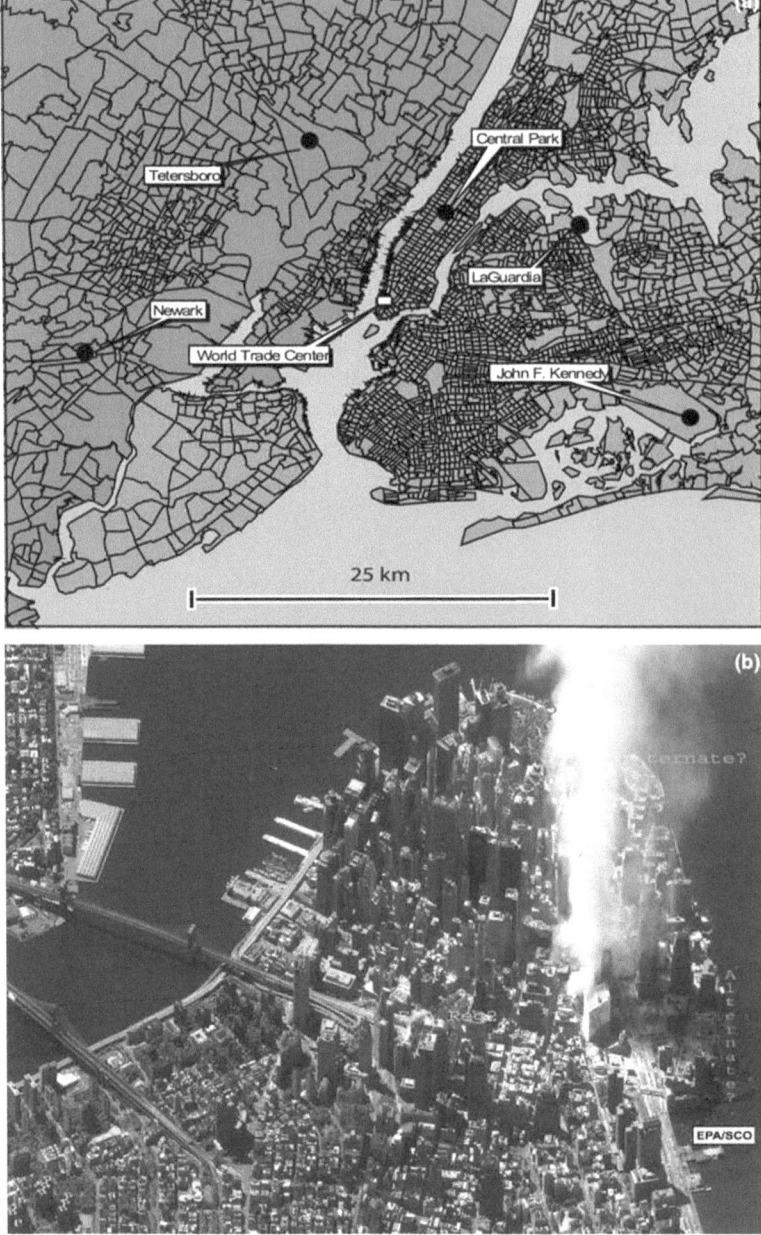

Figure 2
a A plan view of the NWS ASOS stations depicted by solid black circles and the WTC observation site, labeled at the southern tip of Manhattan Island. Figure 2b High-resolution photograph of lower Manhattan on 12 September 2001. The instrumentation cluster used in this study was located on Pier 25 (labeled EPA/SCO on the map).

3. Numerical Model

ARPS (Advanced Regional Prediction System) is a mesoscale meteorological model developed by the Center for Analysis and Prediction of Storms (CAPS), Oklahoma. ARPS was selected for this research because of its advanced physical and numerical schemes. ARPS is a non-hydrostatic, fully compressible primitive equation model suitable for simulating weather phenomena with spatial scales from several meters to several kilometers (XUE et al., 1995). The ARPS uses a terrain following a vertical coordinate system with options for stretched or equal spacing while the horizontal grid spacing is equal in both the x and y directions. Prognostic variables include 3-D wind components, potential temperature, pressure, subgrid scale TKE and moisture related variables (specific humidity, cloud ice, graupel and hail).

The ARPS simulation for this study used a 1.5-order TKE turbulence closure scheme developed by SUN and CHANG (1986). In this scheme a budget equation for subgrid scale TKE is solved, which includes buoyancy, shear production, advection (diffusion and transport) and viscous dissipation. The Lin-Tao 3 Category Ice (LIN et al., 1983) explicit moisture scheme is included along with the Kain Fritsch cumulus parameterization (KAIN and FRITSCH, 1993) for the water cycle in the 32 km domain. Implicit moisture physics schemes are not used in the 5 km and 1 km simulations. Radiation physics are simulated using the atmospheric radiation transfer parameterization developed at NASA/Goddard Space Flight Center, which is tailored for use in the ARPS model. This scheme includes equations for both shortwave (CHOU, 1990, 1992) and longwave (CHOU and SUAREZ, 1994) radiation processes. Details related to the above formulations are discussed in XUE et al. (1995, 2000, 2001). The Noilhan-Planton Land Surface Model is used to represent land surface processes over the region.

The ARPS simulation is initialized from the 32 km ARPS model analysis to generate an intermediate 5 km ARPS domain centered over New York City. For the 1 km simulation domain over New York City, the 5 km simulation domain provides initial and boundary conditions. The inner ARPS domain is shown in Figure 3. The inner domain has 50×50 grid points with 37 vertical sigma levels. The ARPS landuse data are also shown in shaded contours in Figure 3. The data are regridded to evenly fit onto the ARPS grid. Roughness length over the urban region was changed from 0.5 m to 1.5 m to account for the highly urbanized landscape associated with New York City.

4. Results and Discussion

4.1. Observational Analysis

A detailed observational analysis over the New York City area is one of the objectives of this study. The focus of this study period is 10 September 2001 through

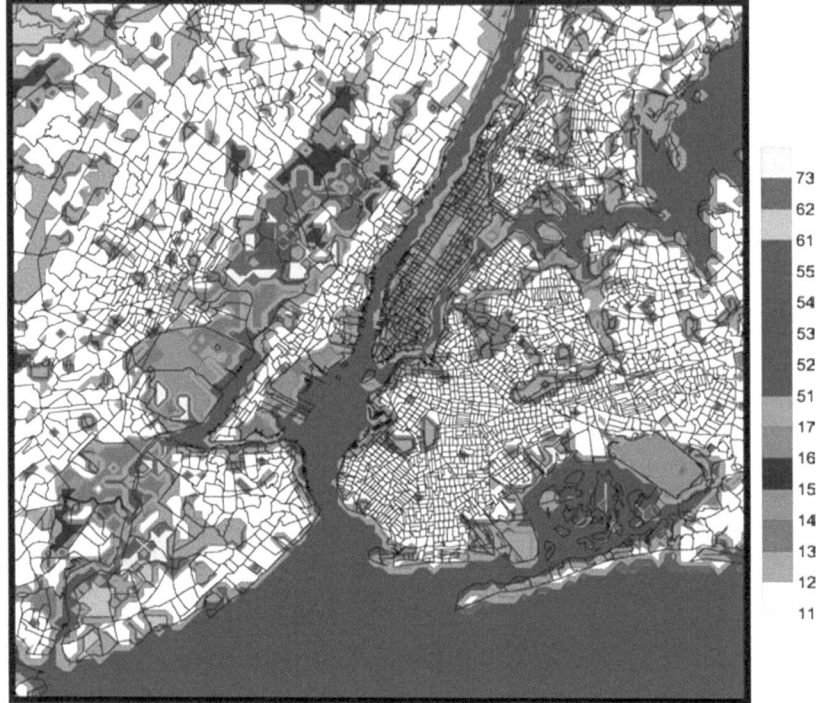

Figure 3
USGS land-use parameterization for the New York City ARPS domain. Land-use data are specified with a grid spacing of 0.9 km.

10 December 2001. Observations from five National Weather Service (NWS) ASOS sites and an independent instrumentation cluster are used in this study as described in Section 2 above.

The synoptic conditions over a three-month study period (September 11, 2001 – December 15, 2001) have been classified for each day into one of seven climatological flow regimes that normally exists during the fall season, or classified as "other" for different synoptic occurrences. Seasonal weather patterns affect the local meteorology and dispersion of pollutants in NYC. The mesoscale boundary layer structure and atmospheric stability vary seasonally and during different synoptic flow situations. Climatologically, a weather system passes on average every 4–6 days (BROWN and SETHU RAMAN, 1981) during the fall season. This cycle, starting after a cold front passage, typically includes a day of moderate to strong (>4–5 ms^{-1}) N-NW winds; followed by a transition day where the wind decreases as it veers from northerly to northeasterly. Next, the region experiences a day during which high pressure is centered near or directly over the area and winds become light and variable. Following this, the high pressure system moves east and winds turn southerly but remain light for a day; then as another frontal boundary approaches from the west, southwest winds increase to moderate levels. Based on this evolution, all days during the study period have been categorized as one of these flow regimes except for a limited few that could not be grouped into the above classifications. These "other" days were mostly situations when either a strong low-pressure system impacted the area or frontal boundaries oscillated over the region, resulting in drastic wind shifts. Figure 4a shows a pie chart illustrating the synoptic flow regimes observed over the New York City region between 10 September 2001 and 10 December 2001. Seven synoptic flow regimes, along with an "other" category for complex flow patterns, are analyzed in Figure 4a. The categories are southerly, westerly and northerly with further divisions by the estimated flow strength (light or strong). The light and strong flow classification was determined by a critical wind speed of $4.0~\text{ms}^{-1}$ that has been linked to the urban heat island formation (BORNSTEIN and JOHNSON, 1977) and sea breeze development (ARRITT, 1993). The flow strength and direction were subjectively determined by examining six-hourly synoptic charts provided by the National Center for Environmental Prediction (NCEP) and surface observations.

The data were then classified based on a daily average of wind speed and direction. The range of wind flow for northerly regimes was defined as flow from 310° to 20°, westerly flow from 250° to 300° and southerly flow from 180° to 250°. Additionally, a light and variable and an "other" classification were included. Four flow regimes dominated: light southerly (18%), strong southerly (18%), strong westerly flow (17%), and light and variable flow (16%). These regimes occurred on 70% of the days. The remaining periods were light westerly (9%), light northerly (6%), strong northerly (7%) and other (9%), respectively. A wind rose valid 10 September 2001 through 10 December is shown in Figure 4b. Hourly observations from Newark, Central Park, LaGuardia and JFK ASOS sites valid 10 September 2001 through 10 December 2001 were used to create the wind rose. Distribution rings are labeled every 5% with wind speed ranges defined as above and below $4~\text{ms}^{-1}$, as discussed above. The wind rose shows that a large percentage (35%) of wind speeds greater than $4~\text{ms}^{-1}$ originated from a direction between southwest and northwest. Additionally, the wind rose indicates that lighter winds were typically observed when the synoptic flow was out of the east and southeast.

4.2. Roughness Length Estimations

Large aerodynamic roughness length variations are often observed over highly urbanized terrain, such as New York City. These variations can significantly affect the surface airflow, causing a reduction in wind speed, turning of the winds, or both. The effect of the urbanized terrain on the surface winds over New York City is highly dependent on the mesoscale flow direction.

In order to quantify this effect, aerodynamic roughness lengths were estimated over lower Manhattan after separating the data into one of the four flow regimes.

(a)

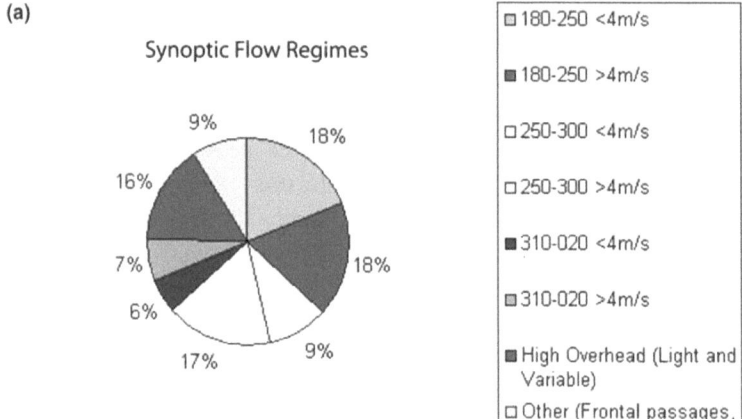

(b)

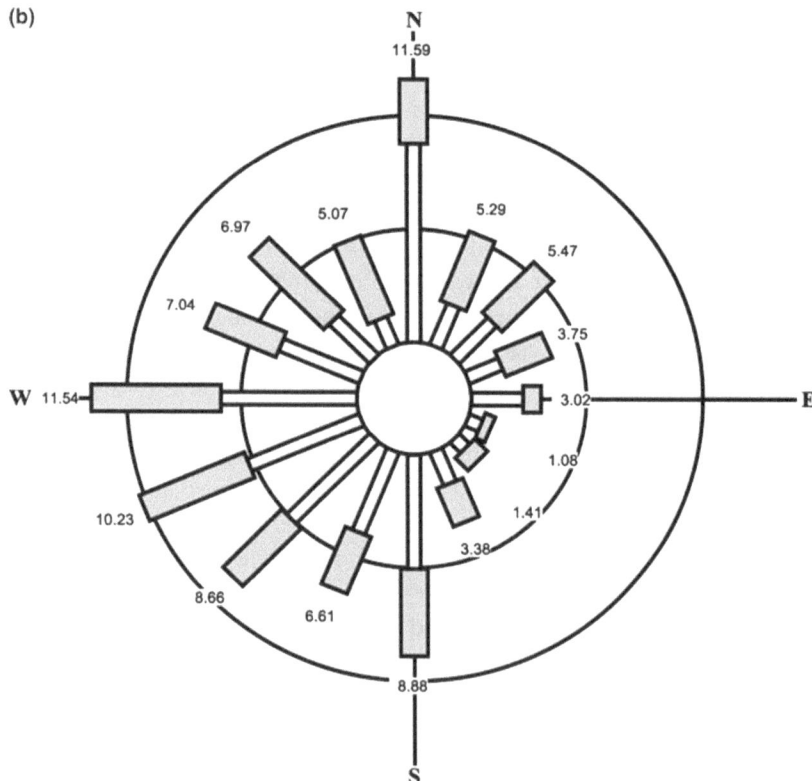

Figure 4

a Pie chart illustrating the frequencies (percentages) of the synoptic flow regimes (wind directions and speed) observed over the New York City region between 10 September 2001 and 10 December 2001. Figure 4b Wind Rose showing the distribution of wind speed and direction over the New York City area valid 10 September 2001 through 10 December 2001. Distribution rings are contoured every 5%.

These regimes include 0–89°, 90–179°, 180–269° and 270–359°. The 0–89° flow moves over the urban core of central Manhattan before reaching the WTC site, while the 90–179° flow moves over the urban core of lower Manhattan before reaching the WTC site. The 180–269° flow moves over Staten Island and the Hudson River before reaching the WTC site while the 270–359° flow moves over the Hudson River and portions of Manhattan Island before reaching the WTC site. Data from the independent 10 m tower over lower Manhattan are used to identify each flow regime. Additionally, the MiniSODAR, located in the vicinity of the 10 m tower, was used in the aerodynamic roughness length estimation. The 10 m tower and the miniSODAR became operational on 8 November 2001. The wind speed data for different flow regimes are averaged over the last month of the study period, November 10 through December 10, 2001. Wind speed and direction are averaged separately over 24-hr periods (00 UTC to 00 UTC), then classified into the appropriate flow regime based on the above conditions. Four 24-hr periods were observed for the 0–89 deg flow regime, while six 24-hr periods were classified in the 90–179 deg flow classification. Ten 24-hr periods were classified into 180–269 deg flow regime, while twelve 24-hr periods were classified into the 270–359 deg flow regime. The data were averaged over 24-hr periods to mitigate the effects of missing data from the miniSODAR.

This period was chosen because high levels of ground level pollutants, including PM-2.5, were observed over the region during this period. Monthly averaged wind speed values were calculated and then broken up into the four ranges of flow directions. The estimated aerodynamic roughness lengths for the 0–89 and 90–179 degree wind directions were approximately 3.8 m. These values seem reasonable as the flow pattern between 0 and 179 degrees is moving directly over the urban core of lower Manhattan. For the 180–269 degree flow directions, the average aerodynamic roughness length was 0.7 m, while for the 270–359 degree flow direction, the average aerodynamic roughness length was 0.9 m. Both of these lower values seem reasonable, as the flow pattern between 180 and 359 degrees moved over the Hudson River before being measured by the instrumentation cluster. Considerably lower values of aerodynamic roughness length, less than 0.01 m, are often observed over the water. However, various near-surface features, including waterfront office buildings, boat depots and even large ships and barges, influence the flow offshore in lower Manhattan. The calculated aerodynamic roughness lengths agree well with the Davenport-Wieringa roughness length classifications (STULL, 1988). This scheme classifies centers of large towns and cities, such as New York City, as chaotic with aerodynamic roughness lengths greater than 2 m.

4.3. Urban Heat Island

This section analyzes the observations during the period 00 UTC (19 LST) 13 November 2001 through 00 UTC (19 LST) 16 November 2001. High-pressure controlled the weather over much of the contiguous United States on 13 November

2001. Given the light synoptic-scale flow, local scale meteorological influences were pronounced on 13 November 2001 over the New York City region. The surface high-pressure center moved slowly off the Mid-Atlantic coast on 14 and 15 November, resulting in a light to moderate southwesterly near-surface wind flow across NYC. This period was selected to study the influences of near-surface wind flow moving off the water on the temperature and wind fields over the WTC site in lower Manhattan. Surface observations from five National Weather Service ASOS sites and a 10 m micrometeorological tower located in lower Manhattan (location of the instrumentation cluster was shown in Fig. 1) have been used in this study. Additional near-surface wind data were obtained from the Model 4000 miniSODAR located in lower Manhattan. The ASOS sites include Central Park, LaGuardia Airport, JFK Airport, Newark Airport and Teterboro (the location of the ASOS sites are shown in Fig. 2a). A surface (2 m) dry bulb temperature time series for the period 00 UTC (19 LST) 13 November 2001 through 00 UTC (19 LST) 16 November 2001 is shown in Figure 5. The daily maximum temperatures appear to be increasing throughout the study period, with an average maximum value of about 12 C observed by the stations on 13 November and a maximum value near 20 C observed on 15 November. Additionally, Central Park and LaGuardia appeared to stay warmer during the nighttime hours, as their temperatures remained nearly 2 C warmer than the other stations, including Newark, Teterboro and JFK. Observations from the 10 m micrometeorological tower in lower Manhattan were in between these extremes. The warmer temperatures observed during the nighttime in Central Park and LaGuardia were likely associated with the urban heat island, as one effect of the urban heat island is to keep surface

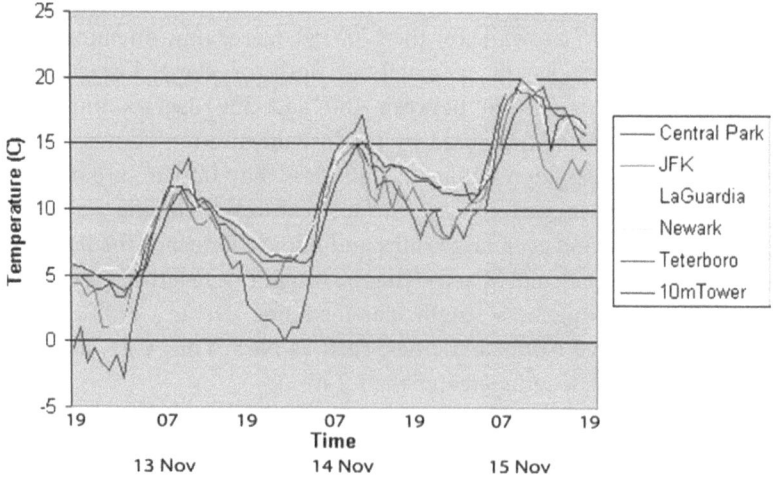

Figure 5
Surface (2 m) dry bulb temperature (c) time series valid 00 UTC (19 LST) 13 November 2001 through 00 UTC (19 LST) 16 November 2001. Five National Weather Service ASOS stations and the 10 m micrometeorological tower data are shown in color shading.

temperatures within the urban core warmer during the nighttime hours. The variability in the minimum temperature between the stations was considerably higher than the variations in maximum temperature. This may be due to light wind conditions, low mixing and differing heat capacities of the buildings and surrounding environment at the different locations. The mean temperature over the study period at Central Park and LaGuardia was between 1 and 2 C higher than the mean temperature over JFK, Newark and Teterboro, respectively. The 10 m micrometeorological tower in lower Manhattan registered a mean temperature of 10.8 C, which was less than the mean temperatures at Central Park and LaGuardia of 11.1 and 11.6 C, respectively although greater than the mean temperatures of 10.2, 9.7 and 9.2 C observed at Newark, JFK and Teterboro, respectively. Central Park, LaGuardia and the 10 m micrometeorological tower in lower Manhattan were located within the highly built-up urban core of New York City, and were likely influenced by the effects of the urban heat island which kept their temperatures warmer at night than surrounding rural locations.

A time series plot from the Model 4000 miniSODAR profile in lower Manhattan (Figure 1) for the period 12 UTC (07 LST) 13 November 2001 through 12 UTC (07 LST) 14 November 2001 is shown in Figure 6. Wind barbs are shown in standard notation. The lowest (15 m) observations are typically unreliable, therefore it should be ignored. The miniSODAR showed west to southwesterly winds at a height of 20 m between 15 UTC (10 LST) and 18 UTC (13 LST) 13 November. At approximately 19 UTC (14 LST) the 20 m winds became more southerly, and were likely associated with the passage of a sea breeze front. Another interesting feature was the vertical profile of nearly uniform wind speed and direction between 03 UTC (22 LST) 14 November and 12 UTC (07 LST) 14 November over lower Manhattan. Such a wind profile is often associated with a daytime convectively mixed boundary

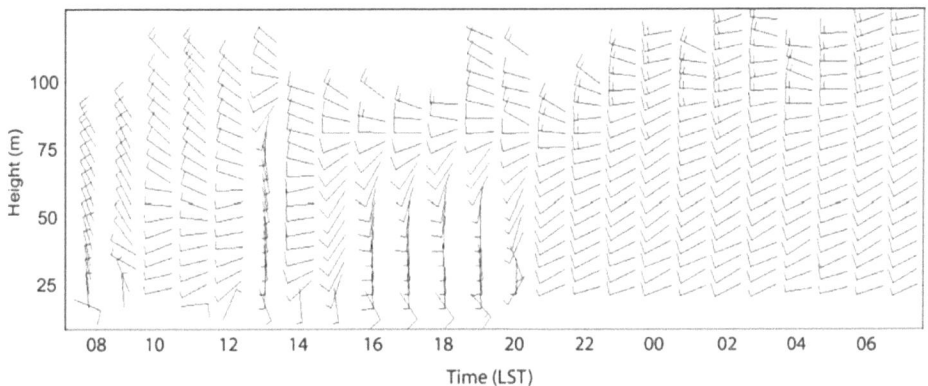

Figure 6

Model 4000-miniSODAR profile at the WTC Instrumentation site in lower Manhattan on 13–14 November 2001. Time is shown in LST.

layer. Near-surface west-southwesterly winds were advecting air into lower Manhattan that previously crossed over Staten Island. This apparent mixed layer may be the result of urban heat island induced static instability, originating over Staten Island, allowing greater turbulent mixing in the nocturnal boundary layer. However, mechanical mixing may also be contributing to this apparent mixed layer.

4.4. Numerical Simulations

Another objective of this research is to study the evolution of the mesoscale boundary layer over the New York City Metropolitan area through numerical simulations. More specifically, the urban heat island effect and sea breeze circulation will be examined in detail. Because of the highly urbanized landscape characteristics of this region, high-resolution numerical simulations are challenging. The ARPS model simulation will be compared and contrasted with surface weather observations taken during a high-pollutant concentration event over New York City in November 2001. Results from the simulation will be used to study the diurnal structure and evolution of the mesoscale boundary layer over the region.

The ARPS model was initialized at 00 UTC 13 November 2001 and integrated over a 60–hr time period until 12 UTC 15 November 2001. This period was chosen because of the formation and propagation of a sea breeze front through lower Manhattan, and also because the synoptic pattern favored the development of the urban heat island. A full synoptic review was presented as part of the Observational Analysis above. The ARPS simulation had a horizontal grid spacing of 1 km with 37 vertical sigma levels.

4.5. Simulated Surface Energy Budget

A detailed map of the New York City Metropolitan area is shown in Figure 7a. Labeled on Figure 7a are the letters B and C, which correspond to the location of surface energy budget time series, shown in Figures 7b and 7c over lower Manhattan and eastern New Jersey between 12 UTC (7 LST) 13 November and 6 UTC (01 LST) 15 November, respectively. Surface latent heat flux is shown in green, surface sensible heat flux in red and ground heat flux in black. All fluxes are plotted in W m^{-2}. There are several interesting features observed on the two time series simulations. The simulated energy budget time series over lower Manhattan, shown in Figure 7b, will be discussed first. Of interest is the occurrence of negative (downward) surface sensible heat flux between 00 UTC and 03 UTC (19 and 22 LST) on 14 November. Sensible heat flux values around negative (downward) 50 W m^{-2} were simulated during this period. By 05 UTC (00 LST), the surface sensible heat flux became positive, and remained positive (upward) until 21 UTC (16 LST). Positive surface sensible heat fluxes during the night over lower Manhattan were likely the result of the formation of urban heat island (SETHU RAMAN and ÇERMAK, 1974). The simulated energy budget time series over New Jersey is shown in Figure 7c. The

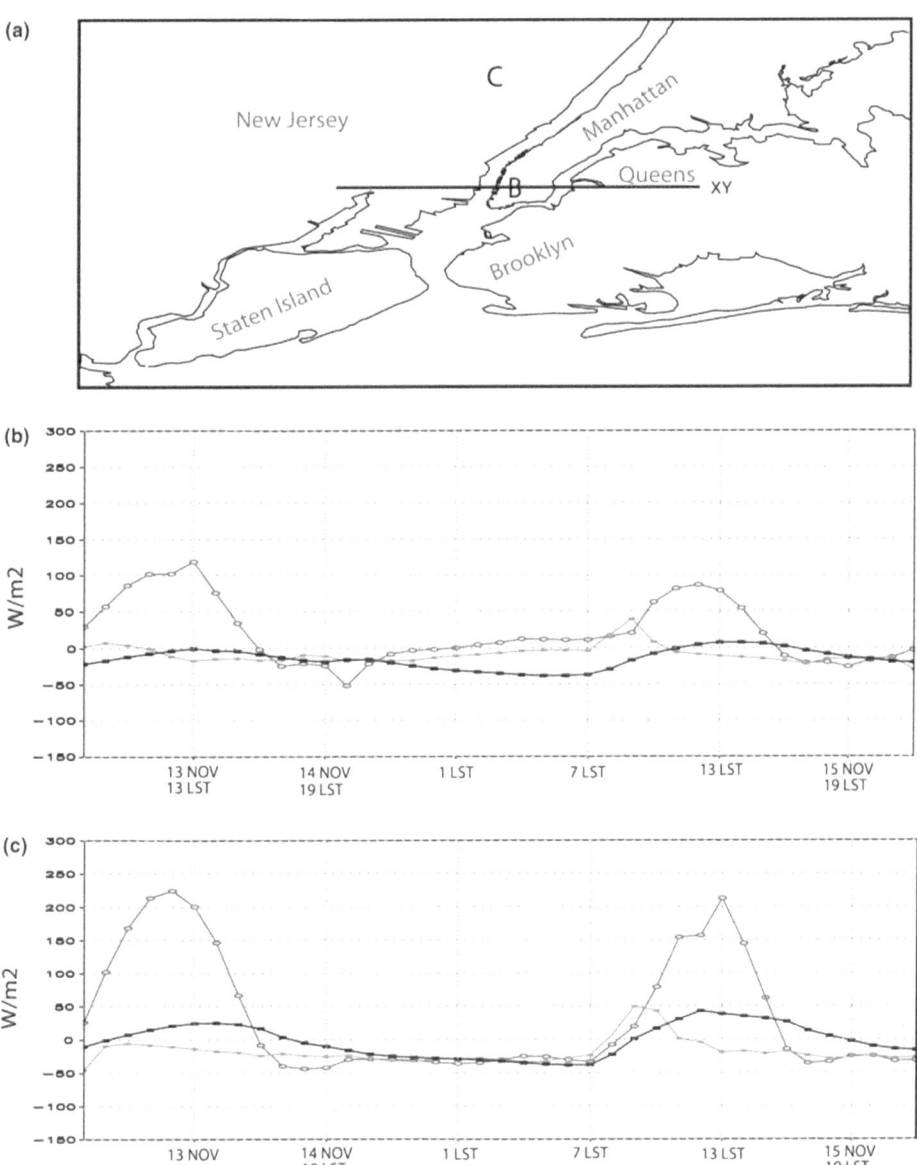

Figure 7
a Detailed map of the New York City Metropolitan area. Figures 7b and 7c show the simulated surface energy budget over lower Manhattan (labeled B on the map) and over New Jersey (labeled C on the map). Surface latent heat flux is shown in green, surface sensible heat flux is shown in red and ground diffusive heat flux is shown in black, respectively. All fluxes are simulated in W/m^{-2}.

simulated surface sensible heat flux values showed a less complex diurnal variation than the simulated surface sensible heat flux values over lower Manhattan. Over New

Jersey, the surface sensible heat flux values were positive during the daytime hours and became negative as nighttime approached (17 LST). Simulated surface sensible heat flux values remained negative throughout the night, as this region is more rural. In addition to the differences in the simulated surface sensible heat flux values during the night, the energy budget time series over lower Manhattan and New Jersey also showed differences between maximum surface sensible heat fluxes. Surface sensible heat flux values of 225 W m^{-2} were simulated over New Jersey at 17 UTC (12 LST) 13 November, while surface sensible heat flux values of 120 W m^{-2} were simulated over lower Manhattan at the same time. Additionally, surface sensible heat flux values exceeding 200 W m^{-2} were simulated over New Jersey at 18 UTC (13 LST) 14 November, while surface sensible heat flux values less than 100 W m^{-2} were simulated over lower Manhattan at the same time. The simulated surface sensible heat flux values were smaller over lower Manhattan because of the southwest flow moving over the Hudson River and Atlantic Ocean causing boundary layer airmass modification over lower Manhattan. Near-surface winds were southwesterly over New Jersey as well, however, these winds were associated with a continental airmass and did not experience any marine airmass modification. Surface latent heat flux simulations were very similar between New Jersey and lower Manhattan, close to zero. This seems reasonable, as surface latent heat flux values over a highly urbanized area are expected to be near zero, and over a residential area, slightly positive.

Several features simulated in the above surface energy budget plots exhibited the signature of an urban heat island. The surface sensible heat flux simulation time series over lower Manhattan was negative briefly during the nighttime hours of 13 and 14 November. Positive surface sensible heat fluxes are often associated with daytime conditions when incoming shortwave insolation is maximized. However, the urbanized structures associated with lower Manhattan act as heat holding materials tending to keep surface temperatures significantly warmer at night (urban heat island). In turn, positive surface sensible heat fluxes are generated over lower Manhattan, and are directly associated with the effects of the urban heat island.

4.6. Simulated Sea Breeze Structure

Wind velocity vectors and vertical velocity (contoured in ms^{-1}) at 100 m above the surface valid 15 UTC (10 LST) 13 November are shown in Figure 8a. The 100 m wind flow generated winds out of the north over much of the domain, becoming more westerly over and just to the east of Staten Island. Observing the vertical velocity contours, an enhancement in upward vertical motion was simulated over and just east of Staten Island where upward vertical velocities were near 0.15 ms^{-1}. To the north of this boundary winds were moving from the north and northeast, while to the south of the boundary, winds were moving from the south and southwest. The independent 10 m tower over lower Manhattan, shown in plan view in Figure 1, validated the model simulation, showing south-southwesterly winds associated with

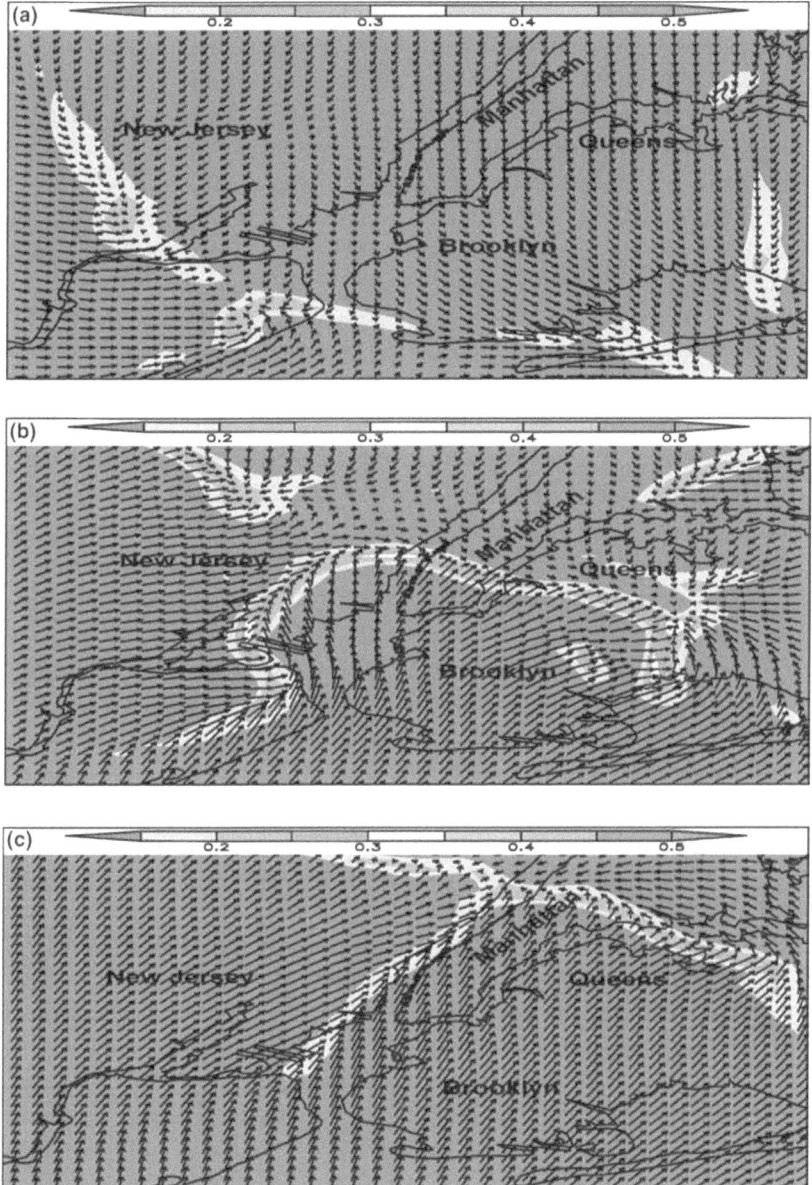

Figure 8
ARPS simulated 100 m wind velocity vectors (m/s) and vertical velocity contours (shown in grayscale defined by the overhead color bar) in m/s. Figure 8a is valid 15 UTC (10 LST) 13 November 2001. Figure 8b is valid 18 UTC (13 LST) 13 November 2001. Figure 8c is valid 21 UTC (16 LST) 13 November 2001.

the passage of the sea breeze front. Observed surface temperatures over land were 12C, while sea-surface temperatures (SST) were observed near 8C. This thermal gradient was strong enough to develop a sea breeze front over the region.

Another interesting feature is the area of enhanced wind speeds simulated over New York Harbor. The ARPS simulation was initialized with the Advanced Very High Resolution Radiometer (AVHRR) SST analyses archived at 1.44 km to allow for a more accurate simulation of boundary layer features along the land-water interface. Figure 8b delineates 100 m wind velocity vectors and vertical velocity (multi-color contoured in ms^{-1}) valid 18 UTC (13 LST) 13 November. A very distinct convergence boundary was observed over the region as an omega-like pattern. The convergence boundary stretched from Staten Island northward into New Jersey, and then spread eastward across lower Manhattan into Queens and Brooklyn before turning southward into Jamaica Bay. Associated with this convergence zone are areas of enhanced upward vertical velocities. Model simulated upward vertical velocities are between 0.2 and 0.35 ms^{-1} as seen in Figure 8b. BORNSTEIN *et al.* (1994) performed numerical simulations over the same area and observed a similar frontal alignment that extended through Staten Island, across lower Manhattan and eastward through Queens and Brooklyn. Figure 8c shows 100 m wind velocity vectors and vertical velocity (contoured in ms^{-1}) valid 21 UTC (16 LST) 13 November. A region of enhanced convergence and vertical motion is simulated over northern Manhattan. This is likely associated with the northernmost extent of the sea breeze's inland propagation. Upward vertical velocity values exceeding 0.2 ms^{-1} were simulated as a result of the low-level convergent forcing.

Figure 9 depicts a model simulated vertical cross section extending from west to east across the NYC Metropolitan area shown as XY in Figure 7a. Wind barbs (ms^{-1}) and vertical velocity (ms^{-1}) are shown as shading for this cross section in Figure 9. Two regions of enhanced upward vertical motion are simulated. The first region is over New Jersey, where a maximum upward vertical velocity of 0.7 ms^{-1} is simulated. This matched the area of enhanced 100 m level convergence simulated over the same region shown in Figure 8b. Another region of enhanced upward vertical motion is evident over Queens and Brooklyn. A well-defined maximum upward vertical motion exceeding 0.6 ms^{-1} was simulated over this region, agreeing closely with the zone of enhanced 100 m convergence simulated in Figure 8b over the same area. With southerly winds simulated in the lowest 250 m of the vertical cross section and westerly to northwesterly winds simulated above 300 m, this frontal feature is shallow in its vertical extent. This simulated feature agrees well with previous research by MICHAEL *et al.* (1998) and BORNSTEIN *et al.* (1994), which showed similar results using Weather Surveillance Radar 88 Doppler (WSR-88D) imagery and numerical simulations, respectively.

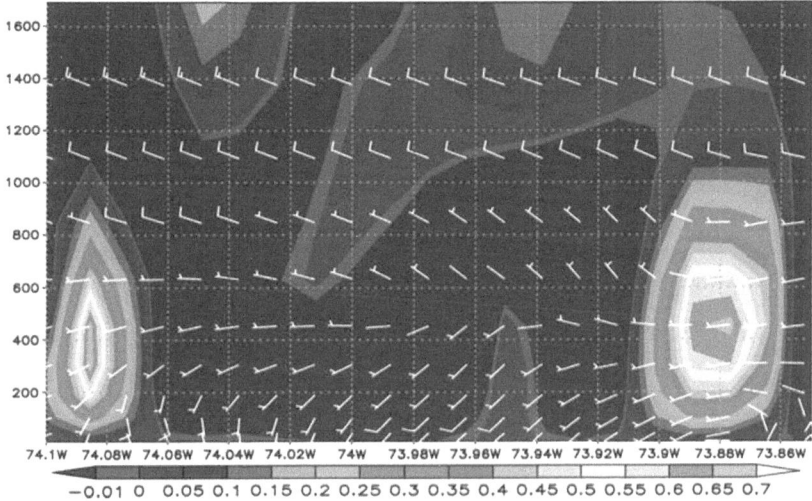

Figure 9
Vertical Velocity (m/s) cross section (m) through a sea-breeze front over extreme southern Manhattan
Island valid 18 UTC (13 LST) 13 November 2001.

4.7. Urban Heat Island

Model estimated 10 m wind direction (vectors), and speed (contoured in ms^{-1}), at 03 UTC (22 LST) 14 November is shown in Figure 10a. Surface wind observations (ms^{-1}) are shown in red. A very complex wind flow pattern is shown in Figure 10a. Southwesterly winds in excess of 4 ms^{-1} were simulated over New York Harbor. As the winds entered the urban core of lower Manhattan they began to slow to less than 2 ms^{-1} and back cyclonically, becoming more southerly over central Manhattan Island. A region of calm winds with speeds less than 0.5 ms^{-1} was simulated over Brooklyn and extreme eastern lower Manhattan. This calm wind was likely a result of the frictional drag caused by the high roughness length associated with lower Manhattan and Brooklyn. Model estimated 100 m wind direction (vectors) and speed (contoured in ms^{-1}) at 03 UTC (22 LST) are shown in Figure 10b. Similar to the 10 m winds, the 100 m winds slowed from 10 ms^{-1} to 6 ms^{-1} as they moved over the more urbanized landscape associated with Manhattan and Brooklyn. However, unlike the 10 m winds, the 100 m winds did not turn cyclonically, remaining southwesterly throughout the entire region. Variations in building height and drag likely resulted in the shallow nature of this cyclonic turning. This feature agrees with observational findings by BORNSTEIN and JOHNSON (1977) that showed nighttime events during stronger flow regimes (>4 ms^{-1}) were associated with distinctive roughness induced cyclonic turning in the winds over the main core of Manhattan and Brooklyn.

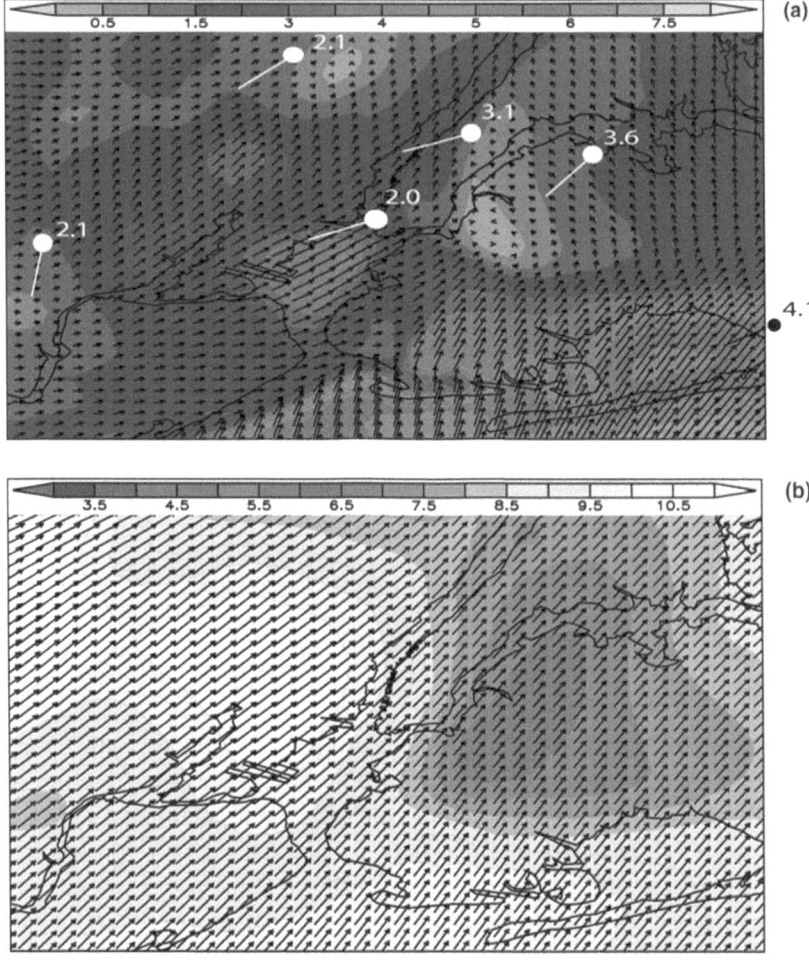

Figure 10

a ARPS simulated 10 m wind velocity (m/s) valid 03 UTC (22 LST) 14 November 2001. 10 m wind observations shown in white with the barb showing wind direction from, and number indicating speed (m/s) Figure 10b ARPS simulated 100 m wind velocity (m/s) valid 03 UTC (22 LST) 14 November 2001.

Vertical TKE $(m^2 s^{-2})$, wind speed (kts, 1 kt = 0.52 ms^{-1}) and potential temperature (K) time series over the World Trade Center (WTC) disaster recovery site valid 12 UTC (7 LST) 13 November through 6 UTC (01 LST) 15 November is shown in Figure 11. A SODAR profile observed at the WTC Instrumentation Site in lower Manhattan from 12 UTC 13 November through 12 UTC (07 LST) 14 November (time is shown in both local and UTC formats) was shown in Figure 6. Wind barbs are shown using the standard notation. From Figure 11, several features are apparent. The first feature analyzed is the region of maximum TKE simulated

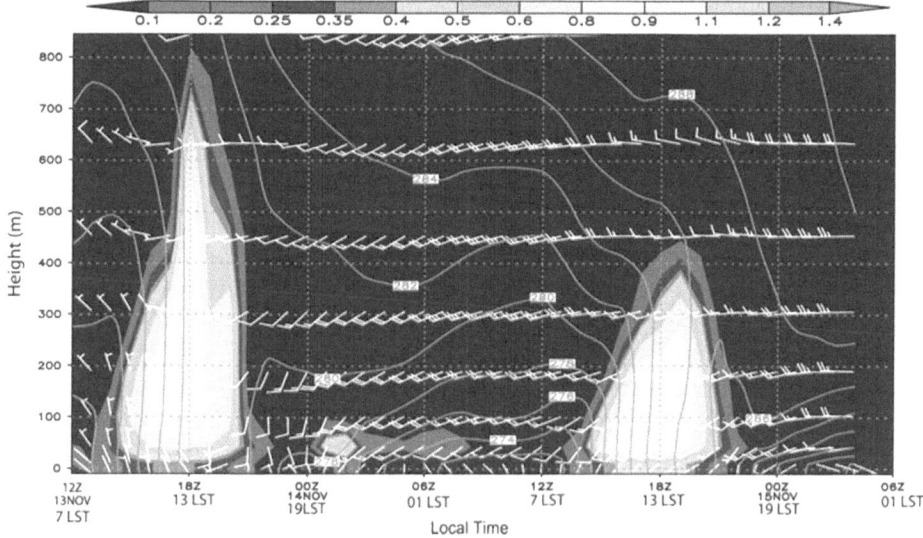

Figure 11

Time series vertical profile of turbulent kinetic energy (m²/s²), horizontal wind (kt) and potential temperature (K) simulated by ARPS at the WTC site. TKE is shaded, potential temperature is contoured and wind is shown by standard barbs notation in knots.

between 16 UTC (11 LST) and 19 UTC (14 LST) 13 November. Boundary layer heights reached nearly 800 m at 18 UTC (13 LST), with TKE values greater than 1.2 m² s⁻² between the altitudes of 300 m and 450 m. Another feature is the boundary layer wind field between 16 UTC and 19 UTC (11 LST and 14 LST). Winds below 300 m were simulated from the north with a magnitude of 5 kts (~ 3 ms⁻¹). Between 18 and 19 UTC (13 and 14 LST), the wind direction changed from northerly to southerly following the passage of the sea breeze front. Data from the SODAR, shown in Figure 6, verifies the ARPS simulation, showing the wind shift from northerly to southerly around 18 UTC (13 LST). Additionally, there is a region of enhanced turbulence and wind speed simulated for the time period between 01 and 08 UTC (20 and 03 LST) 14 November with low-level winds simulated between 10 and 15 kts (5 to 8 ms⁻¹). Analysis from the SODAR data also revealed a region of maximum low-level winds of 15 kts (8 ms⁻¹) observed at 00 UTC (19 LST) 14 November. This feature may be a result of the low-level convergence associated with the urban heat island effect and will be more thoroughly discussed below. After 02 UTC (21 LST) 14 November, the low-level wind flow was simulated and was also observed out of the southwest at 10 kts (5 ms⁻¹). Another region of maximum TKE and boundary layer height was simulated between 16 UTC and 21 UTC (11 and 16 LST) 14 November. This regime differed significantly from the boundary layer structure on 13 November. For example, the maximum boundary layer height was approximately 400 m lower on 14 November than it was on 13 November (800 m). The maximum TKE simulated on 14 November was less than 0.95 m^2 s^{-2}, which was observed at 18 UTC (13 LST). This was significantly less than the 1. 2 m^2 s^{-2} of TKE simulated on 13 November. The decrease in simulated TKE was likely the result of increased static stability seen in the potential temperature distribution. It is apparent from examining the time series and SODAR data that the synoptic scale wind flow overwhelmed any mesoscale and microscale meteorological processes directly related with the urban heat island after 08 UTC (03 LST) 14 November. Southwesterly winds of 10 kts (5 ms^{-1}) and greater were observed during this time period, keeping the boundary layer well mixed and homogeneous over the study area.

Figure 12 presents a model simulated vertical cross section of TKE, displayed from west to east along the line XY shown in Figure 7a. The cross section was centered over the World Trade Center disaster recovery site (40.50°N–74°W). Shown in Figure 12 are wind barbs and TKE (m^2 s^{-2}) valid 18 UTC (13 LST) 13 November, just as the sea breeze front was moving into lower Manhattan. A localized TKE maximum exceeding 1.1 m^2 s^{-2} was simulated over New Jersey, the boundary layer height decreased from 600 m at 17 UTC (12 LST) 13 November, to less than 550 m at 18 UTC (13 LST). The model also simulated a relative minimum of turbulent energy between New Jersey and Manhattan over water. This air column, directly over the Hudson River, registered TKE readings near 0.7 m^2 s^{-2}, and was likely associated with a more stable maritime airmass. A maximum in TKE, exceeding

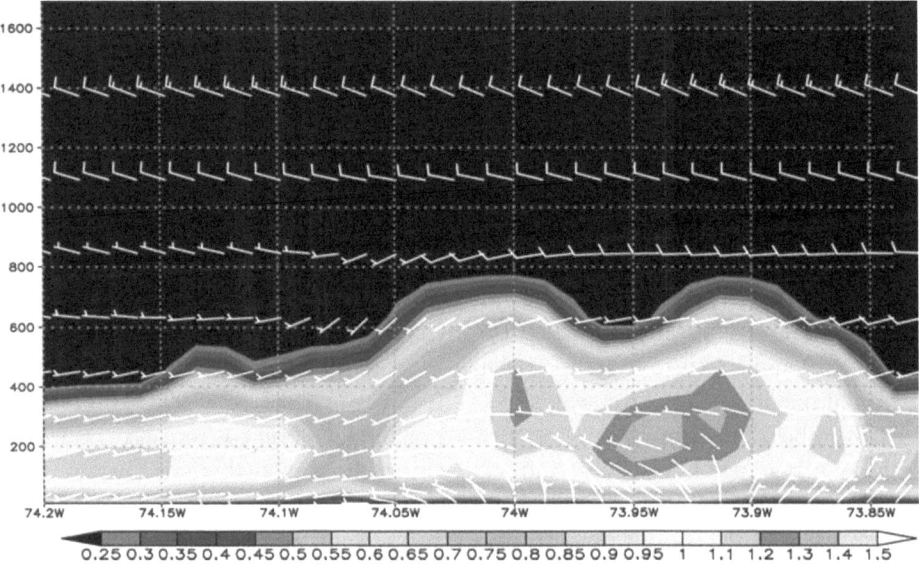

Figure 12
Vertical cross section (m) of TKE (m^2/s^2) and wind barbs valid at 40.50 N at 18 UTC (13 LST) 13 November 2001.

$1.2 \ m^2 \ s^{-2}$, was evident directly over lower Manhattan. The boundary layer height was also maximized in this region, approaching 800 m, experiencing a growth of nearly 400 m between 17 UTC and 18 UTC (12 and 13 LST). This rapid growth was likely associated with greater amounts of surface layer heating which was a direct result of the largely urbanized land use. To the east of this feature was a localized minimum in boundary layer height, approximately 600 m, associated with the more statically stable East River. Boundary layer heights quickly rebounded east of this region, over Long Island, and again approached 800 m. TKE is also maximized over Long Island, with values exceeding $1.3 \ m^2 \ s^{-2}$. The TKE maximum was located at a depth of 200 to 300 m, or approximately one-third the height of the boundary layer, which agreed with the expected region of maximum turbulent energy (STULL, 1988). The wind flow was generally from the north over the lowest 100 m of the simulated cross section, with the exception of a more westerly component developing over New Jersey. This component is a result of the developing convergence zone associated with the strengthening sea breeze front south of lower Manhattan. Above 100 m, $5 \ ms^{-1}$ westerly winds are simulated over New Jersey and Manhattan, while a northwesterly wind associated with enhanced turbulent energy was simulated above 100 m over western Long Island. Over eastern Long Island the wind field above 100 m was out of the north and northeast at $5 \ ms^{-1}$.

5. Summary and Conclusions

Observations from several instrumentation platforms were examined during different synoptic scale flow regimes over NYC. A numerical simulation was conducted to explore the urban heat island, urban roughness effect and sea breeze structure over the NYC region.

Additionally, the effects of wind directions on the roughness lengths observed over lower Manhattan were also analyzed. The roughness lengths varied from 0.7 m with a westerly flow over water to about 4 m with a flow through Manhattan. A nighttime mixed layer was observed over lower Manhattan This apparent mixed layer may be the result of urban heat island induced static instability, originating over Staten Island, allowing greater turbulent mixing in the nocturnal boundary layer.

Simulated surface energy budget also shows the presence of an urban heat island. The ARPS model simulated the development and inland penetration of the sea breeze front over the region. The sea breeze front formed because of strong differential heating between the land of the region and the Atlantic Ocean, in the presence of a light and variable synoptic scale flow. The mesoscale model simulated the sea breeze front moving through lower Manhattan during this period and agreed well with both SODAR and 10 m tower observations from an independent instrumentation cluster maintained by the EPA and State Climate

Office of North Carolina in lower Manhattan. The general structure of the sea breeze front over the region also agreed with previous studies by BORNSTEIN, *et al.* (1994), who performed numerical simulations over the same area and observed a similar frontal alignment that extended through Staten Island, across lower Manhattan and eastward through Queens and Brooklyn.

The nocturnal boundary layer was also studied using surface and 100 m wind simulations, as well as surface energy budget figures and TKE cross sections. Wind simulations revealed a slowing and cyclonic turning of the 10 m wind as the flow moved over Brooklyn, Queens and Manhattan, while 100 m wind simulations reflected a slowing of the wind flow, however no discernable alteration of flow directions. This simulated feature is in agreement with observational findings by BORNSTEIN and JOHNSON (1977) that showed nighttime events during stronger flow regimes (>4 ms^{-1}) were associated with distinctive roughness-induced cyclonic turning in the winds over the main core of Manhattan and Brooklyn.

Vertical profiles of TKE and wind velocity were also examined. Several simulated profiles showed a maximum in TKE over lower Manhattan during nighttime conditions. It appears that this TKE maximum is directly related to the influences of the urban heat island. The simulated location and structure of the nocturnal boundary layer over lower Manhattan are consistent with the results and agreed well with previous research on urban heat islands by SETHU RAMAN and ÇERMAK (1974) and BORNSTEIN and JOHNSON (1977).

Additional near-surface wind and temperature data are needed to further evaluate the numerical model's ability to accurately simulate the mesoscale boundary layer over NYC. These data could also be used for data assimilation into the numerical models.

Acknowledgements

This work was supported by the State Climate Office (SCO) of North Carolina and the U.S. Environmental Protection Agency (U.S. E.P.A.). The SODAR observations were made as part of a joint project with the U.S. E.P.A. The authors thank Dr. Alan Huber (U.S. E.P.A.) for several helpful discussions and also in obtaining the data. Ryan Boyles, Robert Gilliam, and Ameenulla Syed of the SCO also assisted in obtaining the observations. The authors thank Ryan Boyles for technical assistance in preparing this manuscript.

REFERENCES

ANGELL, J.K., PACK, D.H., DICKSON, C.R., and HOECKER, W.H. (1971), '*Urban Influence on Nighttime Airflow Estimated from Tetroon Flights*', J. Appl. Meteor. *10*, 194–205.

ARRITT, R.W. (1993), *Effects of Large-scale Flow on Characteristic Features of the Sea Breeze*, J. Appl. Meteor. *32*, 116–125.

ARYA, S.P., *Air Pollution and Dispersion Meteorology* (Oxford University Press, New York 1999).

ASAI, T. (1970), *Thermal Instability of a Plane Parallel Flow with Variable Vertical Shear and Unstable Stratification*, J. Meteor. Soc. Japan *48*, 129–139.

BORNSTEIN R.D., THUNIS, P., and SCHAYES, G., *Observation and Simulation of Urban-topography Barrier Effects on Boundary Layer Structure Using the Three-dimensional TVM/URBMET Model*. *"Air Pollution and its Application X"* (Plenum Press, New York 1994).

BORNSTEIN R.D. and JOHNSON, D.S. (1977), *Urban-rural Wind Velocity Difference*, Atmos. Environ *11*, 597–604.

BORNSTEIN R.D. (1975), *The Two-dimensional URBMET Urban Boundary Layer Model*, J. Appl. Meteor *14*, 1459–1477.

BROWN, R.M. and SETHURAMAN, S. (1981), *Temporal Variation of Particle Scattering Coefficients at Brookhaven National Laboratory, New York*, Atmos. Environ. *15*, 1733–1737.

CHOU, M.-D. (1990), *Parameterization for the Absorption of Solar Radiation by O_2 and CO_2 with Application to Climate Studies*, J. Climate *3*, 209–217.

CHOU, M.-D. (1992), *A Solar Radiation Model for Climate Studies*, J. Atmos. Sci. *49*, 762–772.

CHOU, M.-D. and SUAREZ, M.J. (1994), *An Efficient Thermal Infrared Radiation Parameterization for Use in General Circulation Models*, NASA Tech. Memo. 104606, 85 pp.

CRESCENTI, G.H. (1998), *The Degradation of Doppler SODAR Performance due to Noise: A Review*, Atmos. Environ. *32*, 1499–1509.

CRESCENTI, G.H. (1999), *A Study to Characterize Performance of Various Ground-based Remote Sensors*, NOAA Technical Memorandum ERL ARL-229, 286 pp.

DRAXLER, R.R. (1986), *Simulated and Observed Influence of the Nocturnal Urban Heat Island on the Local Wind Field*, J. Climate Appl. Meteor. *25*, 1125.

GILLIAM, R.G., CHILDS, P.P., HUBER, H., and RAMAN, S. (2003), *Metropolitan Scale Transport and Dispersion from the New York World Trade Center Following September 11, 2001. Part I: An Evaluation of The CALMET Meteorological Model*, NOAA Technical Memorandum.

GILLIAM, R.G. (2001), *Influence of Surface Heterogeneities on the Boundary Layer Structure and Diffusion of Pollutants*, M.S. Thesis, Department of Marine, Earth and Atmospheric Sciences. N.C. State University, Raleigh, NC.

KAIN, J.S. and FRITSCH, J.M. (1993), *Convective parameterization for mesoscale models: The KAIN-FRITSCH scheme*. In: *The Representation of Cumulus Convection in Numerical Models*, Meteor. Monogr., Amer. Meteor. Soc. 165–170.

LEE, D.O. (1979), *The Influence of Atmospheric Stability and the Urban Heat Island on Urban-rural Wind Speed Differences*. Atmos. Environ. *13*, 1175–1180.

LIN, Y.-L., FARLEY, R.D., ORVILLE, H.D. (1983), *Bulk Parameterization of the Snow Field in a Cloud Model*, J. Climate Appl. Meteor. *22*, 1065–1092.

MELLOR, G.L. and YAMADA, T. (1974), *A Hierarchy of Turbulence Closure Models for Planetary Boundary Layers*, J. Atmos. Sci. *31*, 1791–1806.

MICHEAL, P., MILLER, M., and TONGUE, J.S. (1998), *Sea breeze regimes in New York City region—Modeling and Radar Observations*, Transact. Second Conf. Coastal Atmospheric and Oceanic Prediction and Processes, 78th AMS Annual Meeting, 11–16 January 1998, Phoenix, Arizona.

SAWAI, T. (1978), *Formation of the Urban Air Mass and the Associated Local Circulations*, J. Meteor. Soc. Japan. *56*, 159–173.

SETHU RAMAN, S. and CERMAK, J.E. (1974), *Physical Modeling of Flow and Diffusion over an Urban Heat Island*, Adv. in Geophys. *18B*, 223–240.

SHREFFLER, J.H. (1978), *Detection of Centripetal Heat Island Circulations from Tower Data in St. Louis*, Bound.-Layer Meteor. *15*, 229–242.

SHREFFLER, J.H. (1979), *Heat Island Convergence in St. Louis during Calm Periods*, J. Appl. Meteor. *18*, 1512–1520.

STULL, Roland B, *An Introduction to Boundary Layer Meteorology* (Kluwer Academic Publishers, Norwell, MA 1988).

Sun, W.Y. and Chang, C.Z. (1986), *Diffusion Model for a Convective Layer. Part I: Numerical Simulation of Convective Boundary Layer.* J. Climate Appl. Meteor. *25*, 1445–1453.

Takeuchi, K. and Kimura, F. (1976), *Numerical Simulation of Photochemical Smog in Tokyo Metropolitan Area,* Pap. Meteor. Geophys. *27*, 41–53.

United States Geological Survey, Land Use Land Cover Data (LULC), Web address: http://edc.usgs.gov/products/landcover/lulc.html

Wong, K.K. and Dirks, R.A. (1978), *Mesoscale Perturbations on Airflow in the Urban Mixed Layer.* J. Appl. Meteor. *17*, 677–688.

Xue, M., Droegemeier, K.K., and Wong, V. (1995), *Advanced Regional Prediction System (ARPS) and Real-time Storm Prediction,* Preprint, International Workshop on Limited-area and Variable Resolution Models, Beijing, China, World Meteor. Organ.

Xue, M., Droegemeier, K.K., and Wong, V. (2000), *The Advanced Regional Prediction System (ARPS)- A Multi-scale Nonhydrostatic Atmospheric Simulation and Prediction Tool. Part I: Model Dynamics and Verification,* Meteor. Atmos. Phys. *75*, 161–193.

Xue, M., Droegemeier, K.K., Wong, V., Shapiro, A., Brewster, K., Carr, F., Weber, D., Liu, Y., and Wang, D. (2001), *The Advanced Regional Prediction System (ARPS) – A Multi-scale Nonhydrostatic Atmospheric Simulation and Prediction Tool. Part II: Model Physics and Applications,* Meteor. Atmos. Phys. *76*, 143–165.

(Received May 3, 2004, accepted October 20, 2004)
Published Online First: June 21, 2005

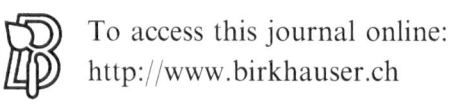

To access this journal online:
http://www.birkhauser.ch

Pure appl. geophys. 162 (2005) 1981–2003
0033–4553/05/101981–23
DOI 10.1007/s00024-005-2701-z

© Birkhäuser Verlag, Basel, 2005

▌Pure and Applied Geophysics

Metropolitan-scale Transport and Dispersion from the New York World Trade Center Following September 11, 2001. Part I: An Evaluation of the CALMET Meteorological Model

ROBERT C. GILLIAM[1,2], PETER P. CHILDS[1], ALAN H. HUBER[2] and SETHU RAMAN[1]

Abstract—Following the collapse of the New York City World Trade Center towers on September 11, 2001, Local, State and Federal agencies initiated numerous air monitoring activities to better understand the impact of emissions from the disaster. A study of the estimated pathway that a potential plume of emissions would likely track was completed to support the U.S. EPA's initial exposure assessments. The plume from the World Trade Center was estimated using the CALMET-CALPUFF dispersion modeling system. The following is the first of two reports that compares several meteorological models, including the CALMET diagnostic model, the Advanced Regional Prediction System (ARPS) and 5th Generation Mesoscale Model (MM5) in the complex marine-influenced urban setting of NYC. Results indicate wind speed, in most cases, is greater in CALMET than the two mesoscale models because the CALMET micrometeorological processor does not properly adjust the wind field for surface roughness variations that exits in a major built-up urban area. Small-scale circulations, which were resolved by the mesoscale models, were not well simulated by CALMET. Independent wind observations in Lower Manhattan suggest that the wind direction estimates of CALMET possess a high degree of error because of the urban influence. Wind speed is on average 1.5 ms^{-1} stronger in CALMET than what observations indicate. The wind direction downwind of the city is rotated 25–34 clockwise in CALMET, relative to what observations indicate.

Key words: Dispersion modeling, CALPUFF, CALMET, Plume modeling, Sea breeze, ARPS, MM5.

1. Introduction

In response to the events on September 11, 2001 at the New York World Trade Center (WTC), a study using the CALMET-CALPUFF (SCIRE *et al.*, 2000a) dispersion modeling system was conducted for a three-month period following the events. Prior to the WTC attack, efforts were already underway to use a similar modeling system for real-time support of air pollution studies in the Raleigh-Durham

[1]State Climate Office of North Carolina, North Carolina State University, Raleigh, NC, U.S.A. E-mail: Gilliam.Robert@epamail.epa.gov
[2]Atmospheric Sciences Modeling Division, Air Resources Laboratory, National Oceanic and Atmospheric Administration. On assignment to the United States Environmental Protection Agency, National Exposure Research Laboratory, Research Triangle Park, NC, U.S.A.

area of North Carolina. After the events at the New York WTC, all efforts were redirected toward the impact of the WTC event on the New York City (NYC) region. The CALMET-CALPUFF simulation was used as a support tool for the US EPA's preliminary assessment of the potential impact of emissions from the WTC site on New York City. In Part I of the study CALMET, the meteorological processor for the CALPUFF (SCIRE *et al.*, 2000b) dispersion model is examined to provide a better understanding of its strengths and weaknesses in representing the meteorology over a complex urban area such as NYC. Additionally, Part I of the study serves to better highlight the uncertainty of the meteorology and inherent limitations of the modeling system before the dependent dispersion results are analyzed in Part II. The study as a whole serves as a documented application for reference by the growing CALMET-CALPUFF user community.

It is well known that atmospheric boundary layer (ABL) heterogeneity is characteristic of a dense urban center like NYC that lies adjacent to an ocean or a large lake, especially when compared to the ABL over rural, inland areas. NYC's landuse is characterized as a highly developed urban core on Manhattan Island, and a sprawling dense suburban area that covers northeastern New Jersey and western Long Island. The NYC urban blocking effect and urban heat island phenomena have been examined in detail by BORNSTEIN and JOHNSON (1977) and BORNSTEIN *et al.* (1994).

Adding to this complex urban surface is a highly variable coastline consisting of many small bays, rivers and sounds (Jamaica Bay, New York Harbor, Hudson River, East River and Long Island Sound). A thermal internal boundary layer (TIBL) often develops along the coast, influencing the boundary layer structure over land. Its variation downwind depends on the surface roughness, upwind atmospheric stability and land-sea temperature contrast (RAYNOR *et al.*, 1979). TIBL exists to some extent during all conditions however is most pronounced during light to moderate synoptic flow cases in which the local temperature and wind variation can dominate the meteorology of the region. FRIZZOLA and FISHER (1963), BORNSTEIN *et al.* (1994), REISS *et al.* (1996) and MICHAEL *et al.* (1998) examine the sea breeze of NYC in detail using numerical models, surface and upper-air observations, and radar imagery. All of these features and their influence on the lower atmosphere make attempts at modeling the region difficult (MICHAEL *et al.*, 1998).

Considering all of these factors, the study first presents an overview of the general weather patterns that occurred during the study period. The synoptic conditions over the three-month study period (September 11, 2001–December 8, 2001) are classified into climatological flow regimes that normally exist during the fall season. Next, independent wind measurements are used for a quantitative evaluation of CALMET near the WTC site. Then, two prognostic models are used to examine the complexity of mesoscale and local scale variations over NYC. The prognostic models are compared and contrasted with the CALMET, as well as with observations, and the historical studies summarized above. It should be stated that the use of the prognostic models is strictly to illustrate that local features, which are important during certain synoptic flow patterns, are not as well represented by diagnostic models like CALMET. Also, observations are rather limited relative to the complexity of the flow patterns. This limitation makes an exact evaluation more difficult consequently some subjectivity is used in the analysis.

2. Methodology

2.1. Synoptic Flow Classification

Weather patterns affect the local meteorology and dispersion of pollutants over NYC. Climatologically, a weather system moves through the region every 4–6 days (BROWN and RAMAN, 1981) during the fall season. This cycle, starting after a cold front passage, typically includes a day of moderate to strong (> 4–5 m/s) N-NW winds; followed by a transition day where the wind decreases as it veers from northerly to northeasterly. Next, the region experiences a day where high pressure is centered near or directly over the area and winds become light and variable. Following this, the high pressure system moves east and winds turn southerly but remain light for a day, then as another frontal boundary approaches from the west, southwest winds increase to moderate levels. Based upon this evolution, all days during the study period have been categorized as one of these flow regimes, except for a limited few that could not be justly grouped into the above classification. These "other" days were mostly conditions when either a strong low pressure system affected the area or frontal boundaries oscillated over the region, resulting in drastic wind shifts.

Table 1

Classification of the synoptic conditions during the September 11–December 8, 2001 study period. Synoptic conditions are classified according to the flow strength and direction observed over the NYC region. Light flow is considered to be less than 4–5 m s^{-1} and strong flow greater than 4–5 m s^{-1}. Percentage of the entire period is shown in parenthesis. Also cited are the modeling case study(s) presented in this research that correspond to the synoptic category.

Synoptic Classification	Number of days (Percentage)	Representative Modeling Case
Light Southerly	17 (19%)	Nov. 14
Strong Southerly	16 (18%)	Oct. 4
Light Westerly	08 (9%)	Sep. 17
Strong Westerly	15 (17%)	Oct. 4
Light Northerly	05 (6%)	N/A
Strong Northerly	06 (7%)	Sep. 11
Light and Variable (High)	14 (16%)	Nov. 13, Sep. 12
Other	08 (9%)	N/A

Table 1 shows the frequency of the various flow classifications that occurred during the study period. The categories are southerly, westerly and northerly with these further divided by the estimated flow strength (light or strong). The light and strong flow classification was determined by the critical wind speed of 4.0 m s^{-1} that has been linked to the urban heat island (BORNSTEIN and JOHNSON, 1977) and sea-breeze development (ARRITT, 1993). The flow strength and direction were subjectively determined by examining six-hourly synoptic charts and surface observations. A light and variable and an "other" classification were included to account for days when the wind was highly variable or a strong storm system affected the region. Four flow regimes dominated: light southerly (19%), strong southerly (18%), strong westerly flow (17%), and light and variable flow (16%). These regimes occurred on 70% of the days. The remaining periods were light westerly (9%), light northerly (6%), strong northerly (7%) and other (9%).

2.2. Calmet Description

To support the EPA's study of the potential impact of airborne pollutants from the WTC tragedy on the populous, a CALMET model domain was designed to cover the central portion of the New York City (NYC) metropolitan area. The grid was centered over lower Manhattan and covered a 50 × 50 km square area at a grid spacing of 0.5 km. Figure 1a shows the extent of the CALMET model domain in relation to the surrounding suburbs of NYC. The vertical grid was stretched in a terrain following a coordinate system with twelve vertical levels: 10, 27, 51, 90, 195, 350, 512, 700, 1000, 1350, 1650 and 2000 meters.

CALMET is a diagnostic, observation-based model that requires three key data sets to function: landuse information, surface observations and upper-air observations. The landuse for the New York City model domain is classified in CALMET according to the United States Geological Survey landuse land cover database. For residential (category #11) the default roughness value was lowered from 1.0 m to 0.75 m (WIERINGA, 1993). The surface properties for water were set to its default values. Built-up urban classification is another landuse that covers a considerable portion of the domain. Both lower and upper Manhattan Island is densely covered by some of the tallest buildings in the world. For this reason, several of the surface properties were changed from their default value to represent this urban anomaly. The surface roughness was increased from 1.0 to 1.5 meters to account for the extremely tall buildings. This follows the high range of aerodynamic roughness length for a "regularly built-up town" proposed by WIERINGA (1993). A roughness of 1.5 m may be too conservative for lower Manhattan, as more recent studies that examine the urban morphology (RATTI *et al.*, 2002) find that extremely built-up areas like Los Angeles or Manhattan are better described by aerodynamic roughness lengths of 5–7 m or more. However, the sensitivity of CALMET-CALPUFF to magnified roughness lengths will be investigated in a later study.

Hourly surface observations are the second required input of the CALMET model. In Figure 1a, the plan view of the model domain is shown along with the locations of the six National Weather Services Automated Surface Observing System (ASOS) stations used in the simulations. The stations are Newark Airport, Teterboro Airport, Central Park in Manhattan, LaGuardia Airport, John F. Kennedy Airport and Islip, located about 100 km to the east of the model domain's center. Quality assured, hourly ASOS data were acquired from the National Climatic Data Center (NCDC) for the study period. These data provide CALMET the near-surface wind speed and direction, temperature, cloud cover and precipitation.

CALMET uses upper-air profile data to estimate the above-surface flow and thermal stratification. Typically, upper-air observations are used from the National Weather Service rawinsonde sites. These observations are available twice daily, at approximately 0700 and 1900 LST (12 and 00 UTC, respectively). Between these morning and evening soundings, considerable boundary layer variation occurs. Additionally, the number of these stations (Islip and Albany) is relatively small in comparison to the region being modeled. For this reason hourly-assimilated model analysis profiles were used (temperature, winds and moisture) from the Advanced Regional Prediction System, Data Assimilation System (ADAS) (ZHANG et al., 1998). Model data, including the ADAS data set, have been successfully utilized in at least one other CALMET-CALPUFF study (LEVY et al., 2002) to provide more physically realistic meteorological fields. The ADAS data were interpolated to the four corners of the CALMET grid, shown as squares in Figure 1a.

After ingesting these data sources, CALMET uses several routines to estimate the winds and stability of the planetary boundary layer. The diagnostic wind field module of CALMET includes several steps. The first step is to apply diagnostic algorithms that account for terrain, kinematic effects, divergence minimization, Froude number adjustment and possible slope flows. In the second step observations are introduced through an objective analysis method based on the inverse distance (R^{-2}) method. A radius of influence (R) limit of 20 km was placed on the interpolation of observations to the CALMET grid. The final step utilizes the O'Brien procedure (O'BRIEN, 1970), which takes the wind field and adjusts the horizontal divergence so that the vertical velocity at the top level of the model is zero. During this procedure, the horizontal winds are adjusted iteratively so that the resulting divergence is lowered to a user-specified value, in this case the default $(5 \times 10^{-6} \text{ s}^{-1})$.

2.3. Instrumentation Description

An instrumentation cluster was deployed near the WTC recovery site in lower Manhattan. The cluster consisted of two Sound Detection and Ranging (SODAR) systems and a three-level micrometeorological tower. The instruments were activated on November 08, 2001 providing approximately one month of independent data during this study period. A plan view showing the location of the instrumentation cluster with respect to lower Manhattan and the WTC recovery site is provided in Figure 1b. The 10-meter wind from the tower and several levels of SODAR data from the Aerovironment Model 4000 miniSODAR are used to evaluate the CALMET-derived winds near the WTC site.

2.4. Dynamic Model Configuration

Two dynamic models, ARPS (XUE, 1998; XUE *et al.*, 2000, 2001) and the fifth generation PSU-NCAR Mesoscale Model (MM5) (GRELL *et al.*, 1995), were used in this investigation to examine the local variation in the meteorology during different synoptic flow scenarios. Of specific interests are the synoptic regimes where ASOS observations may not provide enough resolution to entirely capture the local-scale variations in the meteorology.

ARPS is a nonhydrostatic, fully compressible, primitive equation model, capable of resolving microscale meteorological variations. The turbulence closure scheme in the boundary layer utilizes a 1-1/2 TKE formulation after SUN and CHANG (1986). All moisture processes were activated, an advanced radiation scheme was used and the soil state was integrated using a soil-vegetation model designed according to NOILHAN and PLANTON (1989). A nested ARPS domain was designed to replicate the CALMET grid by centering over the same area and using identical landuse information. The grid spacing was 1.0 km in the ARPS simulation. Boundary conditions were provided by a similarly configured ARPS simulation with a horizontal grid spacing of 5 km. This grid was initialized and its boundary conditions were obtained from the hourly assimilated data of the 32 km ADAS system. Landuse properties of the CALMET domain such as surface roughness and leaf area index were regridded from 500 m to 1000 m for use by the ARPS simulations. Required surface specifications, such as vegetation fraction and soil type, were assigned in ARPS according to the 200 m USGS (United States Geological Survey) CTG (Composite Theme Grid) landuse data set.

The other dynamical model used in this study is the three-dimensional, non-hydrostatic version of the MM5. The mesoscale model incorporates the physics of the Oregon State University (OSU) land surface model (LSM). The OSU model is coupled with the ETA model planetary boundary layer (PBL) scheme. National Center for Environmental Prediction (NCEP) ETA model output was used in this study for initial and lateral boundary conditions. The MM5 default surface variables such as landuse, vegetation type, roughness and topography were acquired from a 900 m USGS database. Default surface roughness value used for the highly urbanized area of NYC in the MM5 model was 1.0 m versus 1.5 m in the ARPS model. For this study, a triple nested version (27, 9, 3 and 1 km) of the MM5 was utilized. The innermost, high-resolution grid was centered over the study area. All four domains had 36 vertical sigma levels (between 1000 hPa and 100 hPa).

3. Results

3.1. Comparison of calmet Winds with Independent 10 m Tower Observations

A statistical analysis is provided of the difference between the WTC tower wind speed and direction observations and the CALMET model simulation. It is realized that comparing point measurements to model output may raise questions. However, the fact that CALMET is a diagnostic model that essentially derives a wind field from observations, and that the WTC tower observations are independent of CALMET; we feel this analysis is a useful exercise. Figure 1b shows the plan view of the lower Manhattan area and the location of the WTC instrumentation cluster, which is important in the following analysis. The statistics are separated according to the observed flow direction at the WTC Tower site. Each of four flow classifications cover a 90° sector that is rotated −30° with respect to north as illustrated by Figure 1b. The flow classifications are northerly (330°–60°), easterly (60°–150°), southerly (150°–240°) and westerly (240°–330°), similar to the classification of the modeling case studies in Table 1. Flow regimes were derived from the plan view to isolate flow directions that are influenced by lower Manhattan (southerly and easterly) from those influenced by the Hudson River (westerly and northerly).

Descriptive statistics of wind speed differences between CALMET and the WTC Tower (CALMET-WTC Tower) in Table 2 are grouped according to the flow direction. The data are hourly wind speeds (m s^{-1}) over the period of November 11–27, 2001. The bias in wind speed is 1.46 m s^{-1} when the flow is from an easterly direction. It is illustrated in the plan view in Figure 1b, easterly flow results in a fetch of up to 8 km over the roughest surface in NYC, so weaker observed winds are not surprising, especially when the wind in CALMET is based on airport observations, taken over considerably smoother surfaces. When the flow is within the southerly quadrant the CALMET bias is 0.98 m s^{-1}, not as pronounced as for the easterly wind flow regime. Westerly flow also has a bias of nearly 1.0 m s^{-1} and the northerly flow was approximately 0.50 m s^{-1}. The average bias over all directions is about 1.0 m s^{-1}.

Wind direction differences between the simulated and observed data are summarized by the statistics in Table 3. A positive wind direction bias indicates that the CALMET wind direction is rotated clockwise relative to the observed wind. Analysis of the data (scatter plot not shown) and compiled statistics in Table 3 reveal

Table 2

Lower Manhattan Tower and CALMET wind speed (m·s^{-1}) comparison statistics grouped by observed wind speed (m·s^{-1}). Bias is calculated by standard (model-observation)

Wind Direction	Direction Criteria	Mean Bias	Mean Absolute Error
Northerly	330°– 60°	0.48	0.82
Easterly	60°–150°	1.46	1.52
Southerly	150°–240°	0.96	1.11
Westerly	240°–330°	0.98	1.15

Table 3

Lower Manhattan Tower and CALMET wind direction (degrees) comparison statistics grouped by observed wind direction. Bias is calculated by standard (model-observation)

Wind Direction	Direction Criteria	Mean Bias	Mean Absolute Error
Northerly	330°– 60°	−4	27
Easterly	60°–150°	34	48
Southerly	150°–240°	25	34
Westerly	240°–330°	−4	20

a few interesting characteristics. First, the mean absolute error (mae) and bias of the wind direction is greatest when the observed wind is from a northerly to southeasterly direction, with the peak differences associated with easterly flow (bias: 34°; mae: 48°). When the observed wind is between southwesterly, clockwise to northerly, the error is considerably less (bias: −4°; mae: 20–27°). The tower location (Fig. 1b) provides insight into these variations. When the flow is from an easterly to southerly direction the air travels over lower Manhattan before reaching the tower. The numerous tall buildings disrupt the flow pattern and increase the wind variability. When the flow is from a westerly to northerly direction, the scatter in the data is considerably less, as the air is traveling across the Hudson River. The lower friction over water and increased stability leads to a less turbulent flow.

A primary question is: How do these CALMET differences relative to a point observation relate to other areas in the model domain? There is at least some evidence from the statistics above that the urban area of lower Manhattan significantly influences the leeside flow in the city. The near-surface wind speed is reduced and the flow direction turns cyclonically as the flow decelerates. Since CALMET does not dynamically generate this general urban flow modification, there is a greater uncertainty in the model results within and directly downwind from the main urban center. This basic statistical analysis would suggest that the flow is lighter and curves cyclonically over the city. BORNSTEIN and JOHNSON (1977) showed similar results with a data set of surface observations. The magnitude of uncertainty may be related to cross-urban flow regimes examined above (i.e., an angle uncertainty of ± 50° in the wind direction and a little over ± 1.0 m s^{-1} in the wind speed). Areas upwind of Manhattan may have less bias in the wind direction, however the wind speed represented in CALMET is likely stronger than reality. Again, the wind speed differences may be related to the footprint of the ASOS stations not being representative of the urban environment. Although evidence of these urban effects is presented, more observations are needed to prove that these processes are occurring.

3.2. CALMET Versus Dynamical Model Simulations

Mesoscale model simulations, both ARPS and MM5 were conducted for a number of cases, each representing a different synoptic flow regime. Three of these

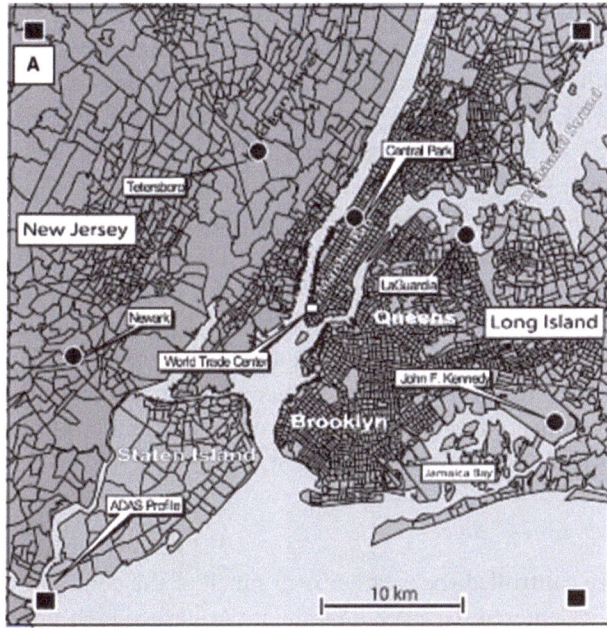

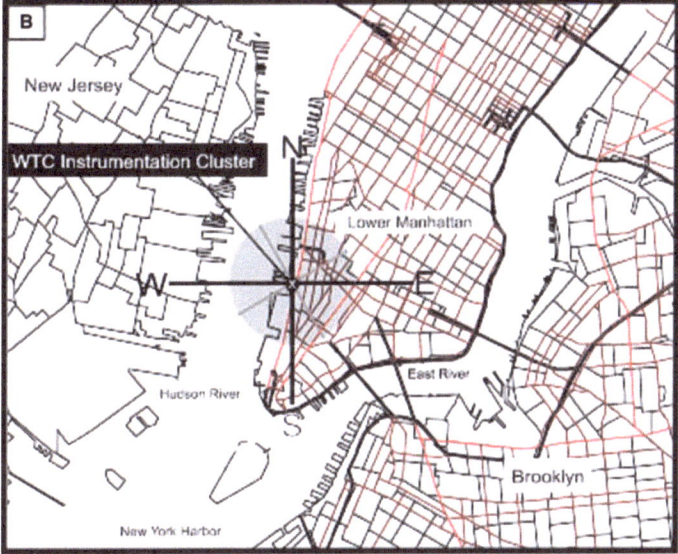

Figure 1
(A) A plan view of the CALMET-CALPUFF, ARPS and MM5 model domain over the New York City area. Surface meteorological observation sites are depicted by solid black circles with a white outline and the ADAS profile locations are shown by the squares. The WTC site as well as other NYC landmarks is labeled. (B) Plan view of the lower Manhattan area showing the location of 10 m tower and two SODAR's. The offset gray axis (-30°) represents the directional grouping (northerly, easterly, southerly, and westerly) of the CALMET-TOWER wind comparison statistics.

cases are presented in this paper. The first case, November 13–14, 2001 represents three of the synoptic classifications: A cool season light and variable case, a cool season light southerly nocturnal and light southerly daytime case. Unlike the other two cases, the WTC instrumentation cluster data were available. Following, a case will be examined in which the large-scale forcing is strong and the wind flow oscillates between a southerly and westerly direction (October 3–5, 2001), satisfying both the strong southerly and westerly flow classifications outlined in Table 1. The final simulation was conducted for September 11–13, 2001 exclusively using the MM5 model. ARPS was not applied because the assimilated data sets used to initialize and provide boundary conditions were not readily available. On September 11, 2001 the flow was brisk out of the north and on the following day, high pressure was centered over the region. This case represents a strong northerly condition and a light and variable warm season scenario. A more detailed description of the synoptic pattern precedes each of the following case studies.

Case I: November 13–14, 2001

High-pressure controlled the weather over much of the contiguous United States on November 13. Centered over West Virginia, the surface high pressure (1036 hPa) resulted in clear skies and calm winds over NYC. Given the light large-scale flow, local influences were more pronounced on November 13, 2001. The surface high pressure moved off the mid-Atlantic coast on November 14–15 resulting in a light to moderate southwesterly flow across NYC. Case I represents the synoptic classifications of a light and variable flow on November 13 and a light southerly flow on November 14.

Model estimated 10 m wind at 1400 LST is shown in Figures 2a (ARPS) and 2b (MM5). Also included are the NWS ASOS 10 m wind observations. The ASOS wind barbs are in knots, however the number to the right represents the wind speed in m s^{-1}. Sea-surface temperature (SST) measured from NOAA's AVHRR (Advanced Very High Resolution Radiometer) satellite was 282 K, while observed land temperatures rose to above 285 K, causing enough temperature differential to induce a weak TIBL at the land-water interface. The ARPS model was initialized with the AVHRR SST that averaged 282 K, but the MM5 used a climatological database that had a much lower SST of 278 K.

Noted in the ARPS wind field at 1400 LST on November 13 (Fig. 2a), there is a distinct shift from a southwesterly to southerly wind along Staten Island and northward along the New Jersey side of the Hudson River. Surface observations agree with this wind variation as the 10 m wind measurements in lower Manhattan display southerly winds while Newark was reporting a southwest wind. The 10 m averaged wind speed from the WTC Tower indicated that this southerly surge of wind simulated by ARPS over the NYC Harbor began at 1400 LST and lasted through the evening at which point the flow veered southwest. FRIZZOLA and FISHER

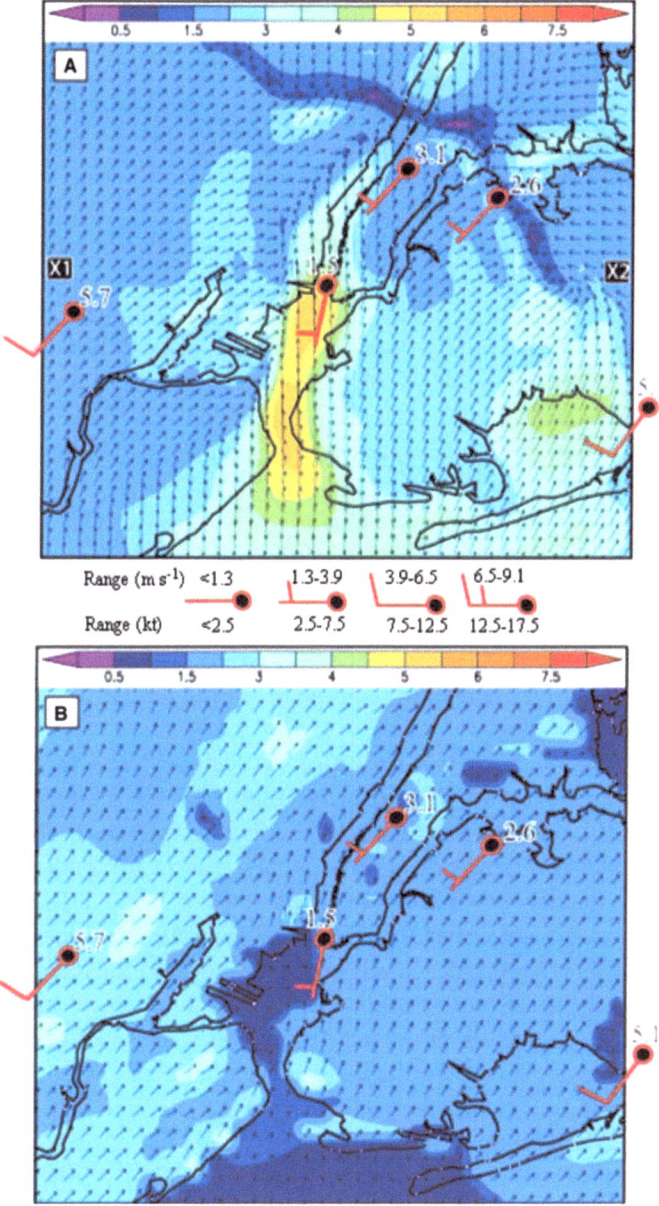

Figure 2
(A) ARPS simulated 10 m wind (m·s^{-1}) at 1400 LST on November 13, 2001. (B) MM5 simulated 10 m wind (m·s^{-1}) at the same time. Wind speed (m·s^{-1}) is shaded with a legend at the top of the plot. Wind vectors are scaled according to the wind speed. ASOS wind observations for the same time are plotted overtop of the simulated fields with the standard barb notation in knots (refer to legend). The observed wind speed values (m·s^{-1}) are shown next to the station plot.

(1963) plotted observations from the same ASOS sites during a similar light synoptic flow pattern that show a nearly identical wind direction variation at the same time of day (1300 LT). BORNSTEIN *et al.* (1994) performed numerical simulations over the same area that showed a similar frontal alignment that extended through Staten Island, across lower Manhattan and eastward through central Long Island. In Figure 2a, the ARPS simulated front extends further north in New Jersey, possibly because the large-scale flow was slightly stronger (4 m s^{-1}) and more northwest in BORNSTEIN *et al.* (1994). Northwest flow would limit the sea breeze from moving inland or northward.

Also of importance is an apparent ARPS simulated Long Island Sound breeze in the far northeastern part of the domain where an easterly flow exists across the northern part of Manhattan Island. Most sea-breeze studies in the area, specifically an observational report by MICHAEL *et al.* (1998) using WSR-88D Doppler imagery, show similar variation in both the radar data along the northern and western shores of Long Island Sound.

The MM5 simulation at the same time is shown in Figure 2b. The general wind flow pattern is similar between the MM5 and ARPS models for a majority of the domain. A comparable but less discrete front is indicated in the MM5 wind field on the west side of the Hudson River. The most noticeable difference between the MM5 and ARPS is the calm winds over water in the MM5, which could be a response to the 4 K cooler sea-surface temperature in the MM5 resulting in a more stable marine boundary layer.

An examination of the ARPS simulated vertical wind and turbulence (m^2 s^{-2}) cross section from X1 to X2 (Figure 2a) at 1400 LST was performed (Figure not shown). The TIBL on the west side of the Hudson River in New Jersey was distinguished in the cross section as a shift of westerly to southerly wind at the surface to approximately 300 m, representing a shallow 300 m TIBL associated with the weak sea breeze front. Ahead of the front were layers of enhanced TKE and deeper mixing, while the turbulence was suppressed behind the front, within the sea-breeze flow. On the same cross section the CALMET estimated mixing depth was plotted. Both models show lower mixing depths over water areas and greater mixing depths over land areas. However, there are substantial differences in the magnitude, with ARPS indicating a mixing layer of 200 m over water and 400 m to 700 m over land while CALMET estimates a depth ranging from 400 m over water to 1200 m over land. The daytime calculation of the mixing height in CALMET, based on the potential temperature lapse rate acquired from 32 km resolution assimilated data, may not accurately resolve the temperature structure of the coastal TIBL over NYC. The dynamical simulation of ARPS that considers the TKE budget, realistic sea-surface temperatures and temperature advection presumably better represents the boundary layer depth.

The CALMET wind field was examined at the same time. Figure 3 illustrates the 10 m wind field from the CALMET. Visual comparisons with the ARPS simulations

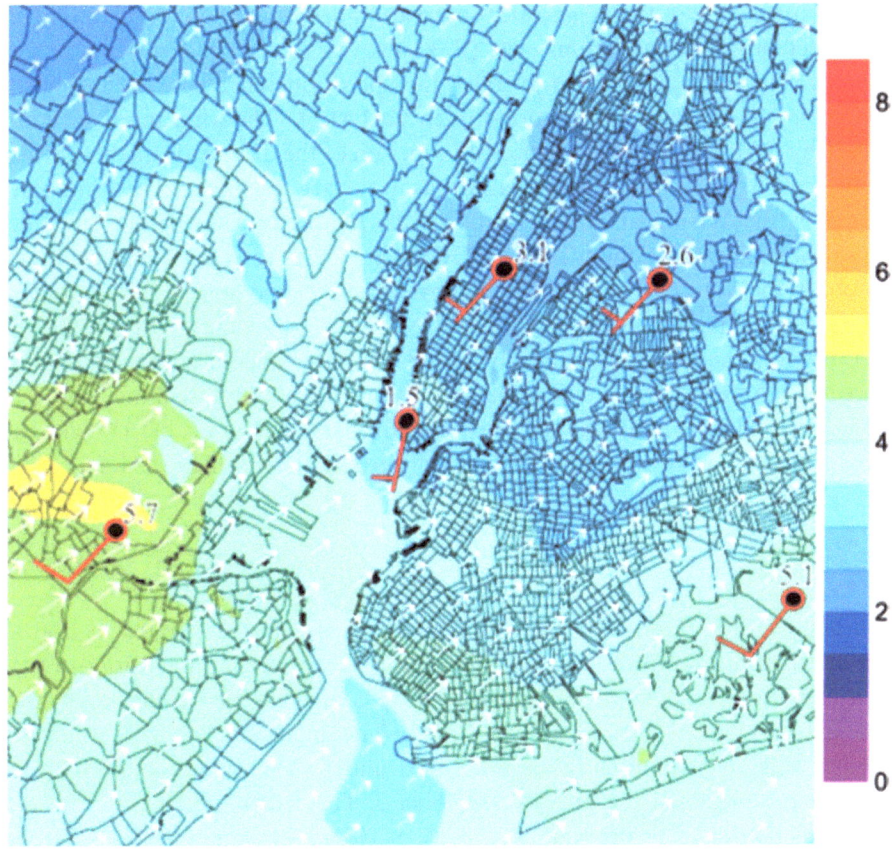

Figure 3
CALMET simulated 10 m wind (m·s^{-1}) at 1400 LST on November 13, 2001. Wind speed (m·s^{-1}) is shaded with a legend at the top of the plot. Wind vectors are scaled according to the wind speed. ASOS wind observations for the same time are plotted top of the simulated fields with the standard barb notation in knots (Refer to wind barb legend in Figure 2). The observed wind speed values (m·s^{-1}) are shown next to the station plot.

in Figure 2a indicate several notable differences. First, the CALMET wind direction is nearly uniform, while the dynamical ARPS and MM5 simulations depict a discrete sea breeze front with an associated wind direction shift. Some may argue that because CALMET is using surface wind observations to derive the wind field, it has to be accurate. However, it was seen in Figures 2a and 2b that the observed ASOS wind directions are consistent with the ARPS and MM5 simulations, as was the wind observation in lower Manhattan. This observed southerly wind at the WTC tower indicates that an abrupt southerly wind shift was not depicted in the CALMET simulation, possibly because the density of surface observations was not great enough to capture smaller scale details in the wind field. This demonstrates an inherent shortcoming of the diagnostic, observation driven method in regions where complex meteorological variations exist (BAKLANOV *et al.*, 2002).

Another feature to note is the overall wind speed difference between the ARPS and MM5 models and the observation-driven CALMET. The wind speed variation generated by CALMET (Fig. 3) is strictly a function of the diagnostic method, which does not explicitly take into account variations in surface roughness. The wind speed simulated by the dynamical models is affected by the prescribed surface roughness. During flow regimes like this, CALMET does not seem to resolve the local-scale variations in the wind flow around Manhattan during the daytime. An evaluation of the prognostic capability of the STEM-FCM model (SILIBELLO *et al.*, 2001) to predict ozone concentrations similarly found that the dynamical model RAMS (PIELKE and COTTON, 1992) provided better meteorological input than the diagnostic CALMET model in coastal regions. The synoptic classification (Table 1) highlights that 16% of the days during the study period were similar to this case study.

Case II: October 2–5, 2001

During October 2–6, 2001 the NYC area experienced a lengthy period of moderate to strong southwesterly flow. This flow dominated a large portion of the eastern United states. A large high-pressure ridge controlled the weather over much of the eastern United States on October 2. The high-pressure system off the southeastern United States coastline intensified under strong upper-level (300 hPa) confluence and remained stationary on October 3. This led to the development of a warm southerly flow over the New York City metropolitan region through October 6. There was increasing concern about the elevated particulate matter concentrations around the city during this three-day period. This episode is examined for this reason and because it fits into the classifications of both strong southerly and westerly flow outlined in Table 1.

The wind field simulated by the high-resolution 1 km ARPS model at 0700 LST on October 4, 2001 indicated that the variation in the wind field was dominated by the surface roughness (figure not shown). The wind speed varied from 5.0 m s^{-1} over water to less than 2.0 m s^{-1} over the rougher urban areas. The simulated wind directions compare well with the ASOS observations, and the simulated wind speeds are about 1.0 m s^{-1} less than observed.

Another important characteristic of the wind field variation is a slight but noticeable cyclonic turning in the low level winds near lower Manhattan; this was also noted in a nocturnal analysis on November 14, 2001. As the simulated west-southwest flow over the open NY Harbor intercepts by the tip of Manhattan Island, the winds slow and become more southwesterly. This feature is in agreement with observational findings by BORNSTEIN and JOHNSON (1977) that showed nighttime events during stronger flow regimes were associated with distinctive roughness

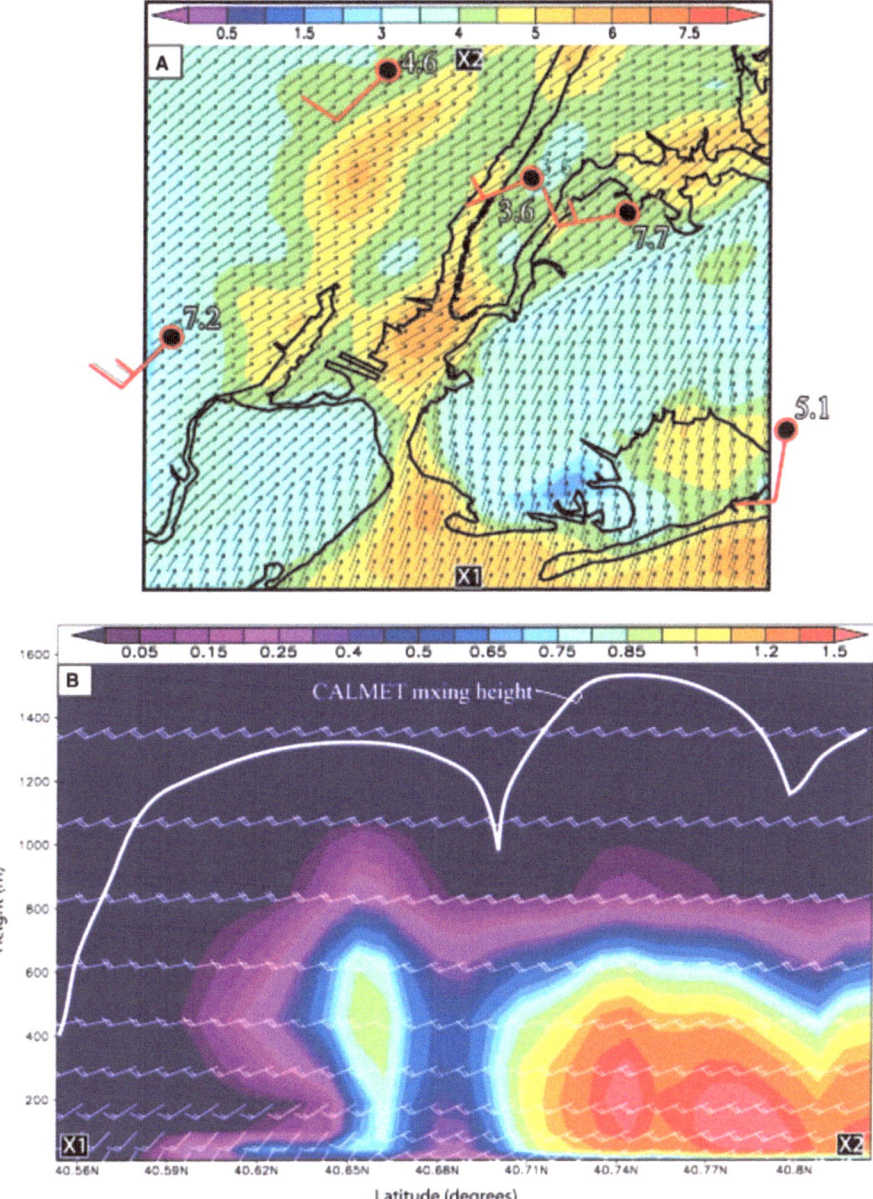

Figure 4

(A) ARPS simulated 10 m wind ($m \cdot s^{-1}$) at 1500 LST on October 4, 2001. Wind speed is shaded with a legend at the top of the plot. Wind vectors are scaled according to the wind speed. ASOS wind observations for the same time are plotted overtop of the simulated fields with the standard barb notation in knots (Refer to wind barb legend in Figure 2). The observed wind speed values ($m \cdot s^{-1}$) are shown next to the station plot. (B) North to south vertical cross-section of TKE ($m^2 \cdot s^{-2}$) and wind from the ARPS model across the model domain from point X1 to X2 in Panel A. TKE is shaded according to legend and wind is shown in standard barb format (kt). For comparison, the white line indicates the CALMET mixing height.

induced cyclonic turning in the winds over the main core of Manhattan. The CALMET wind direction for the same period was generally west-southwest as observed, however the CALMET wind speed was stronger over the land areas compared to the ARPS simulation. Over water, the ARPS simulation had stronger winds (4–6 m s^{-1}) compared to CALMET (2–4 m s^{-1}). The offshore buoy observation 35 km to the south-southeast of Manhattan reported southwest winds 6–8 m s^{-1} between 0630 and 0730 LST.

The second analysis used for evaluating this case occurs on the following afternoon. Temperatures were well above normal during the period with high temperatures reaching almost 300 K while sea-surface temperatures held around 290 K. Figure 4a shows the ARPS simulated 10 m wind field (speed in m s^{-1} shaded) along with the ASOS observations at 1500 LST on October 4, 2001. Evident in the wind field is a wind shift over Staten Island stretching over to the northern portion of Long Island. This is the sea-breeze front that has formed slightly inland. Surface observations agree with the existence of a sea-breeze front somewhere between JFK airport and LaGuardia airport, as the winds are south along the coast and more westerly inland. Other surface observations further inland indicate that the southerly wind associated with the sea breeze has not penetrated to Uptown Manhattan or westward into areas of New Jersey, as these observations show a west-southwest wind. With a stronger opposing wind, relative to the other cases, it is expected that the sea breeze will be held close to the coast along the New Jersey coastline, while the parallel flow to the Long Island coast will generally allow some inland penetration (SIMPSON et al., 1977; ARRITT, 1993; ATKINS and WAKIMOTO, 1997). An examination of the CALMET wind field for the same time (not shown) does not reflect the sharp wind shift of the sea-breeze front, although the general wind direction is represented The interpolation scheme of CALMET allows only gradual changes in wind direction so frontal zones are not explicitly simulated.

The vertical distribution of wind and turbulence from south (X1) to north (X2) through this frontal feature is shown in Figure 4b. The frontal boundary is at 40.66 N as indicated by the south to west wind shift and elevated turbulence associated with an increase in upward motion along the sea-breeze front. The wide variation in boundary layer height determined from the turbulence profile is apparent as the sea-cooled air mass limits the vertical extent of mixing and the land-warmed airmass allows the boundary layer to grow to nearly 1 km. Similar to the previous case study, the CALMET-estimated mixing height (white line) seems to be overestimated when compared to ARPS. The synoptic review indicates that one-third (35%) of the days had strong southerly or strong westerly (Table 1) winds, and could be generally compared to this case.

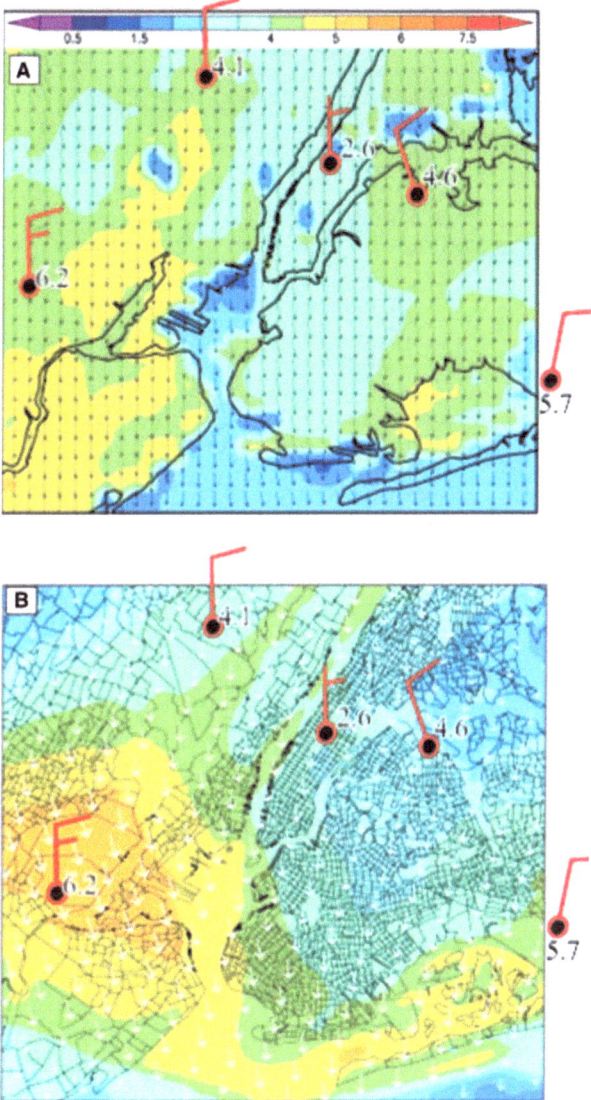

Figure 5
(A) MM5 simulated 10 m wind (m·s⁻¹) at 1400 LST on September 11, 2001. (B) CALMET simulated 10 m wind (m·s⁻¹) at 1400 LST on September 11, 2001. Wind speed is shaded (m·s⁻¹); legend is at the top of the plot. Wind vectors are scaled according to wind speed. ASOS wind observation from the same time period are plotted overtop of the simulated fields with the standard barb notation in knots (Refer to wind barb legend in Figure 2). The observed wind speed values (m·s⁻¹) are shown next to the station plot.

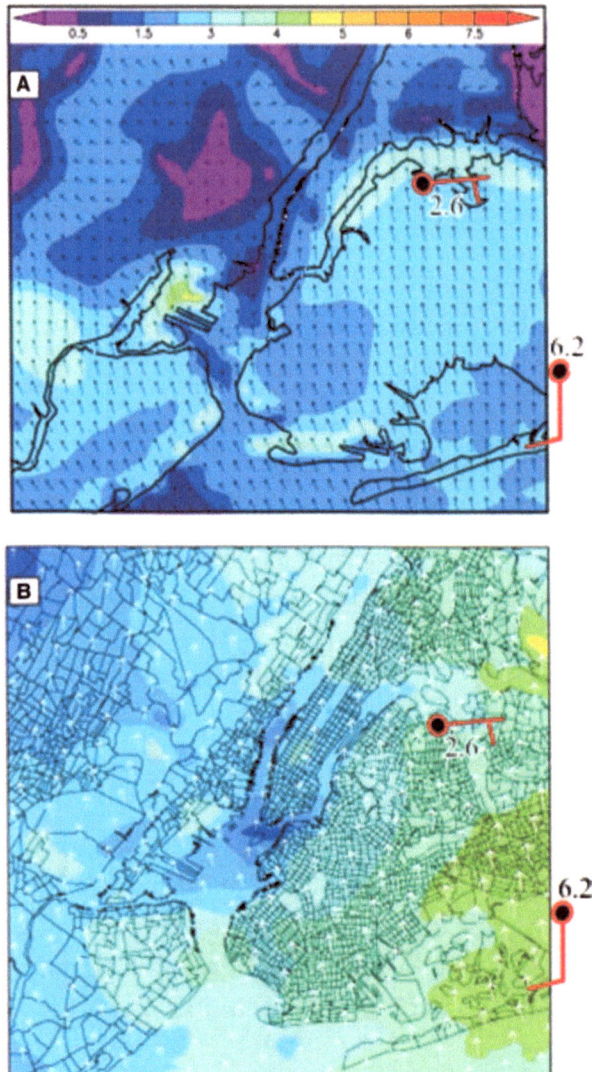

Figure 6
(**A**) MM5 simulated 10 m wind (m·s^{-1}) at 1400 LST on September 12, 2001. (**B**) CALMET simulated 10 m wind (m·s^{-1}) at 1400 LST on September 12, 2001. Wind speed is shaded (m·s^{-1}); legend is at the top of the plot. Wind vectors are scaled according to wind speed. ASOS wind observation from the same time period are plotted overtop of the simulated fields with the standard barb notation in knots (Refer to wind barb legend in Figure 2). The observed wind speed values (m·s^{-1}) are shown next to the station plot.

Case IV: September 11–12, 2001

Late on September 10, a surface cold front moved through New England, and out over the Atlantic Ocean. Behind this surface front, strong high-pressure (1027 hPa) ridged into New England and down through the Mid-Atlantic States. This led to subsidence throughout the atmosphere, which resulted in clear skies and brisk northwest winds over the NYC area on September 11. The strong high-pressure cell became situated directly over the Mid-Atlantic region on September 12. This period contains a case (September 11) that represents a strong northerly flow regime as outlined in Table 1 and also a light and variable, high pressure dominated case (September 12). A MM5-simulated 10 m wind field over the NYC region will be examined on the afternoon of each day.

Strong northerly flow existed on September 11, 2001. The simulated MM5 10 m wind field in Figure 5a shows a homogeneous distribution of the simulated mean wind direction at 1600 LST on the 11th. The simulated wind field over the region remained the same through the afternoon. The simulated wind speed ranges from 3.5 to 5.0 m s^{-1} across the region and seems to be correlated with the surface roughness variations specified by the MM5 landuse data set. The simulated winds are consistent with the overlaid ASOS observations in terms of wind direction, but the overall simulated wind speed is lighter than the observed values. The model simulation depicts an expected flow behavior for stronger northerly flow regimes. The CALMET simulation for the same time period is shown in Figure 5b. The CALMET wind field is similar with respect to wind direction, although the speed is greater. The overall wind speed is approximately 1–2 m s^{-1} stronger in the CALMET simulation.

Figure 6a shows the simulated 10 m wind field using the MM5 on the following day, September 12, 2001 at 1600 LST. High pressure dominated the area so the local effects became more apparent. The wind field manifests a sea-breeze front propagating through the domain. The sea breeze front can be detected as a southerly wind enhancement, inland over northeast New Jersey. The simulated front stretches across Manhattan and then along the East River. Several of the ASOS observations were missing during the afternoon nonetheless at this time the JKF airport ASOS observation agrees with the onset of a sea breeze as the winds have increased from a southerly direction. The LaGuardia airport observation is the only other wind measurement available at this time and it records an easterly wind. The LaGuardia wind observation did turn southerly the following hour, as did the wind at Newark and Teterboro, clearly indicating a sea breeze. The CALMET simulation shown in Figure 6b indicates a southerly wind over the entire area with stronger winds over the eastern domain. Lack of surface observations at this particular time resulted in a simplified CALMET wind field. In such cases the ADAS winds are used to aid in deriving a surface wind field. The 32 km ADAS data work well for upper-level winds that do not vary significantly over small areas, but it cannot resolve local effects closer to the surface.

This case demonstrates the ability of CALMET to represent the wind field over NYC reasonably well during stronger northerly flow regimes. In this case, the wind direction was well represented while the wind speed was overestimated by CALMET.

The synoptic review of the study period in Table 1 indicates that roughly 7% of the days during the study period were similar to the strong northerly flow exhibited in the September 11, 2001 case. Table 1 also indicates that about 16% of the days were similar to the light and variable flow that occurred on September 12, 2001. In these flow regimes, CALMET is less reliable as numerous local effects dominate the meteorology, and observations are not dense enough to provide CALMET an accurate wind field.

4. Conclusions

The goal of this investigation is to evaluate the meteorology of an observational-based CALMET model over the complex region of NYC. The CALMET meteorology is derived from a network of ASOS stations that have inherent bias since the wind observations are taken over open airfields, and are likely not representative of the wind in the built-up urban area. An independent data set taken from an observing system located in lower Manhattan was compared to the CALMET simulation to show any bias or uncertainty in the model. Also, dynamically driven numerical models were used to examine some of the local effects in the region that may not be correctly captured by the CALMET model. Simulations were performed after a careful evaluation of the synoptic weather over the study period, so that the simulations could be representative of not one, but a group of similar days. The following is a summary of conclusions of the results.

- CALMET meteorology is suitably representative of the NYC area during stronger wind flow scenarios, which occurred approximately 42 percent of the study period. We compared the CALMET wind field with several mesoscale model simulations, which showed fairly uniform wind distribution over NYC. The meteorology provided to CALPUFF by CALMET during such flow conditions will likely provide adequate WTC plume transport and dispersion. The CALMET mixing height algorithm may compute a deeper convective mixed layer in coastal areas than that which would be observed. The CALPUFF concentrations, which are influenced by the mixing depth, may therefore be underestimated during the daytime. However, more cases are needed to verify this claim.
- Numerical simulations showed that during light southerly flow regimes, sea-breeze fronts frequently passed over lower Manhattan. In many of these cases the CALMET model experienced problems resolving important details of the frontal evolution including the discrete wind shift, timing and location. A review of the weather patterns after September 11 revealed that during about 35 percent of the days the winds were light and from the south. Past research, observations and the numerical models (both ARPS and MM5) revealed that the wind flow and stability of the sea breeze resulted in complex flow patterns across lower

Manhattan and western Long Island. Typically with the sea breeze, the wind direction in CALMET could be off by as much as 45°. In all cases where CALMET resolves the sea breeze, the front is not discretely represented because of the interpolation method, and the timing is off because all but one ASOS station are located inland. As with strong flow cases, the CALMET mixing depth exceeds simulated by the mesoscale models. If the CALMET model overestimates the mixing height, it will lead to an underestimation of the CALPUFF concentrations. At night, the CALMET-estimated boundary layer depth is close to that from the mesoscale models. These events occurred during 25 percent of the simulation period.

- The observed wind direction downwind of the urban core of Manhattan backed or turned cyclonically 34° relative to the simulated CALMET wind. This wind direction bias in CALMET will influence the plume position estimated by the CALPUFF dispersion model around lower Manhattan.

- CALMET derived wind speeds have a bias as much as 1.5 m s^{-1} in wind speed when compared to observations in lower Manhattan. This bias in the meteorology will influence the dispersion calculation made by the CALPUFF dispersion model, presumably underestimating the concentration.

Overall, the CALMET model was found to provide meteorology that is adequate for driving a plume model most of the time. Naturally, the quality of CALMET is closely related to the quality and representativness of the input observations. For circumstances in which the meteorology is complicated by mesoscale features like the sea/land breeze circulation or a significant urban heat island, and computer resources are available, a full-physics model could be used to provide improved meteorological fields.

Acknowledgements

This work was supported by the United States Environmental Protection Agency, State Climate Office of North Carolina and the Department of Marine, Earth, and Atmospheric Sciences of North Carolina State University. It has been subjected to United States Environmental Protection Agency and National Oceanic and Atmospheric Administration review and approved for publication. Mention of trade names or commercial products does not constitute an endorsement or recommendation for use. Dennis ATKINSON (NOAA) and William BROWN (NOAA) were instrumental in the quick delivery of the quality assured ASOS data from the NOAA National Climatic Data Center. We thank Bob Kelly and Henry Feingersch (US EPA Region 2) for their support of the application of meteorological models and instrumentation following the events of September 11, 2001. We also thank the North Carolina Supercomputing Center for use of computer resources for the ARPS and MM5 simulations. The CAPS group at the University of Oklahoma was instrumental in providing not only the source code of ARPS but also the assimilated data used to initialize the model. We would also like to thank Dr. Gary Lackmann, (associate Professor at North Carolina State University) and James Godowitch (NOAA), for their constructive review of the paper.

REFERENCES

ARRITT, R.W. (1993), *Effects of Large-scale Flow on Characteristic Features of the Sea Breeze*, J. Appl. Meteor. *32*, 116–125.

ATKINS, N.T. and WAKIMOTO, R.M. (1997), *Influences of the Synoptic-scale Flow on Sea Breezes Observed during CAPE*, Mon. Wea. Rev. *125*, 2112–2130.

BAKLANOV, A., RASMUSSEN, A., FAY, B., BERGE, E., and FINARDI, S. (2002), *Potential and Shortcomings of Numerical Weather Prediction Models in Providing Meteorological Data for Urban Air Pollution Forecasting*, Water, Air, and Soil Pollution *2*, 43–60.

BORNSTEIN, R.D. and JOHNSON. D.S. (1977), *Urban-rural Wind Velocity Differences*, Atmos. Environ. *11*, 597–604.

BORNSTEIN, R.D., THUNIS, P., and SCHAYES, G., *Observation and simulation of urban-topography barrier effects on boundary layer structure using the three-dimensional TVM/URBMET model*. In *Air Pollution and its Application X* (Plenum Press, New York 1994) pp. 101–108.

BROWN, R.M. and SETHU RAMAN, S. (1981), *Temporal Variation of Particle Scattering Coefficients at Brookhaven National Laboratory, New York*, Atmos. Environ. *15*, 1733–1737.

FRIZZOLA J.A. and FISHER, E.L. (1963), *A Series of Sea Breeze Observations in the New York City Area*, J. Appl. Meteor. *2*, 722–739.

GRELL, G., DUDHIA, J., and STAUFFER, D. (1995), *A Description of the Fifth-Generation Penn State/NCAR Mesoscale Model (MM5)*, Mesoscale and Microscale Meteorology Division, NCAR/TN-398 + STR, 117 pp.

LEE, D.O. (1979), *The Influence of Atmospheric Stability and the Urban Heat Island on Urban-rural Wind Speed Differences*, Atmos. Environ. *13*, 1175–1180.

LEVY, J.I., SPENGLER, J.D., HLINKA, D., SULLIVAN, D., and MOON, D. (2002), *Using CALPUFF to Evaluate the Impact of Power Plant Emissions in Illinois: Model Sensitivity and Implications*, Atmos. Environ. *36*, 1063–1075.

MICHAEL P., MILLER, M., and TONGUE, J.S. (1998), *Sea-breeze Regimes in New York City Region—Modeling and Radar Observations*, Transac. Second Conf. on *Coastal Atmospheric and Oceanic Prediction and Processes*, 78th AMS Annual Meeting, 11–16 January 1998, Phoenix, Arizona.

NOILHAN, J. and PLANTON, S. (1989), *A Simple Parameterization of Land Surface Processes for Meteorological Models*, Mon. Wea. Rev. *117*, 536–549.

O'BRIEN, J.J. (1970), *A Note on the Vertical Structure of the Eddy Exchange Coefficient in the Planetary Boundary Layer*, J. Atmos. Sci. *27*, 1213–1215.

PIELKE, R.A. and COTTON, W. R. (1992), *A Comprehensive Meteorological Modeling System-RAMS*, Meteor. Atmos. Phys. *49*, 69–91.

RATTI, C., DI SABATINO, S., BRITTER, R., BROWN, M., CATON, F., and BURIAN, S. (2002), *Analysis of 3-D Urban Databases with Respect to Pollution Dispersion for a Number of European and American Cities*, Water, Air, and Soil Pollution *2*, 459–469.

RAYNOR, G.S., RAMAN, S., and BROWN, R.M. (1979), *Formation and Characteristics of Coastal Internal Boundary Layers during Onshore Flows*, Boundary Layer Meteor. *16*, 487–514.

REISS, N.M., KWIATKOWSKI, J., GURER, K., ÇERMAK, J.R., and AVISSAR, R. (1996), *The New Jersey Sea Breeze Experiment (NESBEX): Movement and Structure of the New Jersey Sea Breeze as Diagnosed from Doppler Radar and Other Measurements*, First NARSTO-Northeast Data Analysis Symposium and Workshop, Norfolk, VA, 10–12 December 1996.

SIMPSON, J.E., MANSFIELD, D.A., and MILFORD, J.R. (1977), *Inland penetration of Sea-breeze Fronts*, Quart. J. Roy. Meteor. Soc. *103*, 47–76.

SCIRE, J.S., ROBE, F.R., FERNAU, M.E., and YAMARTINO, R.J., *A User's Guide for the CALMET Meteorological Model* (Version 5) (Earth Tech, Inc. Concord, MA 2000a).

SCIRE, J.S., STRIMAITIS, D.G., and YAMARTINO, R.J., *A User's Guide for the CALPUFF Dispersion Model* (Version 5) (Earth Tech, Inc. Concord, MA 2000b).

SILIBELLO, C., CALORE, G., PIROVANO, G., and CARMICHAEL, G. R. (2001), *Development of STEM-FCM Modeling System: Chemical Mechanisms Sensitivity Evaluated on a Photochemical Episode*, Proc. 2nd Internat. Conf, on *Air Pollution Modeling*, Champs-sur-Marne, April 9–12, 2001.

SUN, W.Y. and CHANG, C.Z. (1986), *Diffusion Model for a Convective Layer. Part I: Numerical Simulation of Convective Boundary Layer*, J. Climate and Appl. Meteor. *25*, 1445–1453.

WIERINGA, J. (1993), *Representative Roughness Parameters for Homogeneous Terrain*, Boundary-Layer Meteor. *63*, 323–363.

XUE, M., *Advanced Regional Prediction System (ARPS) Users Guide, Version 4.0* (Center for Analysis and Prediction of Storms, Oklahoma 1998).

XUE, M., DROEGEMEIER, K.K., and WONG, V. (2000), *The Advanced Regional Prediction System (ARPS)—A Multi-scale Nonhydrostatic Atmospheric Simulation and Prediction Tool. Part I: Model Dynamics and Verification*, Meteor. and Atmos. Phys. *75*, 161–193.

XUE, M., DROEGEMEIER, K.K., WONG, V., SHAPIRO, A., BREWSTER, K., CARR, F., WEBER, D., LIU, and Y. WANG, D. (2001), *The Advanced Regional Prediction System (ARPS)—A Multi-scale Nonhydrostatic Atmospheric Simulation and Prediction Tool. Part II: Model Physics and Applications*, Meteor. Atmos. Phys. *76*, 143–165.

ZHANG, J., CARR, F.H., and BREWSTER, K. (1998), *ADAS Cloud Analysis*, Preprints, 12th Conf. *On Numerical Weather Prediction*, Phoenix, AZ, Am. Meteor. Soc., 185–188.

(Received August 11, 2004, accepted October 20, 2004)
Published Online First: June 21, 2005

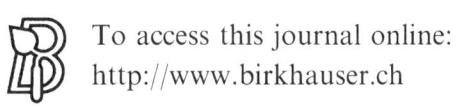

To access this journal online:
http://www.birkhauser.ch

Pure appl. geophys. 162 (2005) 2005–2028
0033–4553/05/102005–24
DOI 10.1007/s00024-005-2702-y

© Birkhäuser Verlag, Basel, 2005

| Pure and Applied Geophysics

Metropolitan-scale Transport and Dispersion from the New York World Trade Center Following September 11, 2001. Part II: An Application of the CALPUFF Plume Model

ROBERT C. GILLIAM[1,2], ALAN H. HUBER[2] and SETHU RAMAN[1]

Abstract—Following the collapse of the New York World Trade Center (WTC) towers on September 11, 2001, Local, State, and Federal agencies initiated numerous air monitoring activities to better understand the ongoing impacts of emissions from the disaster. The collapse of the World Trade Center towers and associated fires that lasted for several weeks resulted at times in a noticeable plume of material that was dispersed around the Metropolitan New York City (NYC) area. In general, the plume was only noticeable for a short period of time following September 11, and only apparent close to the World Trade Center site. A study of the estimated pathway which the plume of WTC material would likely follow was completed to support the United States Environmental Protection Agency's 2002 initial exposure assessments. In this study, the WTC emissions were simulated using the CALMET-CALPUFF model in order to examine the general spatial and temporal dispersion patterns over NYC. This paper presents the results of the CALPUFF plume model in terms of plume dilution and location, since the exact source strength remains unknown. Independent observations of $PM_{2.5}$ are used to support the general dispersion features calculated by the model. Results indicate that the simulated plume matched well with an abnormal increase (600–1000% of normal) in $PM_{2.5}$ two nights after the WTC collapse as the plume rotated north to southeast, towards parts of NYC. Very little if any evidence of the plume signature was noted during a similar flow scenario a week after September 11. This leads to the conclusion that other than areas within a few kilometers from the WTC site, the $PM_{2.5}$ plume was not observable over NYC's background concentration after the first few days.

Key words: Dispersion modeling, CALPUFF, CALMET, plume modeling, sea breeze, particulate matter.

1. Introduction

In response to the events on September 11, 2001 at the New York World Trade Center (WTC), a study utilizing the CALMET-CALPUFF (SCIRE *et al.*, 2000) dispersion modeling system was conducted for the three-month period following the

---

[1]State Climate Office of North Carolina, North Carolina State University, Raleigh, NC, USA
[2]Atmospheric Sciences Modeling Division, Air Resources Laboratory, National Oceanic and Atmospheric Administration. On assignment to the United States Environmental Protection Agency, National Exposure Research Laboratory, Research Triangle Park, NC, USA

events. Prior efforts were already underway to use a similar modeling system for real-time support of local air pollution studies in the Raleigh-Durham area of North Carolina. After the events at the New York WTC, all efforts were redirected toward the WTC impact on the New York City (NYC) region. The results of the study as a whole were used to support the U.S. Environmental Protection Agency's (EPA) preliminary assessments of the WTC emissions impact on the surrounding New York City region.

The purpose of the research reported herein (Part II) is to examine the metropolitan-scale pathway and dilution of airborne emissions from the WTC recovery site after September 11. We examine the first few weeks after the attack using $PM_{2.5}$ observations and the air-dispersion simulation of the WTC plume. Additionally, longer temporal averages of the modeled plume are analyzed to elucidate areas that may have been influenced more regularly because of seasonal flow patterns.

Part I of the study presents an evaluation of the CALMET model results and a summary of the synoptic conditions during the study period. Part I highlighted biases and uncertainties in the meteorological fields derived by CALMET. CALMET has a bias to calculate stronger winds and greater mixing heights over land. Another bias that was observed in the CALMET model was a consistent clockwise wind direction downwind of Manhattan relative to observations. The biases in wind speed and daytime mixing height might lead to less actual plume dilution. There may also be a more northerly (inland) transport of WTC material because of the wind direction bias. The paper also indicated that CALMET tended to poorly define the properties of the sea-breeze circulation, specifically the rapid wind shift along the front and thermal stratification of the marine boundary layer.

More refined studies of the meteorology and plume pathways of the WTC plume are ongoing by several groups, including the US EPA, NC State University, and the Environmental and Occupational Sciences Institute of Rutgers University (EOHSI). The US EPA and EOHSI have an ongoing project to study potential human exposures within lower Manhattan by using models (Computational Fluid Dynamics and Wind Tunnel models) of the detailed flow about urban buildings and detailed environmental concentrations observed near the WTC site. While these detailed refined modeling methods may be critical near the WTC site, application of a CALMET-CALPUFF type system may be useful for human exposure and epidemiological studies over the larger metropolitan region.

CALMET-CALPUFF was used in part because the system is openly available, widely used (GODFREY, 1998; HONAGANAHALLI *et al.*, 2000; LEVY, *et al.*, 2002; BARNA and GIMSON, 2002; ZHOU *et al.*, 2003), and has performed well compared to similar models (U.S. EPA, 1998) and observations (IRWIN *et al.*, 1996, 1998). It is believed the system is adequate to support this application to generally characterize the temporal and spatial area of impact by potential emissions from the collapse of the World Trade Center and associated fires. The model was reviewed (ALLWINE *et al.*, 1998) and recommended by the U.S. EPA, Appendix A of the Guideline on Air Quality by the U.S. EPA (U.S. EPA, 2004), in both near- and long-range transport applications.

2. Methology

2.1. CALPUFF

A CALMET-CALPUFF model domain was designed to cover the central portion of the New York City (NYC) metropolitan area. The domain encompasses a 50×50 km area centered over lower Manhattan which has a horizontal grid spacing of 500 m. Figure 1 shows the area covered by the model domain along with

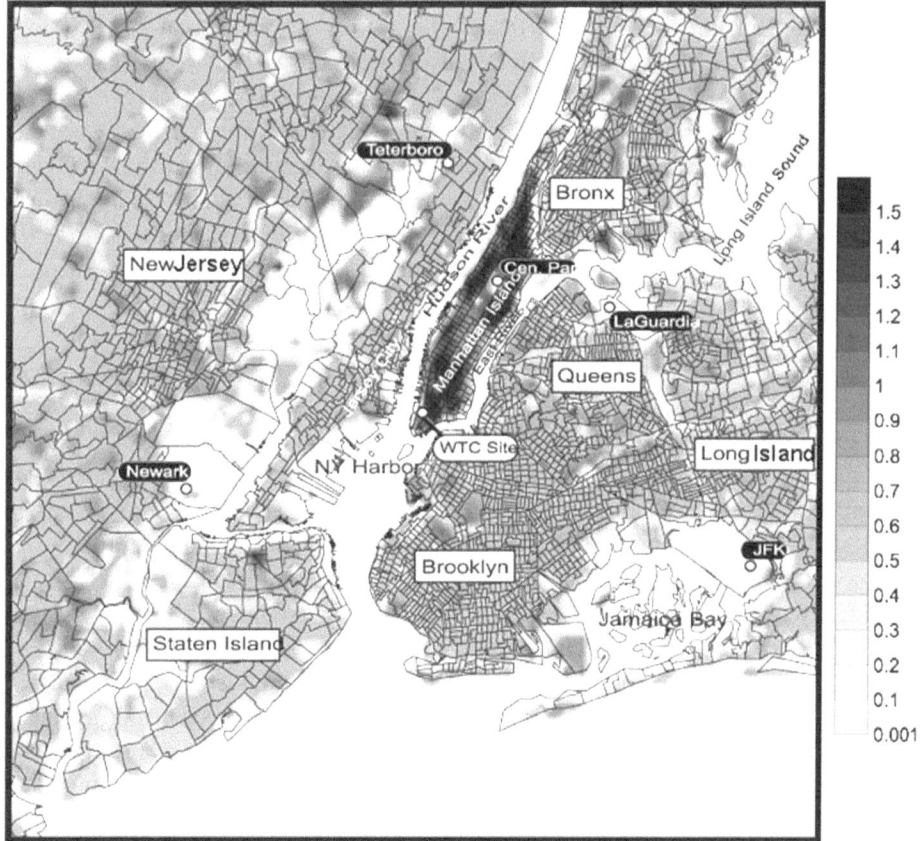

Figure 1

Plan view of the CALMET-CALPUFF model domain. Gray shading represents the parameterized surface roughness length (m). Meteorological observation sites are indicated by the black ovals. Other locations, which are used as reference points in the analysis, are labeled.

prominent Municipalities and bodies of water that will be used as reference points in the analysis. CALPUFF is a Lagrangian puff model that uses the hourly meteorological fields calculated by CALMET (refer to Part I of the study for CALMET configuration) to estimate the growth and transport of released puffs. Transport of the modeled material is governed by the wind diagnosed by CALMET, while the growth or diffusion is determined by turbulence-related boundary layer variables. These variables include the friction velocity (u^*), convective velocity scale (w^*), Monin-Obukhow length (L) and boundary layer height (h). CALPUFF also has algorithms that estimate processes such as dry deposition, wet deposition and various chemical transformations. CALPUFF includes methods to account for influences such as plume rise, stack/building downwash (not used), wind shear and plume penetration of the inversion layer.

Emission at the World Trade center recovery site was specified as a volume source with dimensions of 500 by 500 meters in the horizontal and 50 meters (V_s) in the vertical. The main rationale behind this configuration was to simulate a well-mixed initial volume source consistent with aerial photographs of the void left after the terrorist attack. The hourly emission rate (s_{1hr}) was set to 100 kg of $PM_{2.5}$ for model operational purposes only, since reliable estimates were not known. Hourly averaged ground-level concentrations (c_{1hr}) from the model simulation were normalized by the concentration of $PM_{2.5}$ within the assumed uniformly mixed volume source ($8000\ \mu g\ m^{-3}$) to derive a dilution factor (d, see Equation 1). In the future, this dilution factor can be applied to estimate ground-level concentrations when the source emission rate is determined.

$$d = \frac{c_{1hr}}{s_{1hr}} * V_s. \tag{1}$$

In order to calculate the spatial variation of $PM_{2.5}$ dilution at the ground level, a set of gridded receptors was prescribed with a spatial resolution of 500 meters. The wet and dry deposition options were considered in the simulations. A sensitivity analysis indicated the deposition removed at most 1–2% of the simulated $PM_{2.5}$ thus this was considered to have a negligible effect on the simulated air concentration presented herein. The potential temperature profile provided by the Advanced Regional Prediction System, Data Assimilation System (ADAS) (ZHANG *et al.*, 1998), is taken into account when the vertical distribution of mass is calculated. Additionally, the wind variation with height is considered as a puff splitting algorithm was activated. The dispersion coefficients, which influence the simulated puff growth, were calculated using the micrometeorological variables from CAL-MET.

CALPUFF does have an option to model the material release as a slug or tight series of overlapping puffs. This formulation, in theory, yields a more exact near-source concentration distribution but requires significant computational time. Test examples showed little difference between the puff and slug options beyond a distance of a few kilometers. For this reason, the puff formulation was prescribed with an average puff release rate of one per minute. The spatial and temporal dependent dispersion coefficients were calculated, using the CALMET estimated scaling parameters u^*, w^*, L and h, based on similarity theory. All other parameters were assigned using the recommended defaults, including the deposition and scavenging rate for $PM_{2.5}$. For complete documentation on the model formulation refer to A Users Guide for the CALPUFF Dispersion Model (SCIRE et al., 2000).

2.2. Advantages and Limitations

The NYC CALMET-CALPUFF modeling system has several known limitations that must be considered. Primarily, this approach is not able to resolve the near-source concentration distribution accurately in lower Manhattan. The 3-km area surrounding the World Trade Center site is an extremely non-homogeneous surface consequently similarity relations do not apply well. A complex urban canopy flow is the dominant factor in pollution dispersion within the city and can only be modeled explicitly with a suitable computational fluid dynamic model that can explicitly resolve the relevant turbulent processes. CALMET uses a surface roughness length to describe the surface of the urban canopy. This lower boundary parameterization does not apply well in this situation, as the plume height might be lower than the roughness elements in lower Manhattan. In spite of this, it is believed that the simulated concentration distribution further away from the source (i.e., beyond distances where the plume dimension is considerably larger than the building scale) can provide useful information.

An additional limitation is the relatively coarse time step (1-hour) in both model input and output. Over such a complex region, the time scale of significant variations in the boundary layer structure is in many circumstances considerably less than one hour. Additionally, the NWS ASOS surface observations used by CALMET are only representative of a brief period ($\sim$2 min) before the top of each hour. It was illustrated in Part I of this study that during certain synoptic flow regimes, changes in wind speed and direction can dramatically shift over time periods of less than 1 hour. Other drawbacks with regards to the meteorology are errors in handling certain mesoscale phenomena like sea-breeze fronts and urban heat island effects. Part I of the study also found that the near-surface wind speeds and mixing heights were overestimated by CALMET. These biases are likely to lead to an underestimation of the concentration by CALPUFF. CALMET also tended to poorly define the properties of the sea-breeze circulation, namely the rapid wind shift and thermal stratification.

Advantages of a CALMET-CALPUFF type modeling system are that a complete simulation over a seasonal period may be completed in less than twelve hours using a standard PC. The long study periods that are feasible with this type of model allow for a large pool of modeled concentrations which can be used to derive statistically significant results. This procedure provides a decisive advantage over the application of a model with full physics, which for many applications are only practical for short simulations. Additionally, by incorporating ADAS model analysis profiles in CALMET rather than 12-hourly soundings, we gain better temporal resolution of the meteorology as indicated by LEVY *et al.*, (2002). Also, many other types of observations are injected via the assimilated data (i.e., Doppler radar, satellite and aircraft observations).

3. Results

A series of CALPUFF plume plots is used in the following WTC $PM_{2.5}$ dispersion analysis. As highlighted in the methodology section, the absolute concentrations are not examined, but rather a normalized concentration in the form of a dilution factor which is defined as the plume concentration at any location divided by the concentration of the well-mixed volume source. The plume dilution shading is partitioned quasi-logarithmically with a separate color for each of the following dilution amounts: 10^2, 500, 10^3, 10^4, 10^5 and 10^6. $PM_{2.5}$ observations (New York State Department of Environmental Conservation and New Jersey Department of Environmental Protection) are compared with the simulated plume dilution to ascertain if abnormally high $PM_{2.5}$ is correlated in space and time with the simulated location of the plume. There were a total of 17 permanent sites that were operational on September 11, 2001 and 5 sites that were deployed several weeks after the event. Figure 2 displays the locations of each site. The dense arrays of sites in lower Manhattan, with the exception of *ps64* that was already operational, were deployed in response to September 11. Figure 3 presents the measured concentration averaged over the September 11 through December 8, 2001 period as a function of time of day. A separate average was calculated for the $PM_{2.5}$ sites in lower Manhattan. It is evident in Figure 3 that higher concentrations are measured in lower Manhattan at all times during the day. In both time series, the diurnal component of commuter traffic is the dominant mode. These mean concentrations over the entire study period will be used in the following analysis to sense if the concentrations observed during each case are abnormally high.

3.1. Average Wind Flow and Plume Dilution

Following the dramatic dust cloud generated by the building collapse, widespread fires of burning debris continued for a prolonged period, and a visible smoke plume was dispersed around the New York City area. The following figures provide summary information of the simulated WTC wind and dilution patterns for the three-week period following September 11th. This information is presented using a wind rose plot overlaid on the averaged plume dilution. Contours are plotted for the 10^2, 500, 10^3, 10^4, 10^5 and 10^6 dilution zones relative to the source at ground zero.

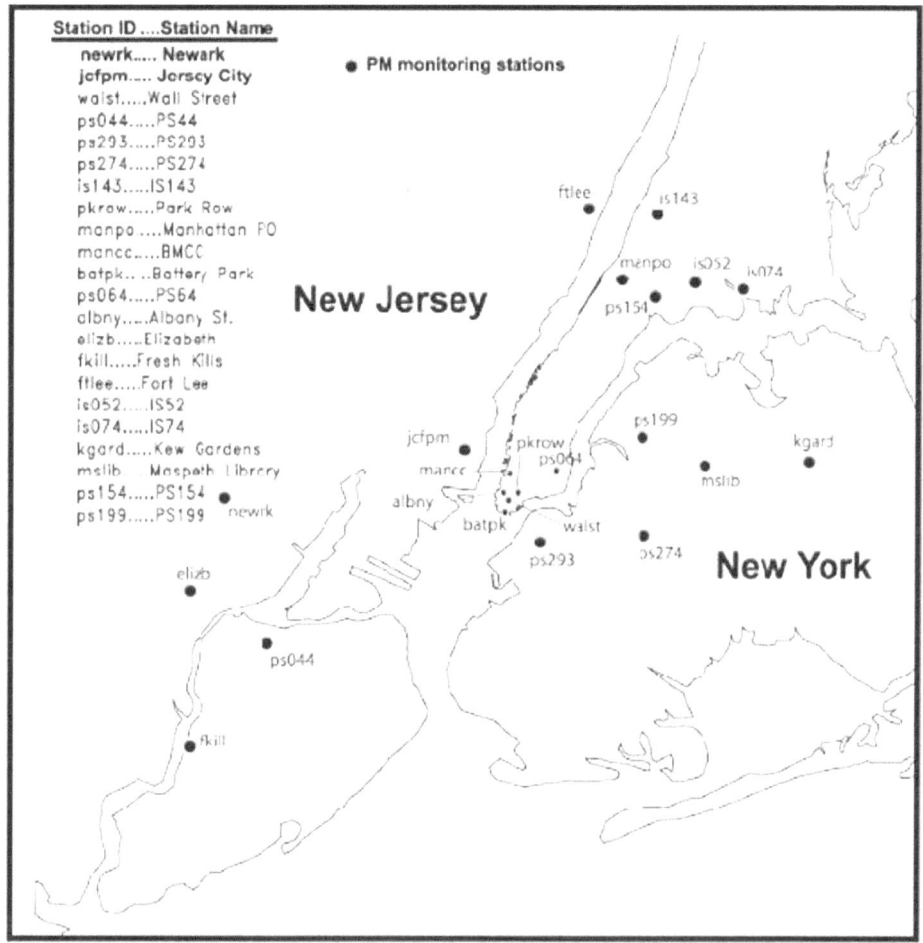

Figure 2

Map showing the name and location of PM$_{2.5}$ monitoring sites in the New York City area. The New York State Ambient Air Monitoring System and New Jersey Department of Environmental Protection manage most of the monitoring sites. Sites in lower Manhattan, marked with small dots (except *ps64*), are temporary monitoring stations deployed after September 11, 2001.

The wind rose shows the direction and speed distribution of hourly wind estimates at lower Manhattan from the CALMET simulation.

The simulated flow strength and direction distribution and average plume dilution for the period from September 11–13, 2001 is shown in Figure 4. During the initial 12–18 hours the wind was a moderate 4–6 m s^{-1} from a northerly direction as shown on the wind rose. Northwest to northeasterly wind was simulated for about 45 percent of this three-day period. Because of this flow direction, the WTC plume was directed over the far western tip of Long Island, specifically the Brooklyn borough of NYC. The CALPUFF simulation estimates that on average the particulate

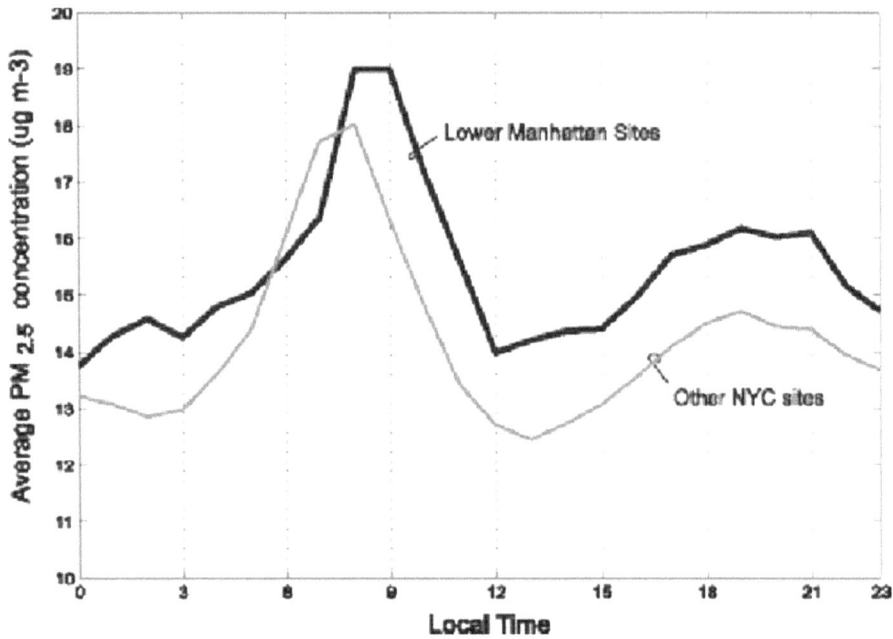

Figure 3

Average PM$_{2.5}$ (μg m^{-3}) concentrations from September 01, 2001 to January 31, 2002 as a function of time of day (LST). The average is partitioned between sites located in lower Manhattan and those farther away. Considered in the calculations are hourly averaged data from 25 combined monitoring stations managed by the New York State Ambient Air Monitoring System and New Jersey Department of Environmental Protection.

concentration from the WTC plume was diluted 1000 to 10000 times before it was advected over this region. The other preferred wind direction over this period was southwesterly, which accounts for more than 50 percent of the wind direction distribution. It is noticed that the winds were lighter on average from this direction (2–4 m s^{-1}). Lower wind speeds are normally linked to less dispersion and higher concentrations. This is seen in the average dilution pattern. Both the 1000 and 10000 dilution contours extend out further from the source as compared to the northerly flow regime. Also plotted are the average concentration measurements at all available PM$_{2.5}$ sites. Most values are in the 10–15 μg m^{-3} range with the exception of *ps64* in lower Manhattan, which reported a three-day averaged concentration of 22 μg m^{-3}. The elevated PM$_{2.5}$ concentration at *ps64* is likely a result of the WTC plume being directed towards the northeast by southwest winds on September 12–13, 2001 as indicated by the CALPUFF simulation.

Several other periods in the weeks following September 11 were examined in a similar manner, although the figures are not shown. Here is a brief summary. During the September 14–16 period the dominant wind direction ($\sim$85%) for the period was from the north to northeast. As a result, the plume dilution pattern was biased to the

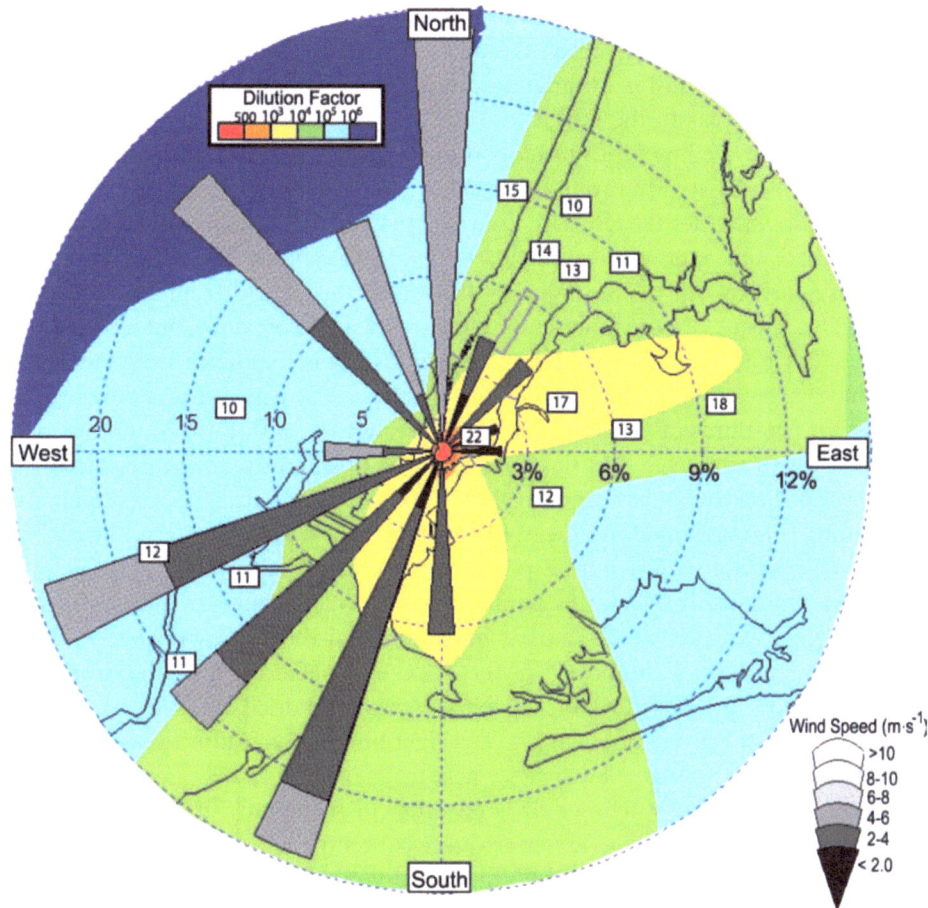

Figure 4

Wind rose of simulated wind at the grid point closest to the World Trade Center (September 11–13, 2001). Also, plotted in the background are the dilution contours of PM$_{2.5}$ from the CALPUFF simulation, averaged over the same period as the wind rose. The distance between concentric percentage rings is approximately 5 km as noted by the numbers left of the center. The plotted numbers (black outlined, white rectangles) represent the average observed PM$_{2.5}$ (μg m^{-3}) concentrations for the active monitoring stations.

southwest of lower Manhattan, specifically over New York Harbor and eastern Staten Island. The 1000-times dilution contour actually extended away from the source considerably farther than the previous three-day period. One factor that reduces the modeled dispersion is the lower mixing depth and increased stability over water. During the September 17–23, 2001 period, the principal wind component was southerly approximately 75% of the period, and generally less than 5 m s^{-1}. Over this time span, the diluted plume influenced the areas north and east of lower Manhattan. The flow direction during the third week after September 11 (September

24–30, 2001) was mostly west-southwest to westerly as well as a brief period of northeasterly winds. Consequently, the dilution contours indicate that the areas to the north and east of lower Manhattan are influenced by higher concentrations as compared to the areas to the west.

Overall, the preceding series of plots provide insight into the dispersion patterns over the New York City Metropolitan area during the most critical period, from an air quality standpoint, after the terrorist strikes on the WTC. Noticeable but diminishing fires and particulate emissions continued until mid-December. This study developed hourly estimates of the plume dilution factor for the period from September 11 through December 8. Figure 5 presents a similar plot as the previous, but for the entire three-month simulation period. More conspicuous in the long-term wind rose is the climatological signature of southwesterly to northwesterly winds. These flow regimes account for two-thirds of the simulated wind directions over this three-month period. Consequently, it is noticed that the average plume position over the three months is more biased towards the east and northeast of lower Manhattan. For example, the 10^4-dilution contour extends out nearly 17 km from the WTC to western Long Island and the Queens Borough. Another potential reason for the higher average concentration over this area is the marine boundary layer. This region is frequently affected by a stable marine layer which restricts horizontal and vertical dispersion. A similar effect is observed to the southwest of the WTC site where the 10^4-dilution pattern seems to follow the coastline of the New York Harbor. This is likely a direct consequence of the lower boundary layer height and limited diffusion of the plume over the water. Over inland New Jersey where the marine layer less frequently resides, the average dilution over the period is an order of magnitude greater. The lowest dilution contours very near the source are more concentric, although slightly biased to the northeast, which is consistent with the climatological wind direction preference.

Overall, it seems that a landuse influence lower on average mixing heights and dilution over the water bodies and a large-scale climatological influence (predominant southwest to northwest flow) can be discerned from the data. Average observed $PM_{2.5}$ concentrations from all available stations are plotted in Figure 5. Over the period, the average concentration at all sites is nearly identical, even sites located in lower Manhattan. This suggests that the longer-term observation records do not distinguish a signature of the WTC plume. Further evidence to support this thinking is presented in the following section through a more rigorous inspection of the data.

3.2. Case Studies

In the following, we present the results of the modeled WTC plume and observed $PM_{2.5}$ concentrations for three of the same cases presented in GILLIAM *et al.*, (2004). Each of the cases was a period highlighted by the US EPA where measured $PM_{2.5}$ concentrations were well above normal. The first case spans several days immediately after the WTC incident (September 11–12, 2001). The second case looks at another

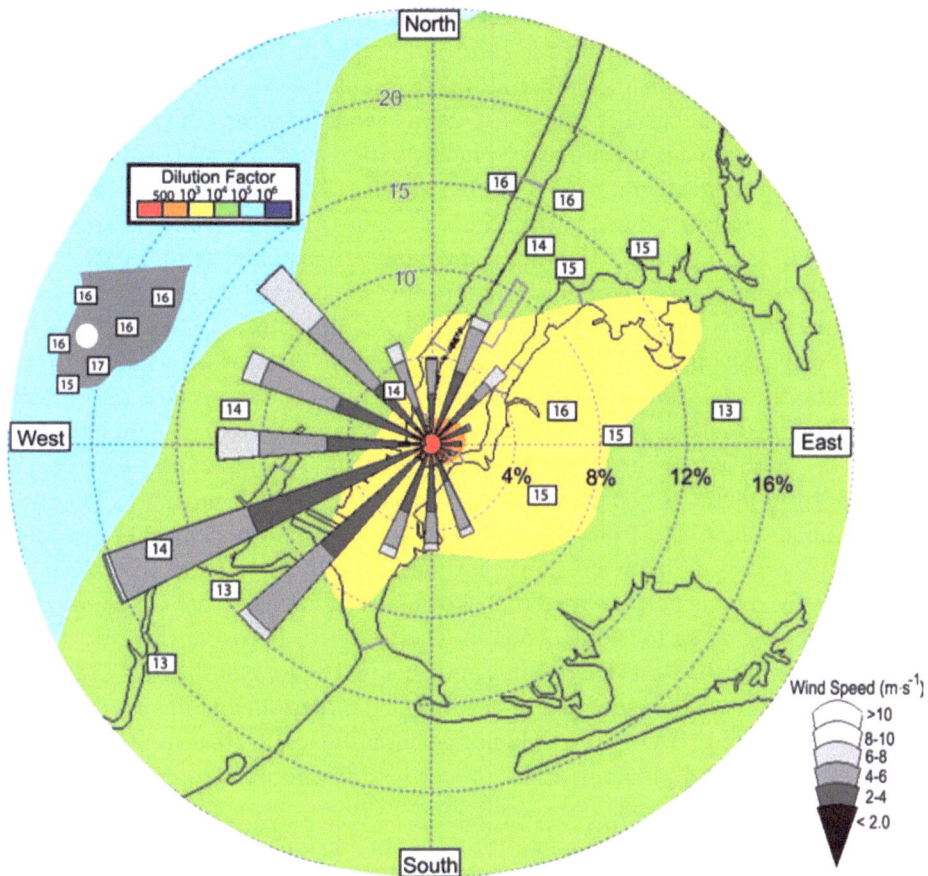

Figure 5

Wind rose of simulated wind at the grid point closest to the World Trade Center (September 11 to December 8, 2001). Also, plotted in the background are the dilution contours of PM$_{2.5}$ from the CALPUFF simulation, averaged over the same period as the wind rose. The distance between concentric percentage rings is approximately 5 km, as noted by the numbers left of the center. The plotted numbers (black outlined, white rectangles) represent the average observed PM$_{2.5}$ (μg m^{-3}) concentrations for the active monitoring stations.

similar flow event about one week after the disaster, September 17–18, 2001. The final case is another period (October 3–5, 2001) during which widespread concentrations measured around the city were abnormally high.

September 11–12, 2001 (Strong northwest to light southerly flow case)

Immediately after the WTC collapse, the emission of dust, smoke and toxic gases into the atmosphere was greatest. A surface high-pressure system was building into the area from the west on September 11. This resulted in moderate to strong northerly winds through the early morning of September 12. During the daytime hours of the 12th, high pressure settled over the region, at which point the winds decreased. High pressure remained in control through the following day (September 13). The northerly flow on September 11 was favorable for efficient mixing and rapid southerly transport of the airborne WTC debris away from the NYC area.

Shown in Figure 6a is the simulated plume dilution from the WTC volume source at 1100 LST on September 11. Also plotted are the $PM_{2.5}$ observations around the region. Figure 3 suggests that a typical concentration of $PM_{2.5}$ around the city at this time of day is 13–14 μg m^{-3}. The sampled concentrations indicate that $PM_{2.5}$ levels are well below the average for this time of day, as observed concentrations range from 50 to 75% of normal. Moderate northwest winds and the fact that a "fresh" air mass was being advected into the region from the northwest are likely reasons for the low pollution levels. The WTC plume spreads across the far western tip of Long Island with ground–level concentrations quickly dropping to 10^3–10^4 times less than that of the specified volume source, within 3 km downwind of the source. The modeled WTC plume is rapidly dispersed because CALMET mixing heights over western Long Island rise from 100–200 m in the early morning to 600–800 m by noon. Another noted characteristic is that the plume is not symmetrical about the centerline. The mixing height varied (not shown) substantially across the plume path, about 400 m over water to 800 m overland. This results in a broader eastward mixing of the plume while the west side remains more compact. None of the $PM_{2.5}$ sampling sites is in the path of the estimated plume, but aerial and satellite images agree well with the simulated CALPUFF plume position. Overall, the moderate wind, convective mixing and plume-path over water help to effectively transport airborne WTC material away from the recovery area.

On September 12, high pressure settled over the region and the winds lightened. In the morning at 0700 LST (Fig. 6b), observed concentrations across the area remained well below normal levels (normal: 16–17 μg m^{-3}) with most values below 75% of average. The simulated WTC plume is advecting material south and west of lower Manhattan over New York Harbor, towards Staten Island. Plume dilution is much less during this time with the 10^2 to 10^3 contours extending farther ($\sim$15 km) from lower Manhattan than the previous mid-day case. The mixing height calculated by CALMET was about 300–400 m over the plume path. The lower mixing height estimate is responsible for the compact, higher concentrated plume. Satellite images on this morning appear to be in good agreement with the plume position.

▶

Figure 6

Simulated CALPUFF dilution (100–10^6 dilution) of a volume source located at the World Trade Center recovery site. The plotted numbers indicate the hourly-averaged $PM_{2.5}$ (μg m^{-3}) concentrations for the active monitoring stations. Also shown is the estimated 10 m wind from the CALMET diagnostic model. Time periods shown are: (**A**) September 11, 2001 at 1100 LST, (**B**) September 12, 2001 at 0700 LST, (**C**) September 12, 2001 at 1500 LST, (**D**) September 12, 2001 at 1800 LST, (**E**) September 13, 2001 at 0100 LST, and (**F**) September 13, 2001 at 0900 LST.

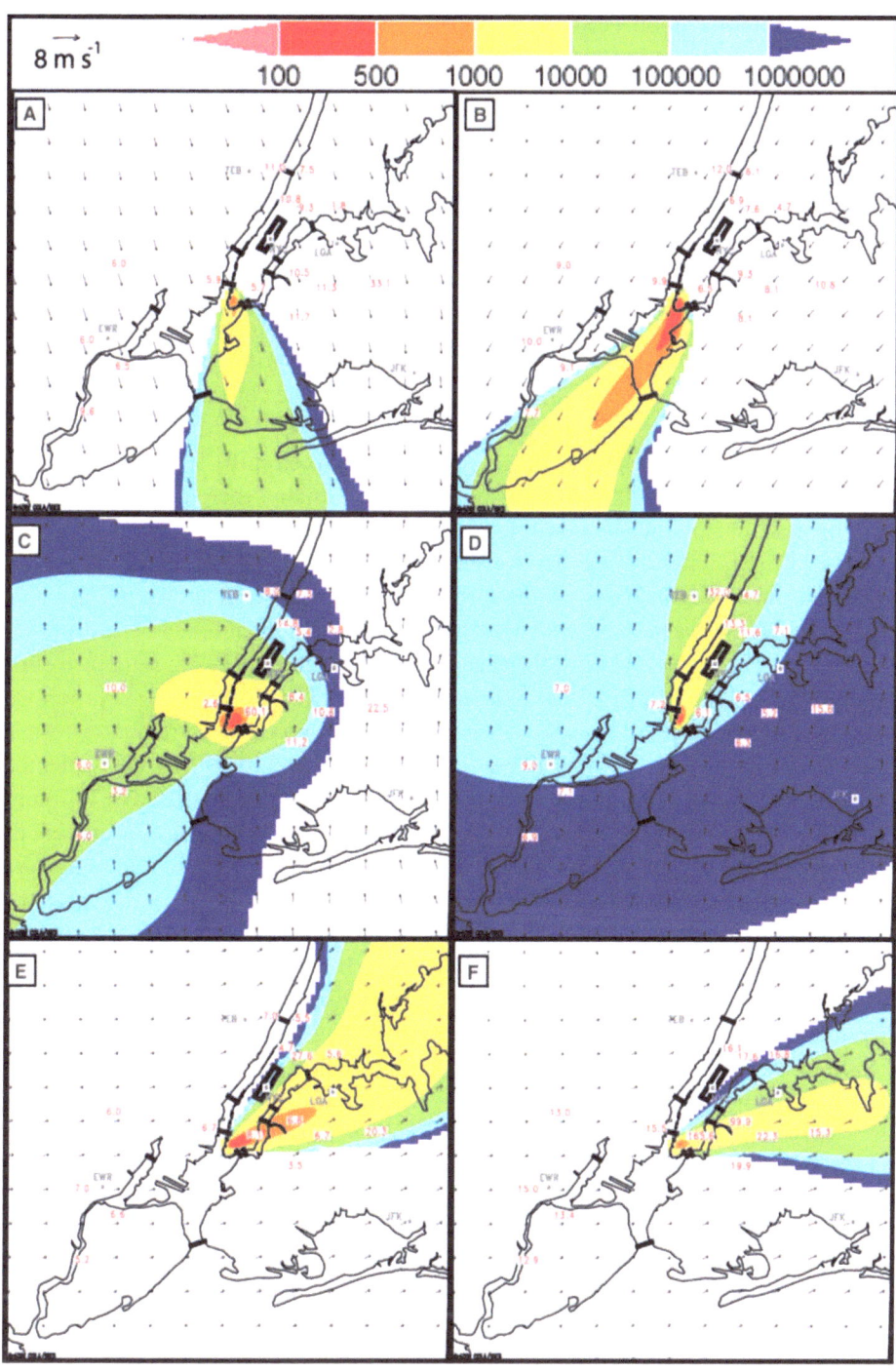

During the afternoon (1500 LST, Fig. 6c) of the same day, a weak sea-breeze boundary developed and moved inland. As a result, the flow becomes southerly and the simulated WTC plume rotated clockwise, passing over areas of New Jersey and Manhattan. Figure 6c shows the simulated plume when the abrupt south to north shift occurred. CALPUFF estimates that the plume is diluted by no less than 10^3 times outside of lower Manhattan. The reason for this dilution, even though winds are light, is the deep convective boundary layer, estimated by CALMET to be in the 1300–1500 m range. The shift in wind transports material over a large portion of New Jersey. However, in this figure the observed concentrations do not reflect a measurable component of $PM_{2.5}$ that can be directly attributed to the WTC plume outside of lower Manhattan. The only available observation in lower Manhattan (*ps64*) does increase to 480% ($\sim$60 μg m^{-3}) of normal. The concentration of $PM_{2.5}$ at (*ps64*) in lower Manhattan jumped from below normal to well above the average in temporal agreement with the simulated plume's northward shift. This may imply that the simulated WTC plume shift is reasonable. Additionally, there is a satellite image showing a similar plume shift on this afternoon.

Several hours later, in the early evening (1800 LST), the simulated plume was directed up the Hudson River on the east side of Manhattan as shown in Figure 6d. Plume dilution was generally above 10^3 within a few kilometers of the source. The plotted $PM_{2.5}$ observations, with the exception of one, do not show a signature of the WTC plume as all values are below the normal level. The Fort Lee observation was actually 213% (32 μg m^{-3}) above the normal average measurement (normal: 15 μg m^{-3}) for this time of day. Subsequent times (0100 and 0900 LST, September 13, 2001) that evening and through the following morning the simulated WTC plume rotated from north to east (Figs. 6e and 6f). A pattern was noticed between the hourly-simulated plume position and the hourly observations. One to two hours after the simulated plume rotates and passes over the observation site, the measured concentration increases to 200–350% of the normal values while those to the west and east remain in the normal or below normal range. It is believed that the more stable conditions allow the plume to remain more concentrated over a greater distance, and therefore is easier to identify by the more distant $PM_{2.5}$ samplers. It is also noted that the simulated plume remains relatively more concentrated farther away from the source when the estimated mixing height and turbulence are low. This agrees with the observations indicating a general increase in $PM_{2.5}$ concentrations around the city.

The next morning (0900 LST on September 13, 2001), the WTC plume in Figure 6f is positioned towards the east-northeast of lower Manhattan. Observations indicate that most sites are reporting near-normal $PM_{2.5}$ values, except for the well-above normal concentrations along the simulated plume centerline, specifically the two observations (*ps64* and *ps199*) that are 600–1000% above normal (normal: 15–19 μg m^{-3}). This indicates that the plume position is well simulated by the dispersion model. The *ps199* observation, in fact, remained well above normal through the entire morning of September 13, 2001 while surrounding stations were at normal levels, in agreement with the simulated plume position. This is evidence that at least initially there was some signature of the WTC plume in the concentration measurements outside of lower Manhattan.

Another more detailed look at the time variation of the modeled CALPUFF plume and the $PM_{2.5}$ observations around the area is presented in Figures 7a and 7b. These figures provide a unique view of the spatial and temporal modeled WTC plume together with the observed $PM_{2.5}$ time series. The shaded values in the upper panel correspond to the plume concentration normalized by the maximum value over the period at a distance of 13 km surrounding the WTC site. The observation sites are noted on the y–axis corresponding to the angle from the WTC (direction from the site is noted on the right y-axis). Below the CALPUFF plume plot are the time series of $PM_{2.5}$ observations from all the stations that are approximately 13 km away from the WTC. Also designated are horizontal lines on the CALPUFF plume variation portion of the plot, which correspond to the direction of the observation station from the WTC site as labeled.

Of particular interest is the period from noon on September 12 through the following day. The CALPUFF plume as shown before and in Figure 7a is directed towards the south at the beginning of the day on September 12. Around mid-day, the shift to west and then north is noticed. Furthermore, the shift corresponds to a decrease in the simulated concentration during the day as the mixing depth increases. In the late day and evening hours of September 12 and early hours of September 13, the plume shifted from north to east-northeast. The observed concentrations indicate an abrupt increase in $PM_{2.5}$ over this exact time span. Additionally, the increase occurs first at Forth Lee (*tlee*), then Manhattan PO (*manpo*), then Public School 154 (*ps154*) and then at Independent School 74 (*is074*). Preceding this, the Newark airport observation actually indicates an increase that is correlated with the simulated WTC plume earlier in the day. However, later that week, from September 14–17, there is little evidence that the WTC plume was observed in the area except possibly at the Manhattan PO (*manpo*) station. There are several spikes in the reported concentration time-series that have some correlation to the simulated emissions from the WTC rubble. The most notable is midnight on September 16. The CALPUFF plume is directed towards the north towards the *manpo* site. However, none of the other stations in the vicinity indicates an increase in the $PM_{2.5}$ so it cannot be conclusively stated that this signature is related to the WTC plume.

A similar look at the simulated WTC plume and observations is provided in Figure 7b. The simulated plume plotted for a ring 3 km from the WTC and the observation time series are from stations in the 3–7 km range. The CALPUFF concentration is of similar variation to the previous figure. The reported observation time series show similar peaks, which correspond well with the simulation. The first peak noted in both the Public School 64 and 199 (*ps64* and *ps199*) time series occurred in the afternoon of September 12 as the sea breeze rotated the plume from

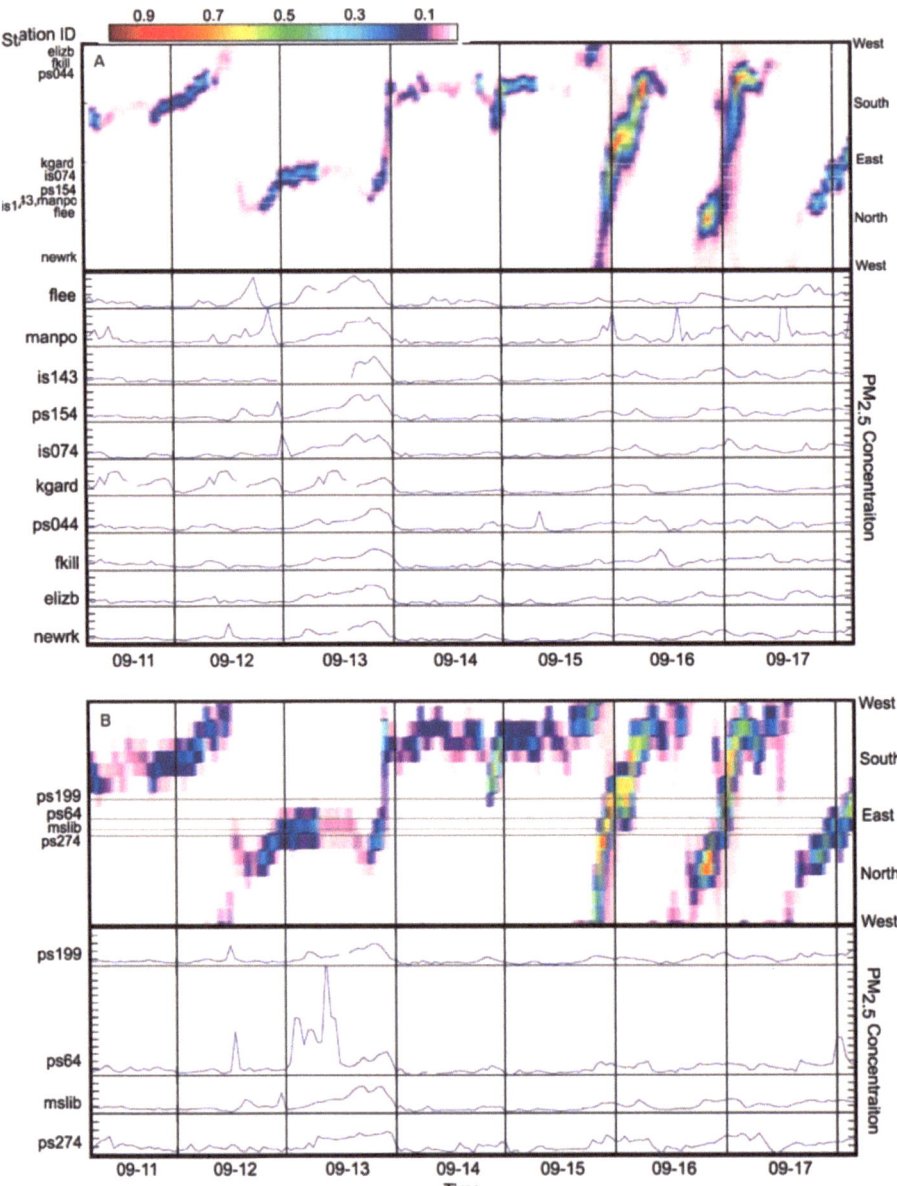

south to north. The second and more dramatic signature occurred the morning of September 13 when the *ps64* observation reported extremely high $PM_{2.5}$ levels over a span of 6–12 hours as discussed in prior results. The simulated plume is in agreement with this incident. The stations that were farther away like *mslib* and *ps274* did not indicate a defined signature; however, *ps199* does show a minor increase. In

◀

Figure 7

The upper portion of the figure panels is the patterns of the CALPUFF simulated World Trade Center (WTC) plume at a distance of 13 km (Panel A), and 3 km (Panel B) from the WTC site (September 11–17, 2001). The shaded values correspond to the simulated $PM_{2.5}$ normalized by the maximum value over the time series. The normalized concentration plume pattern is plotted with respect to the time and direction from the WTC as indicated by the right y-axis label. The stations are labeled on the left y-axis according to the direction from the WTC site (e.g., *flee* is approximately north of the WTC, *kgard* is east of the WTC site). The lower panels are the observed $PM_{2.5}$ (μg m^{-3}) time series for each of the monitoring sites. The station id is located on the y-axis of each time series. Each tick mark on the y-axis represents 10 μg m^{-3} above a zero baseline.

agreement with the previous analysis, there is little indication that the WTC plume influenced the $PM_{2.5}$ observations after this initial period.

September 17–18, 2001 (light and variable flow case)
The next case occurred almost 1 week after the WTC collapse, during a synoptic pattern similar the September 11–13 case. High pressure was positioned over the area on September 17th and remained in control of the weather for several days. During this time the winds were light and variable so local effects dominated the flow patterns. Most noticeable is a diurnal rotation of the winds. During the day the wind turned southerly in response to an inland propagating sea breeze boundary, and then turned northerly at night in response to a weak land breeze. Starting at noon on September 17th, Figure 8a shows the simulated plume dilution/location. The plume is estimated to be diluted 10^3 over the harbor to 10^4 over land areas. The influence of a spatially variable mixing height is observed as the 10^3 contours follow the land-water interface where the mixing height and stability are drastically different. The mixing height over New York Harbor was in the 300–400 m range while increasing to 1200–1400 m over the adjacent landmass. The observed concentrations across the NYC area were all well below normal (normal: 13–14 μg m^{-3}) except at Fort Lee which was slightly above the normal levels. There is no apparent signature in the observations that the WTC plume is significantly increasing $PM_{2.5}$ observations in the surrounding areas.

Several hours later in the afternoon (1800 LST) the plume redirected northward as a sea-breeze front passed over the region. Figure 8b depicts the simulated plume in a similar position as the previous case. Mixing heights are lower in this case compared with the prior, thus the dilution of the plume decreases. CALMET estimated mixing heights across the plume path are in the 800 m (Hudson River) to 1100 m (over New Jersey) range. The 100–500 dilution contours extend up the east side of lower and midtown Manhattan, some 10–12 km away from the source. The 10^3 contour extends even further (20 km) from the source. As for the previous time period, the concentration observations do not show a noticeable signature of the WTC plume as values are mostly in the normal to slightly above normal range (normal: 15 μg m^{-3}). An exception is Fort Lee, which is about 200% of normal.

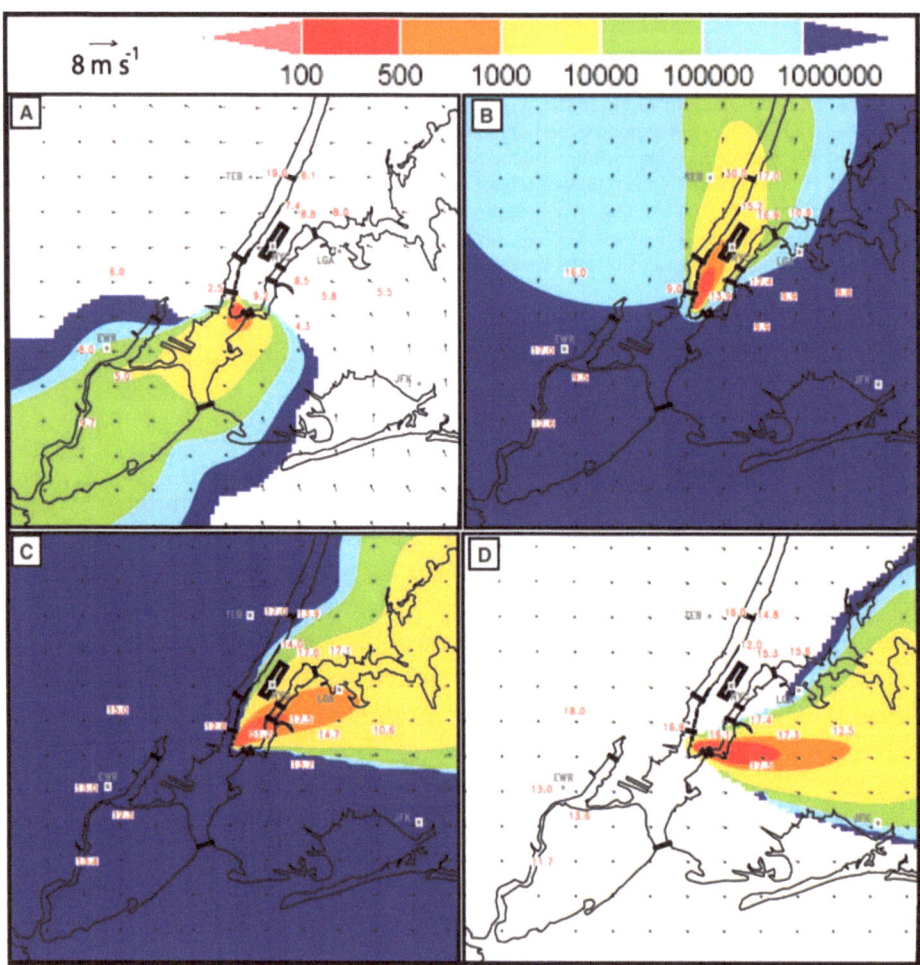

Figure 8

Simulated CALPUFF dilution (100–10^6 dilution) of a volume source located at the World Trade Center recovery site. The plotted numbers indicate the hourly-averaged PM$_{2.5}$ (μg m^{-3}) concentrations for the active monitoring stations. Also shown is the estimated 10 m wind from the CALMET diagnostic model. Time periods shown are: (**A**) September 17, 2001 at 1200 LST, (**B**) September 17, 2001 at 1800 LST, (**C**) September 18, 2001 at 0100 LST, and (**D**) September 18, 2001 at 0300 LST.

Figures 8c and 8d show the same type of plots at 0100 and 0300 LST the following morning. The modeled plume rotated as winds turn from southerly to northerly. The mixing heights decreased overnight to 150–200 m, as a response the simulated dilution value decreases significantly. Most of the PM$_{2.5}$ observations, opposite from the previous case (September 11–12), do not manifest a clear signature of the WTC plume away from lower Manhattan as the reported values are similar over the entire area. The observations do indicate slightly above normal (normal: 12–13 μg m^{-3})

concentrations, which may only be a result of the stable atmospheric conditions. Figure 8c does show that as the plume turned to the northeast, the reported concentration at the *ps64* in lower Manhattan increased to 445% of normal. Several hours later Figure 8d indicates that as the plume turned clockwise, the reported concentration at *ps64* significantly decreased back to near normal (normal: 14 μg m^{-3}). This is certainly evidence that the WTC plume did influence the *ps64* observation and that the turning/placement of the simulated plume is a fair representation of what actually took place. The fact that the meteorological conditions are almost identical between this case and the overnight period of September 12, 2001, and the signature of the WTC plume beyond lower Manhattan is not defined in the observations, would lead one to believe that there was a considerable decrease in the source strength between September 12–13 and September 17–18. Hence there was slight impact from PM$_{2.5}$ emissions on areas outside of lower Manhattan after the first week post September 11, 2001.

October 3–6, 2001 (southwest flow case)
In the early part of October (2–6) attention was placed on an extended period of high PM$_{2.5}$ concentration measurements around the NYC area. A strong surface high-pressure system was anchored off the eastern U.S. for the period; this resulted in a wind flow from the southeast region of the U.S to the northeastern U.S. In this case study the simulated WTC plume and observations are examined during the morning and afternoon of October 3–4.

Figure 9a shows the simulated WTC plume and PM$_{2.5}$ observations on October 3, 2001 at 0500 LST. The plume is directed towards the east-northeast of lower Manhattan. Dilution of the simulated plume material is low with the 100–500 diluted zone extending 15–20 km away from the WTC site. The plume width, loosely defined by the 10^4 dilution contour, is greater relative to other times, spreading out as much as 5 km from the centerline. The very low mixing heights (100 m) and low simulated wind speeds (1–2 m s^{-1}) are the main reason for the relatively high plume concentration away from the source. An animation of the plume over several hours indicates that the wind oscillated from southwest to west, resulting in the wide plume coverage. Observations plotted with the WTC simulated plume demonstrate that the reported concentrations across the entire area are well above normal by 200–300% (normal: 14–15 μg m^{-3}). Even those concentrations upwind of the simulated WTC are well above normal (~300%). PM$_{2.5}$ measurements closest to the WTC site are all slightly higher than the other surrounding sites. Hence it is reasonable to infer that the augmented concentrations during this period are a result of a regional increase in pollution levels and not the WTC site.

Later this day at 1500 LST (Fig. 9b) the simulated plume becomes considerably less concentrated because mixing heights rise to nearly 1500 m. The dilution contour less than 1000 is limited to the area directly over the WTC volume source. Another factor that contributes to the rapid dilution is the

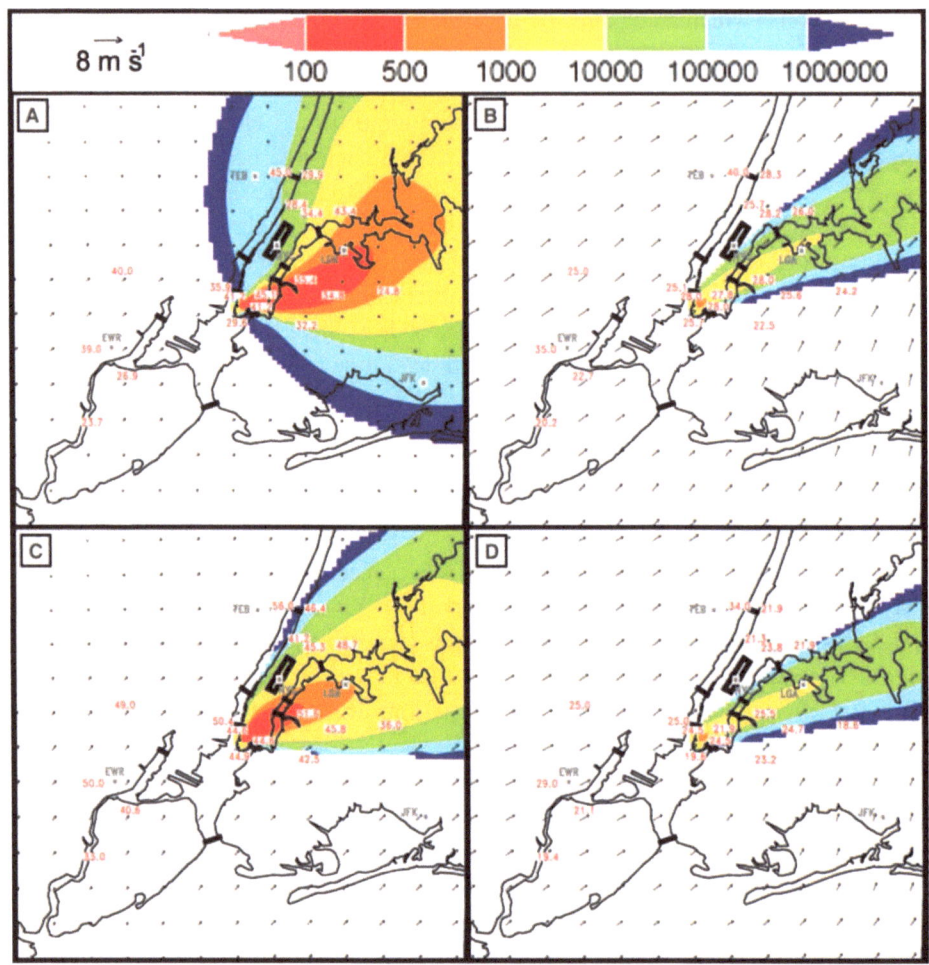

Figure 9
Simulated CALPUFF dilution (100–10^6 dilution) of a volume source located at the World Trade Center recovery site. The plotted numbers indicate the hourly-averaged $PM_{2.5}$ ($\mu g\ m^{-3}$) concentrations for the active monitoring stations. Also shown is the estimated 10 m wind from the CALMET diagnostic model. Time periods shown are: (**A**) October 3, 2001 at 0500 LST, (**B**) October 3, 2001 at October 4, 2001 at 0500 LST, and (**D**) October 4, 2001 at 1500 LST.

stronger southwest wind, which increases mechanical mixing near the surface and quickly disperses the material and transports it away from the source. Several $PM_{2.5}$ observations near the WTC recovery site are noticeably greater (360% of normal) than those more distant. All observed $PM_{2.5}$ values well away from the WTC site remain above 200% of normal.

The following morning at 0500 LST (Fig. 9c) the simulated plume becomes more concentrated again as the mixing heights lower to less than 200 m, similar to the previous night. The winds also lighten as a result of the more stable boundary layer.

The dispersion pattern of the plume is similar to the previous night in that the overall plume widens because of the light winds that meander between southwest and west. The observed concentrations are very similar to the previous night, with consistent 300% above normal values across the region. Observed $PM_{2.5}$ air in and around the WTC site is negligibly different from other areas of NYC. The following afternoon Figure 9d shows that the near-surface winds, boundary layer heights and, hence, the simulated plume is almost a replica of the previous day at 1500 LST. Reported concentrations show $PM_{2.5}$ levels ~150–200% of normal, slightly lower than the previous day but not significantly different.

Although the early October period raised concerns among local officials, as stated above, we believed the synoptic conditions were the main cause, and not the still smoldering WTC rubble. This increase in measured $PM_{2.5}$ is typical of what occurs when upstream winds arrive from a path over areas with substantial $PM_{2.5}$ emissions. Once a front passed on October 6, the observed concentrations dropped to below normal levels as the air mass was replaced with "clean" air from the north.

4. Summary

In this study the dispersion patterns of a simulated WTC plume are analyzed for a three-month period post September 11, 2001. $PM_{2.5}$ measurements from a network of samplers located around the NYC metropolitan area were analyzed to evaluate the potential impact of the WTC plume. These data sets are examined for signatures of the WTC plume during several case studies and over averaged periods ranging from several days to months. A primary goal was to determine if a signature of the plume was evident in the observation records so a conclusion can be reached regarding the air quality impact of the WTC disaster on surrounding areas outside of lower Manhattan. The other objective is to find out how well the plume position and dilution variations are simulated by application of a CALMET-CALPUFF modeling system.

In general, the simulated WTC plume was diluted quicker as material traveled away from the source during the daytime. Estimates from CALPUFF indicated that the material was diluted more than 10^3 times once the plume of material moved out of lower Manhattan. At night the plume became less diluted with the 10^3-contour stretching away from lower Manhattan to areas of Long Island and the 10^4-contour more that 30 km from the source. The higher observed concentrations occurred at night, and are direct responses to the lowered mixing depth and increased atmospheric stability. Although validation of the simulated concentrations is not possible, the diurnal behavior of the simulated plume dilution followed trends similar to those of the $PM_{2.5}$ measurements.

Observations of $PM_{2.5}$ around the NYC region did show some signature of the WTC plume during the first few days after September 11, 2001. During the initial few hours the plume was simulated to travel towards the southeast and did not pass over any of the air monitoring stations, thus no signature was recorded. A plume signature was observed on the night/early morning of September 12–13, 2001 when the simulation showed a concentrated plume northward over several air monitoring stations. Additionally the simulated plume position matched up well with increases in observed concentrations both near and away from lower Manhattan. The simulated plume also matched up with photographs of the plume (not shown) over the metropolitan area that was visible during these first few days.

In the two other cases that were examined (September 17–18 and October 3–4), there was no clear signature of the particulate matter from the WTC site outside of lower Manhattan. The September 17–18 and September 11–13 cases had similar meteorological conditions, however no clear WTC plume signature was evident at the monitoring sites because the source strength at ground zero was substantially lower. The October 3–4 cases had substantially above average particle concentrations at all monitoring sites with no clear signature of the particle matter from ground zero. Overall, the hourly observations combined with the hourly plume modeling indicate it is likely that the plume did not substantially impact areas more than a few kilometers away from the WTC except possibly for the first few days. This is also supported by the uniformity in the 3-month averages of the $PM_{2.5}$ measurements among sites in lower Manhattan, including those sites within several kilometers of the WTC recovery site and those sites outside of lower Manhattan. A comparison of the diurnal variation of average concentrations between the distant and nearby station in lower Manhattan show that the lower Manhattan sites, on average, measure higher concentrations, nonetheless the difference was not very significant, only a few μg m^{-3}.

The main purpose of the plume simulations was to provide general guidance on the likely pathway for pollutant emissions from the WTC site, both temporally and spatially. These simulations are applicable to pollutant emissions near the ground that are well mixed by the buildings surrounding the WTC "ground zero" area. Except for periods during the first few days, particularly the first few hours, the ground-level emission assumption is valid. The plume before the building collapse and during the initial open fires immediately following the collapse was buoyed by the heat of the fire to elevations above the ground. The plume dilution values and locations presented in this paper are representative of the added dilution by atmospheric processes during transport away from the WTC site. It is important to note that the initial dilution from a point or small area source within the "ground zero" area is not being modeled. Also, local wind flow patterns caused by the influences of buildings of lower Manhattan are not being modeled. These limitations are most significant close to the WTC, especially during nighttime periods when the winds are light and the atmosphere is stable. During these periods, emissions from "ground zero" may not become well mixed before they are transported away. For some periods, emissions may even be caught in wind currents "snaking" through the area dominated by large buildings within lower Manhattan. Nevertheless, even for these situations, emissions would eventually be caught within the atmospheric transport that is being modeled by the methods used in this paper. For these reasons the plume simulations are only considered herein appropriate at distances beyond 2 km downwind from the WTC site. Applications such as the CALMET-CALPUFF system have their advantages and limitations because of their known simplifications in characterizing meteorology, and plume transport and dispersion processes. A CALMET-CALPUFF based modeling system, while probably not providing as precise an estimate of pollution levels as mesoscale models, might be adequate for many applications as has been demonstrated in this research. Knowing generally where the plume may have been while not knowing the concentrations precisely can still help determine where to conduct additional monitoring or more refined modeling of human exposures and where to study the population in epidemiological studies. Having such results rapidly as a forecast or screening tool could be valuable to groups that do not have access to sufficient computing resources to operationally run fuller physics models.

Acknowledgements

This work was supported by the United States Environmental Protection Agency, State Climate Office of North Carolina and the Department of Marine, Earth, and Atmospheric Sciences of North Carolina State University. It has been subjected to United States Environmental Protection Agency and National Oceanic and Atmospheric Administration review and approved for publication. Mention of trade names or commercial products does not constitute an endorsement or recommendation for use. We thank Bob Kelly and Henry Feingersh (US EPA Region 2) for their support of the application of meteorological models and instrumentation following the events of September 11, 2001. Gratitude is expressed to the constructive peer review by John IRWIN (NOAA), Tom Coulter (EPA), and Joseph Pinto (EPA). Comments by John Irwin lead to substantial changes in the original draft, including the addition of the companion paper to provide an assessment of the CALMET modeled meteorology. Discussions with Haluk Ozkaynak (EPA) were helpful in guiding the potential applications and limitations of the methods used in this paper for supporting human exposure assessment studies. Dennis Atkinson (NOAA) and WILLIAM Brown (NOAA) were instrumental in the quick delivery of the quality assured ASOS data from the NOAA National Climatic Data Center. The CAPS group at the University of Oklahoma was instrumental in providing the hourly ADAS data set. Matt Darcangelo (NY State Department of Environmental Conservation) and James Oxley (NJ Department of Environmental Protection) were instrumental in providing the air monitoring data needed to support this modeling study.

References

ALLWINE, K.J., DABBERDT, W.F., and SIMMONS, L.L. (1998), *Peer Review of the CALMET/CALPUFF Modeling System*, U.S. Environmental Protection Agency, Research Triangle Park, NC. Web address: http://www.epa.gov/scram001/7thconf/calpuff/calpeer.pdf

BARNA, M.G. and GIMSON, N.R. (2002), *Dispersion Modeling of a Wintertime Pollution Episode in Christchurch, New Zealand*, Atmos. Environ. *36* (21), 3531–3544.

GODFREY, J.J. (1998), *Air Quality Modelling in a Stable Polar Environment - Ross Island, Antarctica*, Atmos. Environ. *32* (17), 2899–2911.

HONAGANAHALLI, PUTTANNA, S., SEIBER, and JAMES, N. (2000), *Measured and Predicted Airshed Concentrations of Methyl Bromide in an Agricultural Valley and Applications to Exposure Assessment*, Atmos. Environ. *34* (21), 3511–3523.

IRWIN, J.S., SCIRE, J.S., and STRIMAITIS, D.G., *A comparison of CALPUFF modeling results with CAPTEX field data results*. In *Air Pollution Modeling and Its Application XI* (edited by Gryning, S.E. and Schiermeier, F.A.) (Plenum Press, New York 1996) pp 603–611.

IRWIN, J.S., *A Comparison of CALPUFF modeling results with 1977 INEL field data results*. In *Air Pollution Modeling and Its Application XII* (edited by Gryning, S.-E. and Chammerhac, N.) (Plenum Press, New York 1998) pp 143–153.

LEVY, J.I., SPENGLER, J.D., HLINKA, D., SULLIVAN, D., and MOON, D. (2002), *Using CALPUFF to Evaluate the Impacts of Power Plant Emissions in Illinois: Mode Sensitivity and Implications*, Atmos. Environ. *36* (6), 1063–1075.

SCIRE, J.S., ROBE, F.R., FERNAU, M.E., and YAMARTINO, R.J., *A User's Guide for the CALMET Meteorological Model* (Version 5) (Earth Tech, Inc. Concord, MA 2000).

SCIRE, J.S., STRIMAITIS, D.G., and YAMARTINO, R.J., *A User's Guide for the CALPUFF Dispersion Model* (Version 5) (Earth Tech, Inc. Concord, MA 2000).

U.S. ENVIRONMENTAL PROTECTION AGENCY (2004), Technology Transfer Network Support Center for Regulatory Air Models. Retrieved on September 22, 2004 from http://www.epa.gov/scram001/.

U.S. ENVIRONMENTAL PROTECTION AGENCY (1998), *A Comparison of CALPUFF with ISC3*. EPA-454/R-98-020. Office of Air Quality Planning and Standards, U.S. Environmental Protection Agency, Research Triangle Park, NC 27711, 50 pp.

ZHANG, J., CARR, F.H., and BREWSTER, K. (1998), *ADAS cloud analysis*, Preprints, 12th Conference *On Numerical Weather Prediction*, Phoenix, AZ. American Meteorological Society, 185–188.

ZHOU, Y., LEVY, J.I., HAMMITT, J.K., and EVANS, J.S. (2003), *Estimating Population Exposure to Power Plant Emissions Using CALPUFF: A Case Study in Beijing, China*, Atmos. Environ. *37* (6), 815–826.

(Received August 11, 2004, accepted October 20, 2004)
Published Online First: June 21, 2005

To access this journal online:
http://www.birkhauser.ch